Power Systems

Electrical power has been the technological foundation of industrial societies for many years. Although the systems designed to provide and apply electrical energy have reached a high degree of maturity, unforeseen problems are constantly encountered, necessitating the design of more efficient and reliable systems based on novel technologies. The book series Power Systems is aimed at providing detailed, accurate and sound technical information about these new developments in electrical power engineering. It includes topics on power generation, storage and transmission as well as electrical machines. The monographs and advanced textbooks in this series address researchers, lecturers, industrial engineers and senior students in electrical engineering.

More information about this series at http://www.springer.com/series/4622

Radu-Emil Precup · Tariq Kamal ·
Syed Zulqadar Hassan
Editors

Advanced Control and Optimization Paradigms for Wind Energy Systems

Springer

Editors
Radu-Emil Precup
Faculty of Automation and Computers
"Politehnica" University of Timişoara
Timişoara, Romania

Syed Zulqadar Hassan
School of Electrical Engineering
Chongqing University
Chongqing, China

Tariq Kamal
Department of Electrical
and Electronics Engineering
Sakarya University
Serdivan, Sakarya, Turkey

ISSN 1612-1287 ISSN 1860-4676 (electronic)
Power Systems
ISBN 978-981-13-5994-1 ISBN 978-981-13-5995-8 (eBook)
https://doi.org/10.1007/978-981-13-5995-8

Library of Congress Control Number: 2018966859

This Springer imprint is published by the registered company Springer Nature Singapore Pte Ltd.
The registered company address is: 152 Beach Road, #21-01/04 Gateway East, Singapore 189721, Singapore

Preface

As a kind of sustainable energy source, wind energy systems have received a substantial jump in power industry and currently the fastest-growing (about 30% annually) energy source worldwide as compared to other renewable energy sources. The main concern regarding wind energy systems is the major difference between the highly intermittent nature of the primary source (wind speed) and the desired demands concerning the electrical energy quality and system stability. This leads to challenging control problems because of several types of disturbance inputs. Therefore, wind energy conversion within the standard parameters imposed by the energy market and power industry is unachievable without the essential involvement of optimization and control. Control and optimization techniques have already exposed their importance in all areas of engineering including energy and sustainability. This book uses the rapid growth of control and optimization paradigms (i.e., adaptive control, fuzzy control, artificial neural networks, modified neural-fuzzy control, predictive control, genetic algorithms, and swarm intelligence algorithms) to increase the conversion efficiency, mechanical reliability, dynamical stability, harmonics mitigation, power regulation, and quality in wind energy systems.

The material of the book is organized in the following ten chapters. All chapters are included in this book after a rigorous review process. Special importance is given to chapters offering novel control and optimization techniques in wind energy systems. The contributed chapters provide new ideas and approaches, clearly indicating the advances made in modeling, analysis, and simulation with respect to the existing state-of-the-art.

Chapter "Nonlinear Modeling, Analysis and Simulation of Wind Turbine Control System With and Without Pitch Control as in Industry" of this book provides nonlinear modeling, and simulation of wind turbine generator dynamics and control with and without pitch control. The modeling part is a comprehensive time-domain layout of the model currently considered by industry, such as General Electric and National Renewable Energy Laboratory. The chapter also summarizes some of the most recent and important observations, such as parameter sensitivity, dynamical stability, and multiple timescales structure found in wind turbine generator system. A data validation for the model versus real measured data of the power–wind curve is also discussed and magnifies the findings of this chapter.

Chapter "Distributed Cooperative Control of Wind Farms with On-site Battery Energy Storage Systems" presents research on output power regulation in wind farms consisting of doubly fed induction generator wind turbines equipped with on-site battery energy storage systems. A novel distributed control strategy based on the leader–follower consensus theory is proposed where a virtual leader is embedded in the wind farm supervisory controller to provide the demand information. A small-signal model of a wind turbine and battery energy storage systems is derived, and eigenvalue analysis is conducted to investigate the stability of the combined system.

Frequency regulation in a power system is always critical for the better-quality power supply to the end user. The growth of wind generators and the unpredictability and variability associated with the resource increase the difficulty level of the frequency regulation tasks in power systems. Chapter "Sensitivity Analysis of Frequency Regulation Parameters in Power Systems with Wind Generation" is devoted to study the impact of fluctuating different system parameters on the overall performance of the traditional frequency regulation system when including contributions of wind energy mix. A model for the inclusion of variable-speed wind turbines in the frequency control loops is analyzed, and parametric sensitivity functions are established using linearized models. The stability analysis for inertia sensitivity of frequency regulation involving wind generation is also carried out.

Among the various power quality problems, harmonic distortion is another important problem of power quality in wind energy systems. This phenomenon can cause serious effects on the grid connection, which may result hosting capacity limitation to preserve the overall performance of the network. Chapter "Wind Turbines Integration into Power Systems: Advanced Control Strategy for Harmonics Mitigation" provides physical factors responsible for harmonic current emissions by full-converter wind turbines. The chapter also presents an advanced control structure to mitigate the harmonics in a wind power generator. The design is directed toward guaranteeing the integration of a large-scale wind farm, through minor changes to the background harmonic distortion at the busbar common coupling into the existing electrical grid.

Among the various control strategies, finite control-set model predictive control strategy has emerged as a simple and promising digital control tool for electric power conversion systems. The predictive control is a nonlinear control method and provides an approach that is better suited for controlling power converters in wind energy systems. Chapter "Power Conversion and Predictive Control of Wind Energy Conversion Systems" presents power conversion systems and predictive control strategies for variable-speed wind energy systems. Various forms of predictive control techniques such as predictive current control, predictive torque control, and predictive power control are discussed considering variable-speed wind energy systems as case studies. The predictive control strategies fulfill various control requirements such as maximum power point tracking, regulation of DC link voltage, grid synchronization, generation of reactive power to three-phase grid, and fault ride-through operation.

Chapter "Adaptive Guaranteed Performance Control of Wind Energy Systems" discusses an adaptive guaranteed performance controller for wind energy conversion system equipped with a doubly fed induction generator. The proposed controller consists of outer loop control concerning the aero turbine mechanical subsystem, and inner loop control concerning the electrical subsystem. The proposed technique is capable of quantifying and further guaranteeing the system performance on both transient and steady-state stages with the help of error transformation techniques. The stability is guaranteed through standard Lyapunov synthesis.

Chapter "Machine Learning and Meta-heuristic Algorithms for Renewable Energy: A Systematic Review" presents a detailed review on the application of machine learning and meta-heuristic optimization algorithms in renewable energy. The chapter discusses artificial neural networks, back-propagation neural networks, fuzzy logic, adaptive neuro-fuzzy inference systems, genetic algorithms, swarm intelligence algorithms (including cuckoo search, artificial bee colony, and particle swam optimization algorithms), and their application in wind energy systems.

Chapter "Design of a Supervisory Control System Based on Fuzzy Logic for a Hybrid System Comprising Wind Power, Battery and Ultracapacitor Energy Storage System" presents a control strategy for the coordinated operation of a wind power generator, and battery/ultracapacitor. The proposed control scheme is based on the use of fuzzy logic to monitor the state of charge of the storage systems, while defining their power references to comply with an imposed grid demand. The control strategy is evaluated through simulation under different operating conditions, proving a satisfactory regulation of the monitored parameters and an adequate supply of the grid requirements.

Chapter "Neural-Based P-Q Decoupled Control for Doubly Fed Induction Generator in Wind Generation System" introduces the robust decoupled control of active and reactive powers of a wind-driven doubly fed induction generator using artificial neural network under fault conditions and varying wind speed conditions. The power estimators based on neural networks are trained by back-propagation method, and they are divided into five subnets, namely real and reactive power measurement, reference active and reactive computation, reference stator current computation, reference rotor current computation, and reference rotor voltage computation.

Chapter "An Indirect Adaptive Control Paradigm for Wind Generation Systems" describes an indirect adaptive wavelet-based control to acquire maximum power from variable-speed wind turbine. The new developed controller maintains its self-adaptive behavior under uncertainties generating from various load disturbances and wind speed variation. The proposed technique is better in terms of efficiency, output power, and steady-state characteristics as compared to the existing state-of-the-art.

Timişoara, Romania Radu-Emil Precup
Serdivan, Turkey Tariq Kamal
Chongqing, China Syed Zulqadar Hassan
December 2018

Acknowledgements

We would like to express our sincere thanks to all authors and reviewers who have contributed directly and indirectly to this book. Special thanks to Prof. Luis Fernández Ramírez for his advice and help during the writing of this book. Finally, thanks to executive editor Dr. Christoph Baumann and project coordinator Mr. Ramamoorthy Rajangam for their great efforts and support during the implementation of this project.

Contents

Editorial Advisory Board

List of Reviewers

Nonlinear Modeling, Analysis and Simulation of Wind Turbine Control System With and Without Pitch Control as in Industry

Sameh A. Eisa

Abstract This chapter introduces the state-of-the-art modeling, analysis and simulation of the wind turbine dynamics and control. The modeling part is a comprehensive time domain layout of the model currently considered by industry, such as General Electric, National Renewable Energy Lab and other major manufacturers. The time domain modeling allows for nonlinear and optimization studies for the highly nonlinear and complex wind turbine control system. Also, this allows for better understanding and intensive study of the very important Pitch control, which is crucial in wind turbine systems, for building/designing control strategies and for optimization objectives. This chapter also provides a documentation for what have been published recently (2016–2018) regarding important dynamical properties and parameter sensitivities in the wind turbine control system. In this regard, the chapter also provides a possible reduction to the wind turbine control system based on the range of wind speeds the wind turbine is exposed to. This allows scholar to study the wind turbine dynamics and control in three different regions, one of them has the Pitch control activated in the case of higher wind speeds. Moreover, the chapter provides an illustration of the dynamical stability and the possibility of approximating the wind turbine control system by multiple time scales. Additionally, The chapter provides different simulations of the system, which can be helpful for academic studies that intend to run non-autonomous scenarios. Also, we cite in a recently (2018) published work, a data validation for the model versus real measured data of the power-wind curve, which magnify the findings of our study.

S. A. Eisa (✉)
Mechanical and Aerospace Engineering Department,
University of California, Irvine, USA
e-mail: seisa@uci.edu; sameheisa235@hotmail.com

R.-E. Precup et al. (eds.), *Advanced Control and Optimization Paradigms for Wind Energy Systems*, Power Systems, https://doi.org/10.1007/978-981-13-5995-8_1

List of Symbols

P_{wind}	wind power in the airstreams
ρ, A_r, v_{wind}	air density, rotor area (m^2), wind speed (m/s)
C_p, P_{mech}	aerodynamic power coefficient, power extracted by the turbine
w_{ref}, P_{elec}	rotor reference speed, electrical (active) power delivered to the grid
V	the magnitude of the terminal voltage
R, X, E	infinite bus parameters: resistance, reactance, infinite bus voltage
Q_{gen}	total reactive power delivered to the grid
H, H_g	turbine and generator inertia constants
w_0, w_{base}	initial speed, base angular frequency
D_{tg}, K_{tg}	shaft damping and stiffness constants
f_1, f_2	integrals of differences of speeds and powers
P_{stl}, K_{pp}	rated power and Pitch control proportional
K_{ip}, K_{pc}	integral gain and Pitch compensation proportional
θ, K_{ic}	Pitch angle and integral gain
p_{inp}, T_{pc}	power order (subject to modifications) and its time constant
K_{ptrq}, K_{itrq}	torque control proportional and gain
P_{1elec}, T_{pwr}	filtered electrical power and its time constant
V_{ref}, K_{Ql}	reference voltage and its gain
E_{qcmd}, K_{vi}	reactive voltage command and terminal voltage control gain
Q_{droop}, T_{lpqd}	the droop function and its time constant
Q_{inp}	the input to the droop function block
V_{1reg}, T_r	filtered supervisory voltage and its time constant
V_{reg}, T_r	supervisory voltage and its time constant
$Q_{wvl}, Q_{wvu}, K_{pv}, K_{iv}$	two integrals lead to Q_{ord} and their gains

1 Brief Introduction

Humanity future is depending much on advancement and development of renewable energies. There are many reasons of why we need to expand our energy systems. This is due to economic justifications and environmental concerns. No matter what the reasons are, we require additional understanding of the generation of renewable energies if we are to fully utilize them.

Based on the US department of energy reported [1], wind energy is the fastest growing source of renewable energies. Consequently, we need more studies and research and to fully comprehend the dynamics and behavior of Wind Turbine Generators (WTGs) if we are to gain the most from this valuable resource. Both corporations and governments are highly interested in understanding the challenges of integrating

WTGs with other conventional power systems. Because of the complexities involved in the WTGs implementation, researching control systems, optimization, energy storage, and power generation of WTGs has dramatically increased recently. In this regard, this chapter is intended to provide a state-of-the-art comprehensive modeling effort that should guide scholars working in the research areas mentioned earlier in this paragraph.

The provided modeling effort in this chapter is a summary for the state-of-the-art nonlinear modeling of WTGs control system dynamics. The industry publications, namely General Electric (GE) ones [2, 3], have been intensively investigated in the last two years through the publications [4–12]. These studies converted the model found in GE reports into nonlinear system of differential-algebraic equations, followed by a wide range of analysis and simulation results. The resultant time domain nonlinear model can be reduced based on the wind speed v_{wind} range the WTG is exposed to. This important possibility of reduction to the model, will be covered and presented collectively in Sect. 2. Also, we will summarize some of the most recent and important observations these studies have concluded about the WTG system, such as parameter sensitivity, stability and different time scale structure found in the WTG system. In Sect. 3, the Pitch control and its significance will be presented. Additionally, some non-autonomous simulations for the given model under Pitch control, is provided. In the same section, we will provide a Simulink verification of the model and how it compares to National Renewable Energy Lab [13, 14]. In this regard, it is important to mention that our modeling intensive study recognized some other modeling sources such as [15–18]. Also, at the end of this chapter, we will provide and discuss a real data validation for the power-wind curve of our model. These verifications and validations are a supportive evidence that the modeling effort presented in this chapter is reliable. This is essential in any optimization or control study. The reader is recommended to check the Ph.D. dissertation [19] for more detailed information about the topics covered in this chapter.

2 State-of-the-Art Nonlinear Modeling of WTGs

In this section, we provide a mathematical model that is in time domain (can be solved by stiff differential equations solvers such as ODE15s in Matlab). This full scale modeling allow for better and more in-depth control studies. This is especially true because the WTG system is highly nonlinear. Also, a system/model formulated in time domain, usually provide better framework for non-autonomous simulations, keeping in mind that non-autonomous simulations are more practical to present extreme scenarios. We start by explaining the different controls in WTGs and translate them into differential equations. Then, we provide tables that summarize and collect the parameters, C_p coefficients, and limiters (control limits) needed for the model. Also, we give a method to eliminate the algebraic equation resulting from the network equation. This results in a system of differential equations instead of a system of differential-algebraic equations, which allows for simpler implementation in numerical solvers.

The main references used while constructing the model are [2–6, 18]. In [2], the control blocks are consistent of the wind power extraction block, one/two mass block, Pitch compensation control block, and reactive power block (power factor and supervisory voltage cases). In [3], C_p curves are provided and explained. The GE team suggested an extra two optional blocks to, possibly, be added (active power and inertia blocks). The GE team in [3] introduced the so called Q Droop function, which has been intensively studied in [6] and fully analyzed in [11]. The study [18] introduced their model effort citing [20] and GE studies. The reader may ask a legitimate question: Why and how GE models relate to other WTGs? In another words, how building the model is inclusive to the-state-of-the-art modeling efforts if it follows heavily GE modeling? These questions were answered by detail in [4–11]. The answers though can be grouped in the two points below:

1. The GE team made the case in their reports [2, 3] that their model can be used to represent WTG models for other manufacturers/companies. As a matter of fact, they have provided many validation results, as can be found in [21].
2. In [8], it is shown that the GE modeling is equivalent to the NREL [13] if we fix the parameters. The Simulink projects used for this comparison are also given in Sect. 3.4. Additionally, we provide in Sect. 3.4 a discussion regarding the data validation for the proposed model (uses intensively GE) versus the model of [18, 20].

2.1 Main Outline of the Model

- Wind power model: Using basic physics, the wind power in the air streams is given by $P_{wind} = \frac{1}{2}\rho A_r v_{wind}^3$ Per Unit (pu), see [3]. This block models how a WTG extracts power from the air and with what efficiency. The model's main purpose is to introduce the C_p curves such that the power extracted by the WTG is $P_{mech} = \frac{1}{2}C_p\rho A_r v_{wind}^3$. As discussed in the introduction, and as in [22], the ideal C_p is the Betz limit which is approximately 0.59. No WTG can extract more than the Betz limit of the power available in the air-streams. C_p curves of the three bladed wind turbine (type-3) are better other types for some tip ratios (Fig. 1).
- Rotor model: This model represents the dynamics of the generator and turbine speeds due to the electrical and mechanical torques. The two-mass model has been introduced in [2, 3, 18] while in [23] this block was represented by a single-mass rotor. It can be noticed that GE studies [2, 3] hinted that single mass rotor may be used for simplification. Later (in Sect. 2.2.1) we will mention the representative differential equations for both models. Figure 2 shows the transfer function for this block as in [3].
- Reference speed: This block models how the reference speed is calculated. The reference speed dynamics are dependent on the generated electric power such that

at steady state $w_{ref} = f(P_{elec})$. GE studies [2, 3] mentioned that the reference speed should increase slowly with the generated electric power until it reaches the rated speed. This speed is essential to control the generator and turbine speeds. There is a difference between [2, 3, 18] regarding the transfer function of the reference speed. Later (in Sect. 3.4) we will discuss this difference in more detail.

- Pitch control and compensation: This block captures the dynamics of the Pitch. This has been a growing area of research. This control calculates the Pitch angle based on the differences between the rated power and the power order, and between the reference speed and the generator speed. The Pitch angle has direct effect on power extraction efficiency. This is an important control to keep the WTG producing the rated power for a higher range of wind speeds. Figure 3 shows the transfer function for this block as in [3].
- Reactive power control: This control manages the generated reactive power from the WTG. This control can be in the power factor setup or the supervisory voltage setup. The first case occurs when the WTG is treated as one unit by itself, while the second case occurs when the WTG is treated as one unit in a compound of units. These two cases were introduced in [2, 3, 18]. Figure 4 shows the transfer function for this block as in [3].
- Electrical control: Unlike the previous block where the control was for the reactive branch that feeds the generator, the electrical control shows how the active current can be generated and controlled. This block is the same across the references [2, 3, 18] that covered it. Figure 5 shows the transfer function for this block as in [3].
- Active power and inertia controls: Usually these controls are not activated. The function of these two controls is to manage the power order produced by the WTG. This management depends on and corresponds to changes in bus frequency. The two controls provide extra power in the case there is lower than normal bus frequency (reference frequency) and vise versa. The active power control provide extra power by setting up the maximum rated power and cutting out, if needed, the available power to the WTG. On the other hand, the inertia control does the same function, but by providing extra power from the rotor inertia. GE [3] has hinted that most current WTGs have yet to implement these controls as of 2010. Figures 6 and 7 show the blocks as in [3].
- Converter/Generator model: This is the step where the output the WTG is delivered to the power grid. Two branches are considered in this model, active and reactive ones, which deliver the active and the reactive power to the grid respectively. In [2, 3] this model is very similar, with some lower and upper limit differences for the controls, however, in [18] we see that a third branch is added to the model for the phase shift convergence between the resultant components (current and voltage) of the wind turbine and the grid. For more detail about how this difference is insignificant when we have stability, the reader is recommended to read about the convergence between the models in [19]. Figures 8 and 9 show the generator model as in [2, 18] respectively (Fig. 10).
- Terminal voltage and grid model: The terminal voltage is the connection between the converter/generator model and the grid model. In the models we follow, the wind turbine is connected to the grid in order to work. This implies that even for

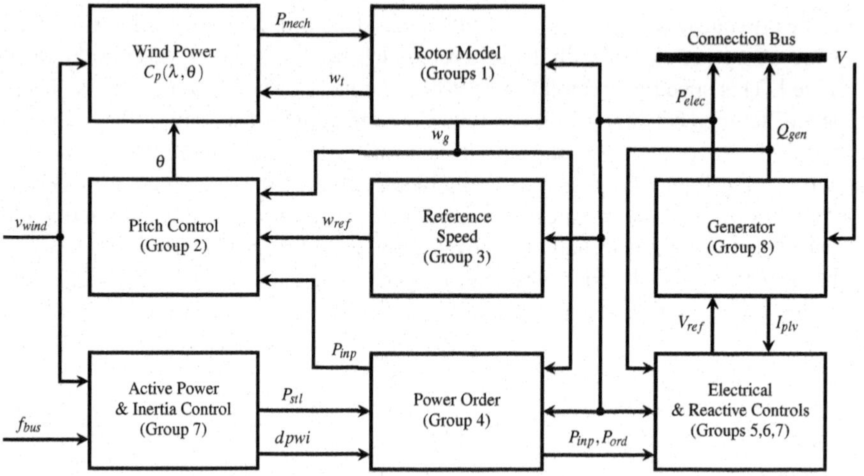

Fig. 1 WTG control blocks and dynamics

theoretical/mathematical studies, the grid should be modeled so we can have an algebraic equation (the network equation) from Kirchhoff's law, that relates the dynamics of the WTG to the grid. In our study, we follow the model used in [18] and suggested in [2, 3] to represent the grid by an infinite bus model, see Fig. 11. Therefore, the terminal voltage will be given by the following equation as in [18]:

$$(V^2)^2 - [2(P_{elec}R + Q_{gen}X) + E^2]V^2 + (R^2 + X^2)(P_{elec}^2 + Q_{gen}^2) = 0 \quad (1)$$

Note that, if the grid model changes to another model other than the infinite bus, a new algebraic constraint will need to be derived and analyzed. Without this part of the grid modeling, the wind turbine is working without load and has undefined inputs to some of the control dynamics. Figure 12 gives the transfer functions of the WTG as in [3].

2.2 *Characteristics and Dynamical Analysis*

2.2.1 Translating the Blocks of Transfer Functions and Controls into a System of Differential Algebraic Equations

Having first reviewed the transfer functions and control blocks in Sect. 2.1, we now begin the process of breaking down the blocks (in every Fig.) into algebraic relations in the transfer function domain. This will be done by deriving the transfer function relations after specifying nodes of variables.

Group 1: Two mass model as in Fig. 2.

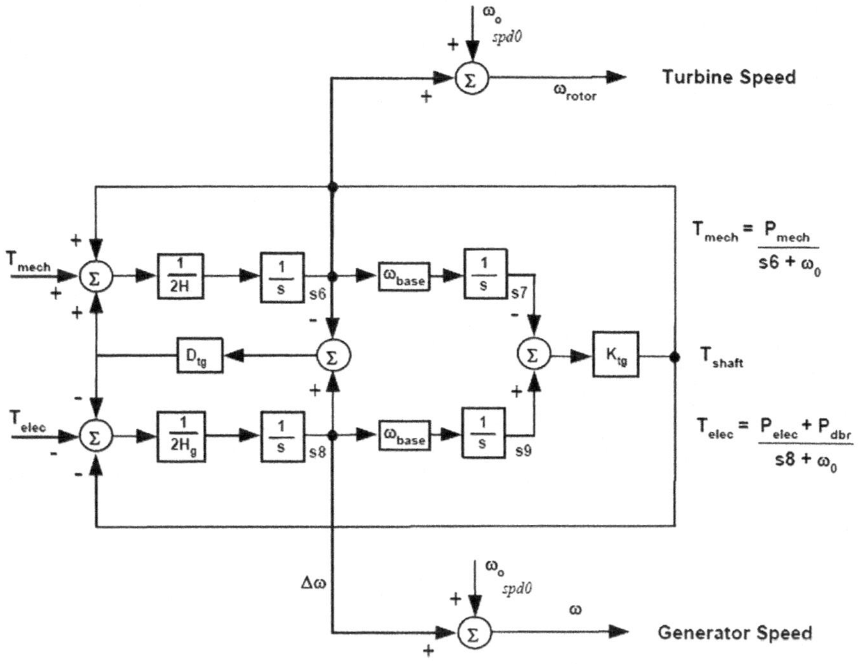

Fig. 2 Two mass model of a WTG as in [3]

In Fig. 2, we let the nodes $s_6 = w_g$ and $s_8 = w_t$, so the turbine speed will be the sum of w_0 and the node s_6. Therefore, $w_{turbine} = w_{rotor} = w_t + w_0$ and similarly the generator speed $w_{generator} = w = w_g + w_0$. Also we let $\Delta\theta_m = s_9 - s_7$, so $T_{shaft} = K_{tg}\Delta\theta_m$. Thus w_t is given by,

$$w_t = \frac{1}{2H} \cdot \frac{1}{s}[T_{mech} + D_{tg}(w_g - w_t) + T_{shaft}]. \tag{2}$$

Similar to Eq. (2) we get,

$$w_g = \frac{1}{2H_g} \cdot \frac{1}{s}[-T_{elec} - D_{tg}(w_g - w_t) - T_{shaft}] \tag{3}$$

and,

$$\Delta\theta_m = \frac{w_{base}}{s}(w_g - w_t). \tag{4}$$

The above equations contain the dynamics of the two mass rotor model.
Group 2: Pitch control as in Fig. 3.

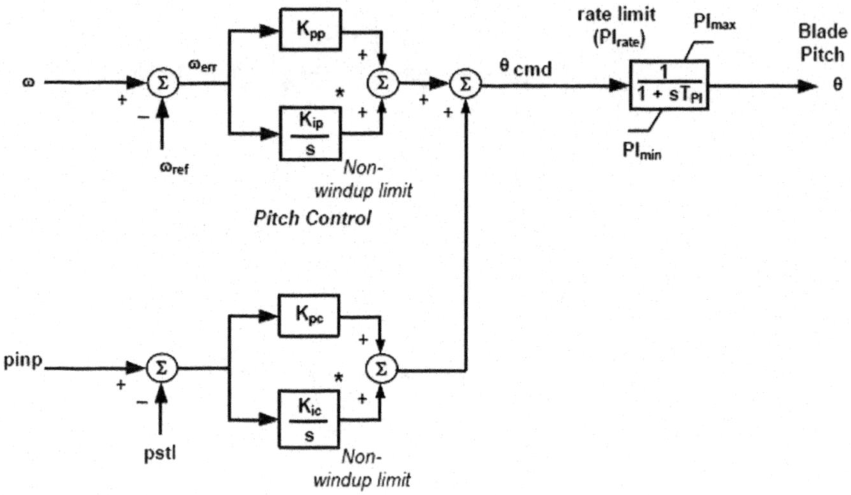

Fig. 3 Pitch control model of a WTG as in [3]

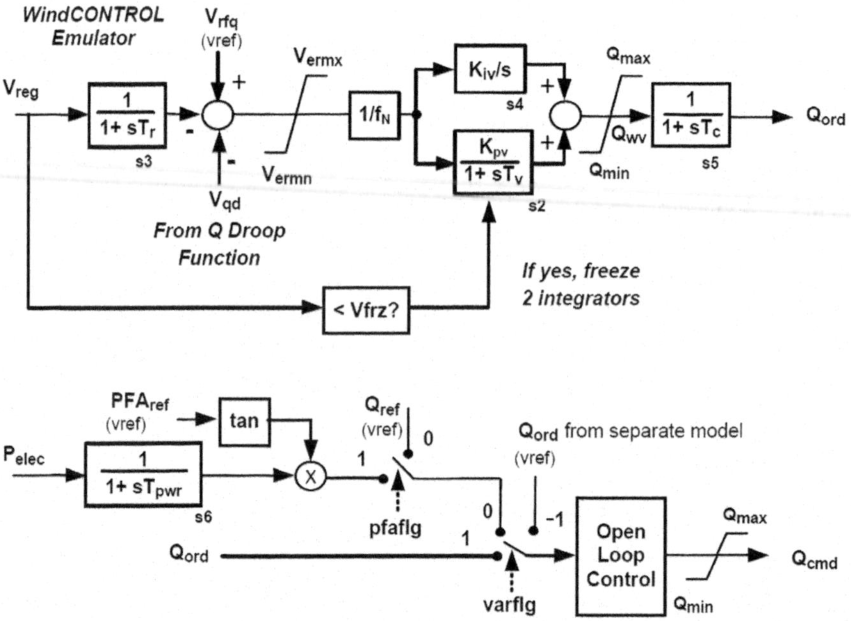

Fig. 4 Reactive power control of a WTG as in [3]

In Fig. 3, we start with the two integrators (branches that have $\frac{1}{s}$). We let f_1 be the output of the transfer function $\frac{K_{ic}}{s}$ and we let f_2 be the output of the transfer function $\frac{K_{ip}}{s}$. Thus,

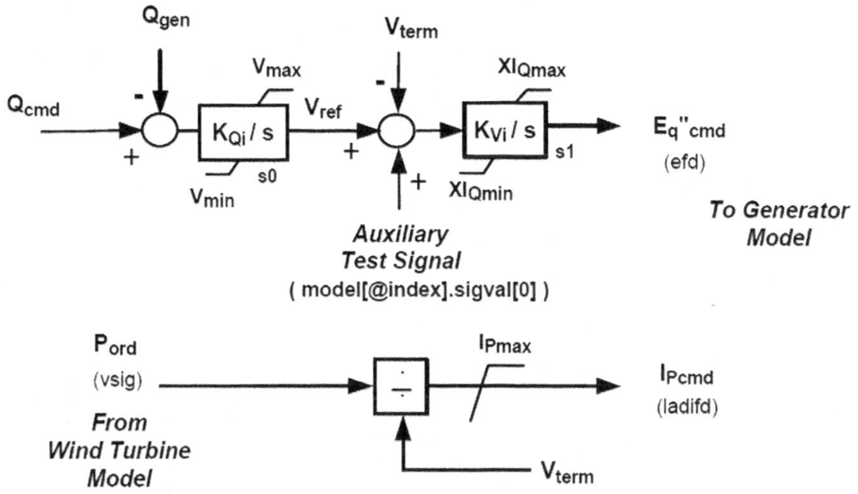

Fig. 5 Electrical control of a WTG as in [3]

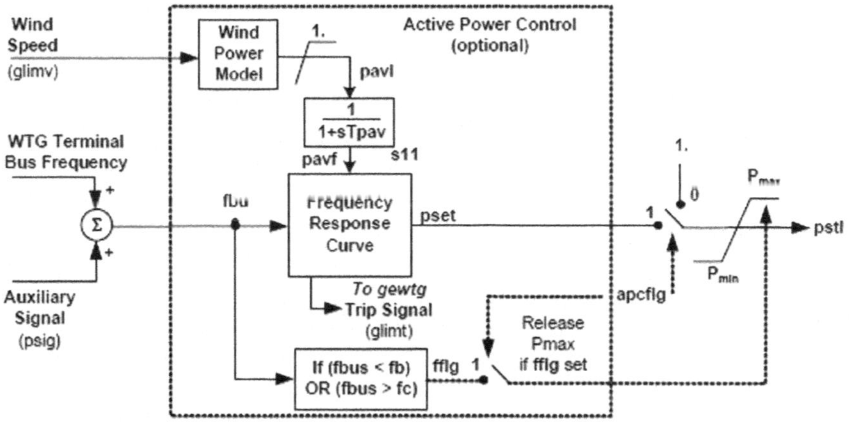

Fig. 6 Active power control of a WTG as in [3]

$$f_1 = \frac{(w - w_{ref})}{s} = \frac{(w_g + w_0 - w_{ref})}{s} \tag{5}$$

and,

$$f_2 = \frac{(P_{inp} - P_{stl})}{s}. \tag{6}$$

The Pitch angle command (θ_{cmd}) is the node after summing the upper and the lower outputs of the Pitch control. Also, it is the node before the transfer function of T_{pl}. Thus θ_{cmd} is given by,

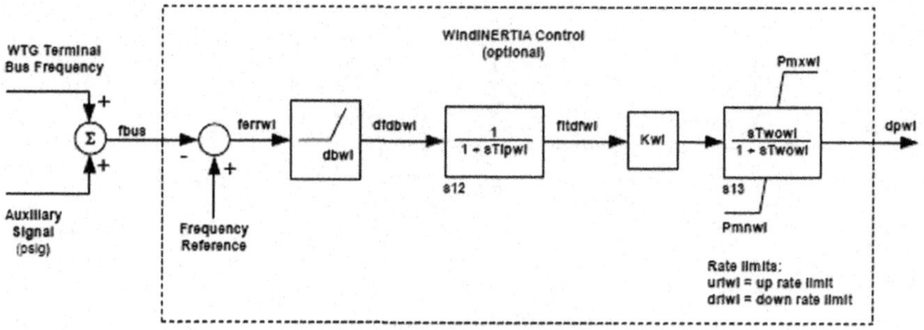

Fig. 7 Inertia control of a WTG as in [3]

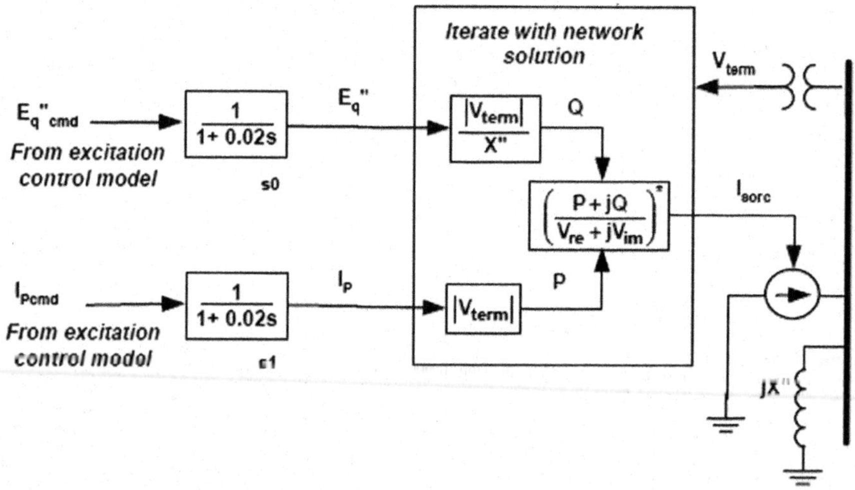

Fig. 8 Converter/Generator model of a DFAG/DFIG WTG as in [2]

$$\theta_{cmd} = K_{pp}(w_g + w_0 - w_{ref}) + K_{ip}f_1 + K_{pc}(P_{inp} - P_{stl}) + K_{ic}f_2. \tag{7}$$

The Pitch angle (θ) is the output of the transfer function of T_{pl}, which has θ_{cmd} as an input. Thus θ is given by,

$$\theta = \theta_{cmd}\frac{1}{1 + s \cdot T_{pl}}. \tag{8}$$

After algebraic re-arrangement we get,

$$\theta = \frac{K_{pp}(w_g + w_0 - w_{ref}) + f_1 + K_{pc}(P_{inp} - P_{stl}) + f_2}{1 + s \cdot T_{pl}}. \tag{9}$$

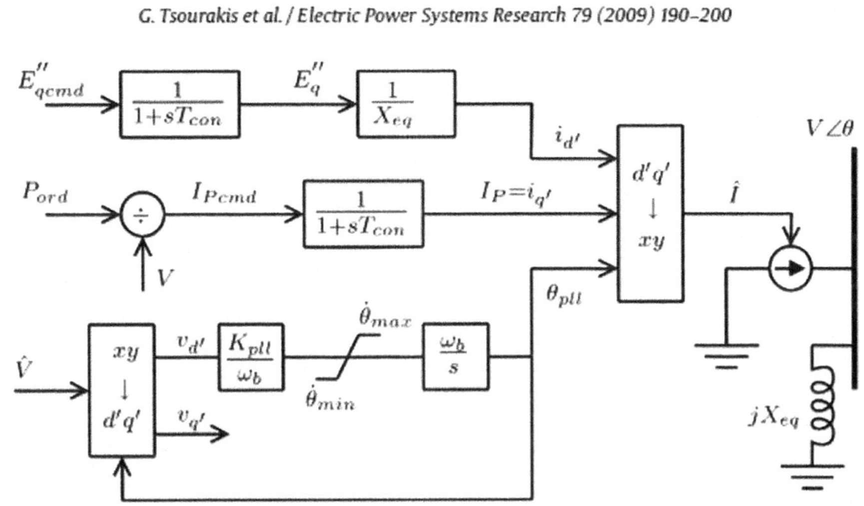

Fig. 9 Converter/Generator model of a DFAG/DFIG WTG as in [18]

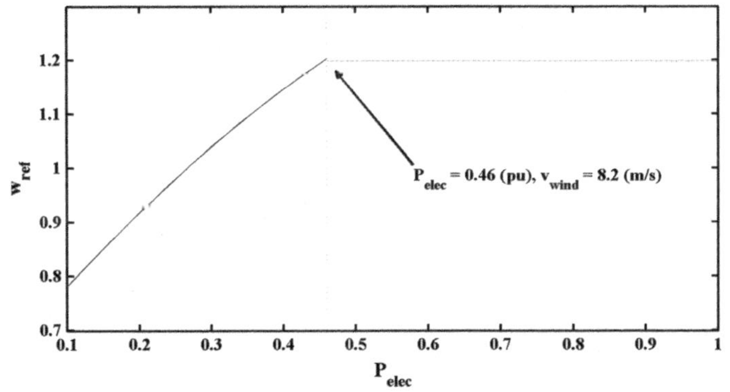

Fig. 10 w_{ref} steady state as a function of P_{elec} as in [4]

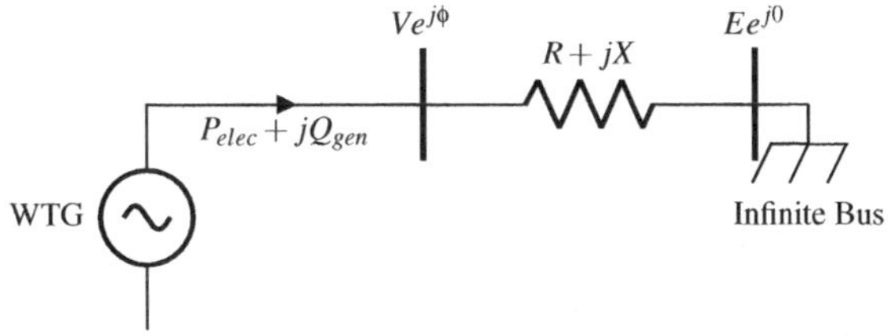

Fig. 11 Single machine infinite bus test system as in [4]

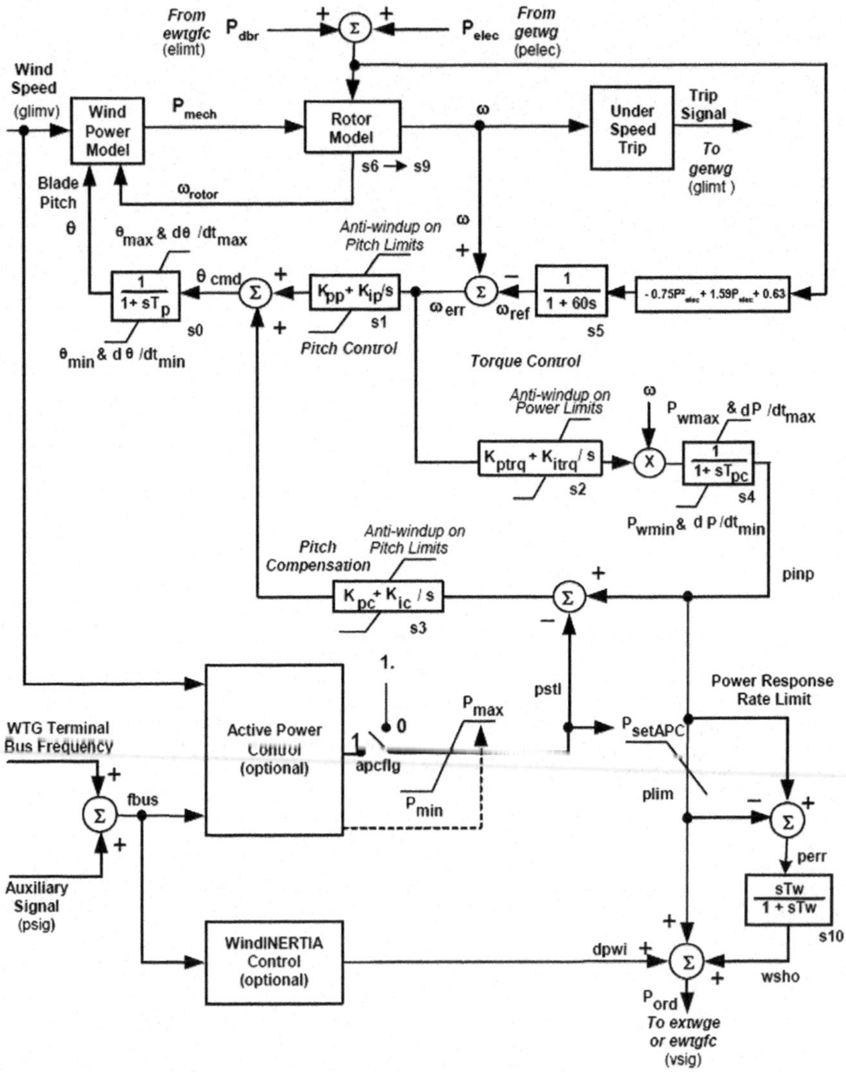

Fig. 12 All of the WTG model transfer functions and controls as in [3]

Equations (5), (6), and (9) contain the dynamics of the Pitch control.

Group 3: Reference speed as in Fig. 12.
The reference speed w_{ref} is the output of the transfer function ($\frac{1}{1+s\cdot 60}$), which has the node symbol s_5 (at the upper part of Fig. 12). The input for this transfer function is $-0.75P_{elec}^2 + 1.59P_{elec} + 0.63$. Thus w_{ref} is given by,

$$w_{ref} = \frac{-0.75P_{elec}^2 + 1.59P_{elec} + 0.63}{1 + s \cdot 60}. \tag{10}$$

Equation (10) represent the dynamics of w_{ref}.

Group 4: Power order as in Fig. 12.
The main power order P_{inp} is the output of the transfer function of T_{pc}, which has the node symbol s_4 (in the middle of Fig. 12). The input for This transfer function is the multiplication of w and the output of the transfer function that has the node symbol s_2. With f_1 in Eq. (5) and $w = w_g + w_0$, P_{inp} is given by,

$$P_{inp} = \frac{(w_g + w_0)(K_{ptrq} + K_{itrq}f_1)}{1 + s \cdot T_{pc}} \tag{11}$$

and w_{sho} (the output of the transfer function of T_w with the node symbol s_{10}) is given by,

$$w_{sho} = \frac{(P_{inp} - P_{lim}) \cdot s \cdot T_w}{1 + s \cdot T_w}. \tag{12}$$

As shown at the sum after the node s_{10} in Fig. 12, the final power order is given by,

$$P_{ord} = P_{lim} + w_{sho} + dpwi. \tag{13}$$

Group 5: Reactive power in power factor setup case and electrical controls as in Figs. 4 and 5.
Since we consider the reactive power control operating in power factor case, then the lower part in Fig. 4 is operating. We let the output of the transfer function of T_{pwr} be P_{1elec} (see Fig. 4). Thus P_{1elec} is given by,

$$P_{1elec} = \frac{P_{elec}}{1 + s \cdot T_{pwr}}. \tag{14}$$

Q_{cmd} is the output of the multiplier in Fig. 4. Therefore,

$$Q_{cmd} = P_{1elec} \cdot \tan(PFA_{ref}). \tag{15}$$

In the electrical control (Fig. 5), V_{ref} is the output of the transfer function of K_{Qi} (upper part of Fig. 5). This transfer function has $Q_{cmd} - Q_{gen}$ as an input. Thus V_{ref} is given by,

$$V_{ref} = \frac{K_{Qi}(Q_{cmd} - Q_{gen})}{s}. \tag{16}$$

Equations (14) and (16) contain the dynamics of group 5.

Group 6: Reactive power in supervisory voltage setup case and electrical controls as in Figs. 4 and 5.
The Q_{droop} function as shown in Fig. 12 in [3] is given by,

$$Q_{droop} = \frac{(Q_{inpt} - Q_{droop})}{1 + s \cdot T_{lpdq}}. \tag{17}$$

Since we consider the reactive power control operating in supervisory voltage case, then the upper part in Fig. 4 is operating. We let V_{1reg} be the output of the transfer function of T_r, which has the node symbol s_3 (see Fig. 4). This transfer function has V_{reg} as an input. Thus V_{reg1} is given by,

$$V_{1reg} = \frac{(V_{reg} - V_{1reg})}{1 + s \cdot T_r}. \tag{18}$$

In Fig. 4, we let $f_n = 1$ or included in the gains (see page 4.7 in [3], second paragraph). We let the outputs of the transfer functions of K_{pv} and K_{iv} be Q_{wvl} and Q_{wvu} respectively. The input for those two transfer functions is $V_{ref} - V_{1reg} - V_{qd}$ (see Fig. 4) Thus Q_{wvl} and Q_{wvu} are given by,

$$Q_{wvl} = \frac{K_{pv}(V_{rfq} - V_{1reg} - V_{qd})}{1 + s \cdot T_v} \tag{19}$$

and,

$$Q_{wvu} = \frac{K_{iv}(V_{rfq} - V_{1reg} - V_{qd})}{s}. \tag{20}$$

As shown in Fig. 4, Q_{wv} is given by,

$$Q_{wv} = Q_{wvl} + Q_{wvu}. \tag{21}$$

The output of the transfer function of T_c is Q_{ord} (see Fig. 4). The input for this transfer function is Q_{wv}. Thus Q_{ord} is given by,

$$Q_{ord} = \frac{Q_{wv}}{1 + s \cdot T_c}. \tag{22}$$

Since the reactive power control is operating in supervisory voltage case, then $Q_{cmd} = Q_{ord}$ from Eq. (22). Equation (16) holds for this group 6 as a representative for the electrical control. Equations (17)–(20), (22), and (16) contain the dynamics of group 6.

Group 7: Active power and inertia controls as in Figs. 6 and 7.
For the active power control (Fig. 6), P_{avf} is the output of the transfer function of T_{pav}. This transfer function has P_{avl} as an input. Thus P_{avf} is given by,

$$P_{avf} = \frac{P_{avl}}{1 + s \cdot T_{pav}}. \tag{23}$$

For the inertia control (Fig. 7), *fltdf wi* is the output of the transfer function of T_{lpwi}, which has the node symbol s_{12}. This transfer function has *dfdbwi* as an input. Thus *dfdbwi* is given by,

$$fltdf\,wi = \frac{dfdbwi}{1 + s \cdot T_{lpwi}}. \tag{24}$$

The final output of the inertia control (*dpwi*) is the output of the transfer function of T_{wowi}, which has the node symbol s_{13}. This transfer function has *fltdf wi* multiplied by the gain K_{wl} as an input. Thus *dpwi* is given by,

$$dpwi = \frac{K_{wl} \cdot dpwi \cdot s \cdot T_{wowi}}{1 + s \cdot T_{wowi}}. \tag{25}$$

Equations (23)–(25) contain the dynamics of group 7.

Group 8: DFAG generator/converter as in Fig. 8.
In order to have equations for E_q and I_{plv} (outputs of the transfer functions with the node symbols s_0 and s_1 respectively), we need to relate E_{qcmd} and I_{pcmd} (inputs of the transfer functions with the node symbols s_0 and s_1 respectively) to other variables we have that represent the dynamics in other controls. Looking at the electric control (Fig. 5), we notice that E_{qcmd} is the output of the transfer function of K_{iv}, which has the node symbol s_1. Similarly, in the lower part of Fig. 5, we find I_{pcmd} as the output of the divider ($\frac{P_{ord}}{V}$). Thus E_{qcmd} is given by,

$$E_{qcmd} = \frac{K_{vi}(V_{ref} - V)}{s} \tag{26}$$

and,

$$E_q = \frac{E_{qcmd}}{1 + s \cdot 0.02}. \tag{27}$$

We have $I_{pcmd} = \frac{P_{ord}}{I_{plv}}$, then I_{plv} is given by,

$$I_{plv} = \frac{\frac{P_{ord}}{I_{plv}}}{1 + s \cdot 0.02}. \tag{28}$$

We note that I_p in [2] is equivalent to I_{plv} (the symbol used in this document) in [3]. Equations (26)–(28) contain the dynamics of the generator.

After applying inverse Laplace transform to the equations above, we derive a system of differential equations as follows:

Group 1: Two-mass rotor.

$$\frac{dw_g}{dt} = \frac{1}{2H_g}\left[-\frac{P_{elec}}{w_g + w_0} - D_{tg}(w_g - w_t) - K_{tg}\Delta\theta_m\right]. \tag{29}$$

$$\frac{dw_t}{dt} = \frac{1}{2H}\left[\frac{P_{mech}}{w_t + w_0} + D_{tg}(w_g - w_t) + K_{tg}\Delta\theta_m\right]. \tag{30}$$

$$\frac{d(\Delta\theta_m)}{dt} = w_{base}(w_g - w_t). \tag{31}$$

As discussed when we were introducing the different controls, a one mass model may be used to replace the two-mass model in group 1. The one mass differential equation was given in [24]:

$$\frac{dw}{dt} = \frac{1}{Hw_{base}}[P_{mech} - P_{elec}].$$

The following relations hold:

$$P_{mech} = \frac{1}{2}C_p(\lambda,\theta)\rho A_r v_{wind}^3 = \frac{1}{2}\left(\sum_{i=0}^{4}\sum_{j=0}^{4}\alpha_{i,j}\theta^i\lambda^j\right)\rho A_r v_{wind}^3$$

and,

$$P_{elec} = VI_{plv}.$$

Group 2: Pitch control.

$$\frac{df_1}{dt} = w_g + w_0 - w_{ref}. \tag{32}$$

$$\frac{df_2}{dt} = P_{inp} - P_{stl}. \tag{33}$$

$$\frac{d\theta}{dt} = \frac{1}{T_p}[K_{pp}(w_g + w_0 - w_{ref}) + K_{ip}f_1 + K_{pc}(P_{inp} - P_{stl}) + K_{ic}f_2 - \theta)]. \tag{34}$$

Group 3: Reference speed.

$$\frac{dw_{ref}}{dt} = \frac{1}{60}[-0.75P_{elec}^2 + 1.59P_{elec} + 0.63 - w_{ref}]. \tag{35}$$

Group 4: Power order.

$$\frac{dP_{inp}}{dt} = \frac{1}{T_{pc}}[(w_g + w_0)(K_{ptrq}(w_g + w_0 - w_{ref}) + K_{itrq}f_1) - P_{inp}]. \tag{36}$$

$$\frac{dw_{sho}}{dt} = \frac{dP_{inp}}{dt} - \frac{dP_{stl}}{dt} - \frac{1}{T_w}w_{sho}. \tag{37}$$

Group 5: Reactive power control in the power factor setup case.

$$\frac{dP_{1elec}}{dt} = \frac{1}{T_{pwr}}[P_{elec} - P_{1elec}]. \tag{38}$$

$$\frac{dV_{ref}}{dt} = K_{Qi}[Q_{cmd} - Q_{gen}] \tag{39}$$

where,

$$Q_{gen} = \frac{V(E_q - V)}{X_{eq}}.$$

Q_{cmd} is explained in detail in Sect. 2.2.2.

Group 6: Reactive power control in the supervisory voltage setup case.

$$\frac{dQ_{droop}}{dt} = \frac{1}{T_{lpqd}}[Q_{inpt} - Q_{droop}]. \tag{40}$$

$$\frac{dV_{1reg}}{dt} = \frac{1}{T_r}[V_{reg} - V_{1reg}]. \tag{41}$$

$$\frac{dQ_{wvl}}{dt} = \frac{1}{T_v}[K_{pv}(V_{ref} - V_{1reg} - V_{qd}) - Q_{wvl}]. \tag{42}$$

$$\frac{dQ_{wvu}}{dt} = K_{iv}(V_{ref} - V_{1reg} - V_{qd}). \tag{43}$$

$$\frac{dQ_{ord}}{dt} = \frac{1}{T_c}(Q_{wvl} + Q_{wvu} - Q_{ord}). \tag{44}$$

Equation (39) is holding in all reactive power groups.

Group 7: Active power control and inertia control.

$$\frac{dP_{avf}}{dt} = \frac{1}{T_{pav}}[P_{avl} - P_{avf}]. \tag{45}$$

$$\frac{d(fltdfwi)}{dt} = \frac{1}{T_{lpwi}}[dfdbwi - fltdfwi]. \tag{46}$$

$$\frac{d(dpwi)}{dt} = \frac{K_{wi}}{T_{lpwi}}[dfdbwi - fltdfwi] - \frac{dpwi}{T_{wowi}}. \tag{47}$$

Group 8: DFAG generator/converter.

$$\frac{dE_{qcmd}}{dt} = K_{vi}[V_{ref} - V]. \tag{48}$$

$$\frac{dE_q}{dt} = \frac{1}{0.02}[E_{qcmd} - E_q]. \tag{49}$$

$$\frac{dI_{plv}}{dt} = \frac{1}{0.02}\left[\frac{P_{ord}}{V} - I_{plv}\right]. \tag{50}$$

Group 9: The algebraic equation resulting from the network (see [18]):

$$0 = (V^2)^2 - [2(P_{elec}R + Q_{gen}X) + E^2]V^2 + (R^2 + X^2)(P_{elec}^2 + Q_{gen}^2). \tag{51}$$

Table 1 represents the model's parameter values as in [3], however the grid parameter values are taken from [18]. Table 2 has the C_p curves' needed coefficients as in [3]. Also, we define the control limits introduced in [3] to be the lower and upper bounds as following in Table 3:

Table 1 The model's parameter value

Parameter	Value
w_0	1 (choice larger than 0)
D_{tg}	1.5 (60 Hz) or 2.3 (50 Hz)
K_{tg}	1.11 (60 Hz, 1.5 MW)
K_{tg}	1.39 (50 Hz, 1.5 MW)
$\frac{1}{2}\rho A_r, K_b$	0.00159 and 56.6 respectively
w_{base}	125.66 (60 Hz) or 157.08 (50 Hz)
$H(two\ mass)$	4.33
$H(one\ mass)$	4.94 (60 Hz), 5.29 (50 Hz)
H_g	0.62 (60 Hz), 0.96 (50 Hz)
K_{pp}, K_{ip}	150, 25 respectively
K_{pc}, K_{ic}	3, 30 respectively
T_p, p_{stl}	0.3, 1 respectively
T_{pc}, K_{ptrq}	0.05, 3 respectively
K_{itrq}, T_w	0.6, 1 respectively
T_{pwr}, K_{Qi}	0.05, 0.1 respectively
T_{lpqd}, T_r	5, 0.02 respectively
T_v, K_{pv}	0.05, 18 respectively
K_{iv}, T_c	5,0.15 respectively
T_{pav}, T_{lpwi}	0.15,1 respectively
K_{wi}, T_{wowi}	10,5.5 respectively
K_{vi}, X_{eq}	40,0.8 respectively
R, E	0.02,1.0164 respectively
$X = X_l + X_{tr}$	$X_l = 0.0243$, $X_{tr} = 0.00557$ respectively

2.2.2 Reduction of the Model

Here we go through a number of possible cases that reduce the system. These reductions are based on the range of wind speeds the WTG is operating on, or on which optional controls, such as active power and inertia controls are deactivated.

Wind Speeds versus Reference Speed: The rated reference speed is $w_{ref} = 1.2$ pu. Physically the WTG can't reach this rated speed with low wind speeds. That is why w_{ref} increases gradually as shown in Eq. (35), until it reaches 1.2 pu. Given the model and the parameter in this study, the rated reference speed (1.2 pu) is reached at $v_{wind} = 8.2$ m/s. Therefore, the differential equation of w_{ref} can be seen as,

$$\frac{dw_{ref}}{dt} = \begin{cases} \frac{1}{60}[-0.75P_{elec}^2 + 1.59P_{elec} + 0.63 - w_{ref}] & v_{wind} < 8.2 \text{ m/s} \\ 0, \; w_{ref}(0) = 1.2 & v_{wind} \geq 8.2 \text{ m/s} \end{cases}$$

Table 2 C_p coefficients $\alpha_{i,j}$

i	j	$\alpha_{i,j}$	i	j	$\alpha_{i,j}$
4	4	4.9686e−10	4	3	−7.1535e−8
4	2	1.6167e−6	4	1	−9.4839e−6
4	0	1.4787e−5	3	4	−8.9194e−8
3	3	5.9924e−6	3	2	−1.0479e−4
3	1	5.7051e−4	3	0	−8.6018e−4
2	4	2.7937e−6	2	3	−1.4855e−4
2	2	2.1495e−3	2	1	−1.0996e−2
2	0	1.5727e−2	–	–	–
1	4	−2.3895e−5	1	3	1.0683e−3
1	2	−1.3934e−2	1	1	6.0405e−2
1	0	−6.7606e−2	0	4	1.1524e−5
0	3	−1.3365e−4	0	2	−1.2406e−2
0	1	2.1808e−1	0	0	−4.1909e−1

Table 3 Control limits to be applied as in [3]

Variable	Lower bound	Upper bound
$V_{1reg} + V_{rfq} - V_{qd}$	$V_{ermn} = -0.1$	$V_{ermx} = 0.1$
Q_{wv}	$Q_{min} = -0.436$	$Q_{max} = 0.436$
Q_{cmd}	$Q_{min} = -0.436$	$Q_{max} = 0.436$
V_{ref}	$V_{min} = 0.9$	$V_{max} = 1.1$
E_{qcmd}	$Xl_{Qmin} = 0.5$	$Xl_{Qmax} = 1.45$
$\frac{P_{ord}}{V}$	$I_{pmin} > 0$	$I_{pmax} = 1.1$
θ	$\theta_{min} > 0$	$\theta_{max} = 27$
P_{inp}	$P_{wmin} = 0.04$	$P_{wmax} 1.12$
P_{avl}	$P_{wmin} = 0.04$	1
$dpwi$	$P_{mnwi} = 0$	$P_{mxwi} = 0.1$
$\frac{dP_{inp}}{dt}$	$dP_{min} = -0.45$	$dP_{max} = 0.45$
$\frac{d\theta}{dt}$	$d\theta_{max} = -10$	$d\theta_{min} = 10$

Therefore based on the above equation, Eq. (35) can be considered as part of the system's dynamics (if $v_{wind} < 8.2$) or eliminated (if $v_{wind} \geq 8.2$) by setting $w_{ref} = 1.2$.

Electric Power versus Pitch Control: Unless mentioned otherwise, the rated electric power generated is 1 pu. The Pitch control gets activated only when the WTG would otherwise generate more power than the rated power. In this case, the Pitch angle increases, so less power is extracted, and the electric power is held at the rated power. When $\theta = 0$ extraction of power is maximized. Therefore the differential equation of θ can be seen as,

$$\frac{d\theta}{dt} = \begin{cases} \frac{1}{T_p}[K_{pp}(w_g + w_0 - w_{ref}) + K_{ip}f_1 + K_{pc}(P_{inp} - P_{stl}) \\ +K_{ic}f_2 - \theta)] & P_{elec} > 1 \text{ pu} \\ 0, \ \theta(0) = 0 & P_{elec} \leq 1 \text{ pu} \end{cases}$$

Therefore, based on the above equation, Eq. (34) can be considered part of the system's dynamics (high wind speeds such that $P_{elec} > 1$) or eliminated by setting $\theta = 0$ (to maximize power extraction for low wind speeds when $P_{elec} \leq 1$). If we set $\theta = 0$, we eliminate f_2 as well in Eq. (33).

Reactive Power Control Q_{cmd} Cases: In the reactive power control, Q_{cmd} is dependent on whether the reactive power control is operating in power factor case or supervisory voltage case. This difference was explained in group 5 and 6 in Sect. 2.2.1. We can summarize that difference in the following relation:

$$Q_{cmd} = \begin{cases} P_{1elec} \cdot \tan(PFA_{ref}) & \text{Power factor case} \\ Q_{ord} & \text{Supervisory voltage case} \\ \text{constant or considered from another model} \end{cases}$$

The nature of the study determines the reactive power setup case (represented by Eqs. (38)–(39) in the power factor case or by Eqs. (40)–(44) and (39) in the supervisory voltage case). As mentioned in [3] it can also be a constant or from a separate model depends on the study and its conditions.

The Power Order P_{ord} Cases: The power order as shown at the sum in the lower part of Fig. 12, has three main parts. Those parts are the regular power order P_{inp}, the effect of the active power control w_{sho}, and the output of the inertia control $dpwi$ (see Eq. (13)). But the active power control and the inertia control can be activated or deactivated. This lead to P_{ord} can be one of the following cases:

$$P_{ord} = \begin{cases} P_{inp} & \text{Active Power and inertia controls (deactivated)} \\ P_{lim} + w_{sho} & \text{Active Power control (activated)} \\ P_{lim} + dpwi & \text{Inertia control (activated)} \\ P_{lim} + w_{sho} + dpwi & \text{Active Power and inertia controls (activated)} \end{cases}$$

WTG Power versus Wind Speed Curve and the Study Cases: Based on the v_{wind} range, the dynamics of the WTG can be divided into regions. Giving the parameter values in Tables 1 and 2, we have the following cases:

- Region 1: Wind speeds between the minimum cut off speed (3 m/s) and 8.2 m/s. Within this range, the Pitch angle θ is fixed at zero in order to extract all possible power from the air, as the rated power is not reached by this range of wind speeds. Also, the rated reference speed, 1.2 pu is reached when $v_{wind} = 8.2$ m/s so, the reference speed should be seen to gradually increases versus the wind speed. Therefore, in this region of dynamics, Eq. (35) is considered, while Eqs. (33) and (34) are eliminated, and we set $\theta = 0$.
  This case can be further modified by taking into account the activated or deactivated

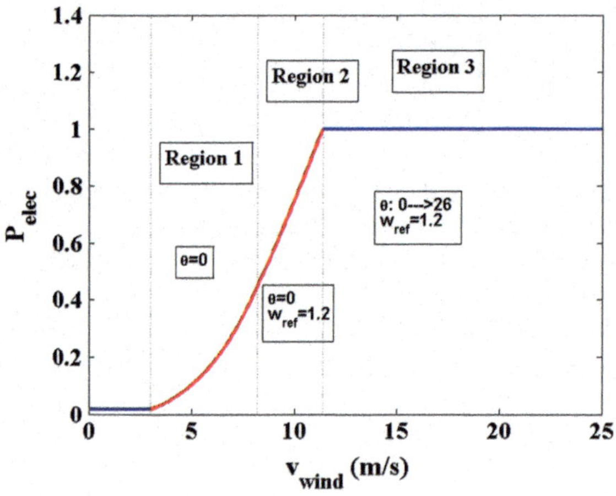

Fig. 13 Power curve profile for the WTG

optional controls (active Power and inertia controls). Also, it can be modified to any of the reactive power cases.

- Region 2: Wind speeds between 8.2 and 11.4 m/s. In this region, the reference speed is at the rated level 1.2 pu, while the power remains below the rated level. The WTG reaches the rated power 1 pu at $v_{wind} = 11.4$ m/s. Therefore, in this region, Eqs. (35), (33), and (34) are eliminated and we set $w_{ref} = 1.2$ and $\theta = 0$. As in the previous case, this case can be further modified.
- Region 3: Wind speeds between 11.4 and 25 m/s (the maximum allowed speed). The dynamics of this region take into consideration Eqs. (33)–(34), while Eq. 35 is eliminated and we set $w_{ref} = 1.2$. Also this case can be further modified as mentioned in the previous cases.

We built a numerical simulator for the dynamical system in the three regions above. The stable steady state of the generated electric power versus wind speeds is as expected for any WTG power curve profile. Figure 13 shows the result of the simulation in the three regions of dynamics, and the power curve profile for the WTG.

2.3 Documented Results and Conclusions About the Model

In this subsection, we present some of the important information and conclusion that have been made about the model derived in Sect. 2.2.1. These points and conclusions are summarized below:

- **Parameters**: The model's parameters can be different based on the sources found in literature. Therefore, scholars are encouraged to determine the conditions in which their model is used. For instance, some few differences can be found between [2, 3, 13, 18, 20]. However, we remark that later in Sect. 3.4 we verified our model versus real data, which suggest credibility for the parameters given in Tables 1 and 2. These parameters are mainly taken from [2, 3] and also some from [18].

- **Stability**: Stability for the model has been significantly studied through eigenvalues as in [5, 25]. It seems however, from [5] that in the case of the power factor set up (see Sect. 2.2.2), only the grid parameters (R and X) can have a large effect on transitioning the system from Stable to Unstable. Figure 14 shows the region where Stability and Instability occurs in the grid parameter space. We note that Fig. 14 has been produced in [5, 8, 10] respectively for the model in all ranges (Regions 1,2 and 3) of wind speeds introduced in Sect. 2.2.2 and Fig. 13. However, it is important to notice that [5] reported the possibility of a Hopf bifurcation for very small value of X. Note that small values of X has been reported by the NREL [26] to also cause the WTG acting funny and break. The interesting part about small value of X that it can lead to a Hopf bifurcation behavior, but also under the control limits given in Table 3 as discussed in [5]. This phenomenon of how allowable oscillations can be allowed by the WTG controls as reported by the NREL [26], has been further investigated in [12] to continue on the work of [5] and use the model provided in this chapter to provide a theoretical explanation for the phenomenon. On the other hand, if the system is in the supervisory voltage reactive power control set up (see Sect. 2.2.2), then it is required to have the Q_{droop} function "activated" to maintain stability (see [6, 11]). As a matter of fact, the Q_{droop} function has to be in a a feedback mode that is feeding a gain of the reactive power delivered to the grid to have stability, not just a specific constant (see [6, 11]).
- **Parameter Sensitivity**: Checking how the system steady states and local trajectories would react (change in response) to small changes in a given parameter has been studied in [4, 8–10]. In these papers, it was concluded that they system is highly sensitive to v_{wind} and sensitive enough to all grid parameters X, R and E for all wind speed ranges (Regions 1,2 and 3, Fig. 13). To determine local trajectories sensitivity to parameters, we need to study eigenvalue sensitivity to parameters as done in [9]. Only the region (Region 3, Fig. 13) has eigenvalue sensitivity towards v_{wind}, which will be included in our study to the Pitch control next section.
- **Boundedness, Existence and Uniqueness**: If we consider the control limits (bounds) given in Table 3, it can be shown mathematically that the network equation in Eq. (51) has a unique solution that allow us to have the system in differential equations instead of the system being in differential-algebraic equations. As proved and introduced in [7], We can use directly Eq. (52) to represent V:

$$V = f(I_{plv}, E_q; X; R; E) = \frac{-B + \sqrt{B^2 - 4AC}}{2A} \tag{52}$$

With $A = 1 + \frac{2X}{X_{eq}} + \frac{R^2+X^2}{X_{eq}}$, $B = -\left[2I_{plv}R + \frac{2XE_q}{X_{eq}} + \frac{2(R^2+X^2)E_q}{X_{eq}}\right]$ and $C = \frac{R^2+X^2}{X_{eq}} + (R^2+X^2)I_{plv}^2 - E^2$. This reduction helps significantly in any nonlinear optimization and/or control study. To the best of our knowledge, no WTG system

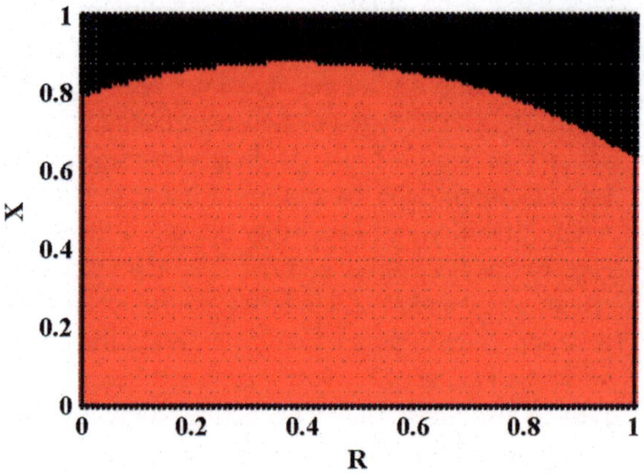

Fig. 14 Stable (red) and Unstable (Black)

has been mathematically analyzed to even conclude existence and uniqueness for the resultant system of differential equations, as have been done in [7]. It is also worth mentioning that in [7], it was mathematically proved and found that there is a safe region in the parameter space of R and X, in which the system always maintains existence and uniqueness of solutions. This region is given in the Fig. 15.

- **Multiple Time Scale Structure**: In [7] and intensively in [8], it has been shown that the WTG system has different time scales in it. In fact, it was shown that the system can reduces into fast-slow (two) time scales or fast-medium-slow (three) time scales. This is the first time in literature, as we think, that some study covered the multiple time structure in WTGs. Reduction to the model (two or three time scales), with guaranteeing eventual convergence (shown in [7, 8]), can expand and enhance the real time domain simulations and provide nonlinear studies with another tool to use while studying the WTGs.
- **Eliminating the control limits**: In [6, 11], it was shown that the attraction limits in the case of stability are larger than the control limits provided in Table 3. In another words, the stable steady states are shown to attract trajectories in a range that is even larger than the control limits. This, as concluded by [6, 11] means that the limiters can be eliminated when we build simulations for the system, at least if we are concerned about small-signal stability studies. The reader then is recommended to take a look at how the block diagrams, in transfer function, for the WTG components/controls, look like without any of the limiters in [11].

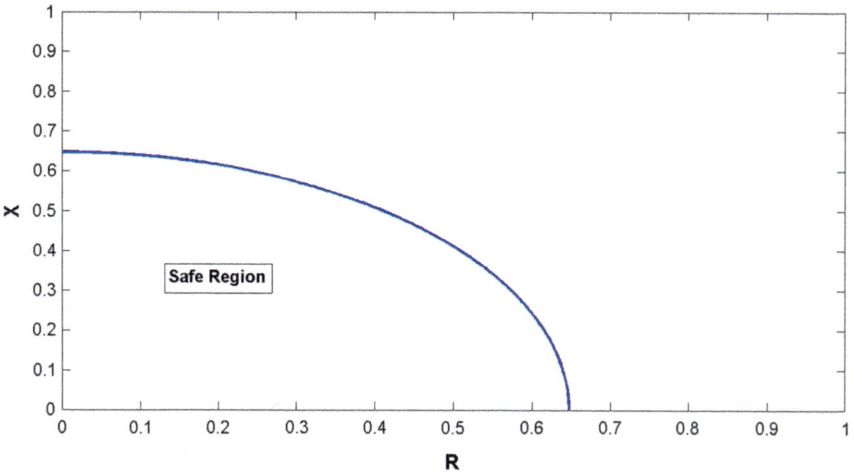

Fig. 15 Safe region in which solutions uniquely exists

3 Pitch Control, Simulations, Simulink Verification and Real Data Validation

3.1 Modeling and Analysis of Pitch Control

In this subsection, we provide a full time domain study for the dynamics in Region 3 (see Fig. 13) where the important Pitch control is activated. We start first by considering the power factor setup for the range of wind speeds in Region 3 (11.4–25 m/s) with the two mass model,the power factor, and without the active power and inertia controls. Next subsection, we expand the analysis to the reactive power control being in the set up of supervisory voltage and, having the Q_{droop} function in effect. The differential equations reduce to Eqs. (29)–(34), (36), (38)–(39), (48)–(50). For this system of 12 nonlinear differential equations, there is no algebraic (network) equation, as we eliminate the algebraic equation using Eq. (52). We find the steady states versus the wind speed, so we can see the Pitch control function that stabilizes the system. We study stability in parameter space.

3.1.1 Pitch Control with Power Factor

The Steady States and Eigenvalues: Once the wind speed approaches 11.4 m/s, the power extracted from the air-streams exceeds the 1 pu (rated power). Consequently, the Pitch control gets into action and enforce less power extraction. Figure 17 shows the Pitch angle in steady state as a function of wind speed. Because of the Pitch dynamics, in the steady state $P_{elec} = P_{mech} = P_{inp} = P_{1elec} = 1$. Figure 16 shows

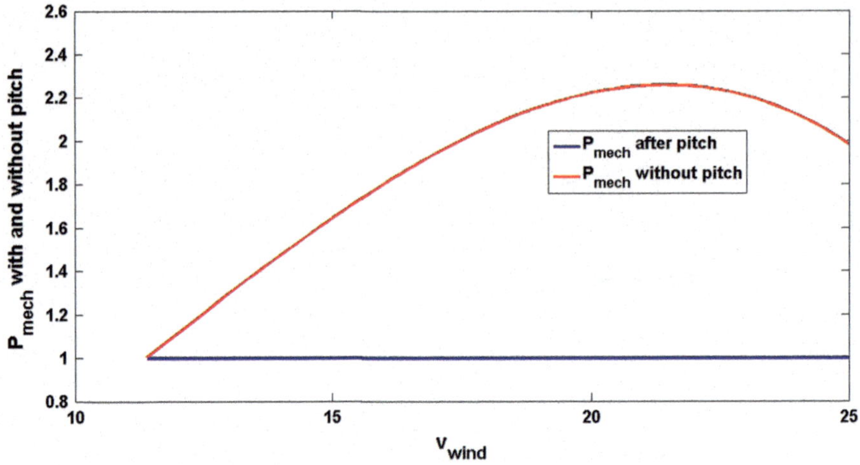

Fig. 16 P_{mech} when the pitch angle is fixed at zero versus the pitch control activated

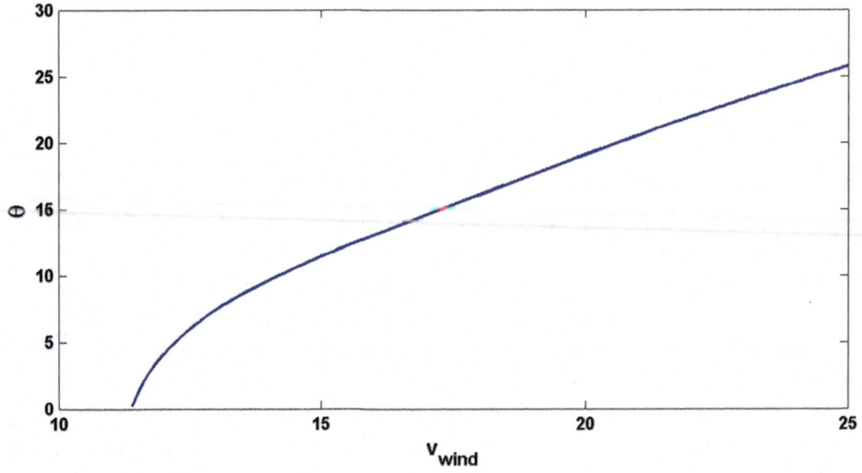

Fig. 17 Pitch angle steady state as a function of v_{wind}

what happens to the power to be extracted if the Pitch are fixed at zero versus when the Pitch control is activated. Since the Pitch control fix the power generation, the physical steady state in Region 3 is constant versus v_{wind}. We computed these values, $w_g = w_t = 0.2$, $E_q = E_{qcmd} = 1.167$, and $V = V_{ref} = 1.039$.

Because the C_p function is forth degree polynomial of θ, the resultant steady states are not necessary unique. We ran tests numerically to see if there are other possible steady states. For every fixed wind speed, we could find two different Pitch angles. One of these steady states is negative and the other is positive. As a result we fixed our codes to only find these Pitch angles that are in the acceptable range.

Table 4 The eigenvalues computed at the steady state

	Real $v_{wind} = 11.4$	Imag $v_{wind} = 11.4$	Real $v_{wind} = 25$	Imag $v_{wind} = 25$	% change
λ_1	−52.25	0	−52.25	0	0, 0
λ_2	−48.94	0	−48.94	0	0, 0
λ_3	−19.99	0	−19.99	0	0, 0
λ_4	−16.11	0	−16.11	0	0, 0
$\lambda_{5,6}$	−1.34	±12.15	−1.35	±11.6	0.7, 4.5
λ_7	−3.14	0	−0.88	0	71.9, 0
$\lambda_{8,9}$	−0.68	±2.16	−0.68	±2.16	0, 0
$\lambda_{10,11}$	−0.18	±0.46	−1.52	±3.2	744, 59.5
λ_{12}	−0.13	0	−0.16	0	23, 0

For instance, $v_{wind} = 20$ is associated with $\theta = -3.775$ and $\theta = 19.115$ as steady states values. Our code then chooses only $\theta = 19.115$. It is important to hint that only the acceptable range of Pitch angles belong to a locally stable set of steady states. When the Pitch control is deactivated in other Regions, the eigenvalues showed no sensitivity to the wind speed (see Sect. 2.3). In Region 3, the eigenvalues are highly sensitive to v_{wind}. Table 4 shows complex pair of the eigenvalues and their change when $v_{wind} = 11.4$ m/s and $v_{wind} = 25$ m/s. The sixth column presents the percentage of change for both the real part and the imaginary part respectively.

Grid Parameters and Stability: The resistance R and the reactance X are the grid parameters of interest (see Fig. 11). Changes in parameter, while fixing $v_{wind} = 12$ in our trial, is expected to have an effect on the terminal voltage. We ran a code that discretized the parameter space of R and X using a high resolution unit square grid, and found the steady states based on the given parameters at each point (Fig. 14). The code computes the Jacobian matrix and linearizes around the steady states. The reader is recommended to see [5] for Figures of the steady states plotted in the grid parameter space.

3.2 Pitch Control and Q Droop Function

By setting up the system to be the supervisory voltage case and two-mass rotor model, the model reduces to Eqs. (29)–(35), (39)–(44), and (48)–(50). Note that the model can reduce again depending on the wind speed Region of interest (Regions 1, 2 and 3 in Fig. 13). Equation (40) has the formulation of the Q Droop function dynamics. Furthermore, the effect of this dynamics is in the term $V_{qd} = K_{qd} Q_{droop}$ (Eqs. (42)–(43)), which eventually lead to Q_{ord} in Eq. (40). The Q Droop function is activated when the gain K_{qd} is larger than 0. The Q Droop function is a slow acting

Table 5 Steady sates versus K_{td}

	w_g	V	Q_{gen}	E_q	I_{plv}
$K_{td} = 0.04$	0.2	1.03	−0.154	0.911	0.97
$K_{td} = 0.05$	0.2	1.032	−0.08	0.97	0.96
$K_{td} = 0.06$	0.2	1.033	−0.05	0.99	0.96

function, and is supposed to help reducing the effective reference voltage in the reactive power control, corresponding the changes in the reactive power control. The result of applying this function is to have improved coordination between multiple integral controllers regulating the same point in the system. This positive result is claimed by GE [3]. As described by more details in [6, 11] the system has to have the Q Droop function activated to maintain stability. Moreover, this activation has to be a feedback mode from the reactive power delivered to the grid, that is $V_{qd} = K_{qd} Q_{gen}$ in Eqs. (42)–(43), and not to have Q_{droop}=constant, as this has shown to cause instability [5]. This is a very important consideration and result both [6, 11]. It is important to mention that in throughout and detailed analysis [11] has shown that Q_{droop}=constant result in an unstable system, which directly contradict the GE claim in [3] that the Q_{droop}=constant is acceptable.

In Region 3 ($11.4 < V_{wind} < 25$), the physical state variables are constant as discussed earlier. As suggested in [3], we have $K_{qd} = 0.04$. We fixed $v_{wind} = 11.4$ m/s and computed the steady states in Table 5 by varying the gain K_{td} to test its sensitivity. Similarly, Table 6 is for the eigenvalues. It is noticeable that changing K_{td} from 0.04 to 0.06 (50% change) have the changes 67%, 0.3%, 8.6%, and 1% in Q_{gen}, V, E_q, and I_{plv} respectively. It can be noted as well that some of the eigenvalues had over a 100% change. In Table 6, the sixth column has the change in percentage (real and imaginary parts respectively). From these results it is clear that the gain parameter K_{dq} has direct effects on both the steady states and the eigenvalues. Eigenvalues sensitivity means that this parameter can affect local trajectories behavior, so it is important to have more parameter estimate studies, especially if tunning is needed based on the application conditions.

3.3 *Simulations*

3.3.1 System Response to a Pulse Wind Profile

Figure 18 shows a v_{wind} profile for a pulse changing from 20 to 21 m/s and again to 20 m/s with the system response. The profile equation is $v_{wind} = 20 + e^{-(t-10)^2}$.

Table 6 Eigenvalues at $v_{wind} = 11.4$

	Real $K_{td} = 0.04$	Imag $K_{td} = 0.04$	Real $K_{td} = 0.06$	Imag $K_{td} = 0.06$	% change
λ_1	−52.25	0	−52.25	0	0, 0
λ_2	−48.94	0	−48.94	0	0, 0
λ_3	−19	0	−19	0	0, 0
λ_4	−16.11	0	−16.11	0	0, 0
$\lambda_{5,6}$	−1.34	±12.15	−1.35	±12.15	0.74, 0
λ_7	−3.14	0	−3.14	0	0, 0
$\lambda_{8,9}$	−0.1	±1.19	−0.002	±1.18	98, 0.84
$\lambda_{10,11}$	−0.18	±0.46	−0.18	± 0.46	0, 0
λ_{12}	−0.13	0	−0.13	0	0, 0
λ_{13}	−8.83	0	−8.82	0	0.11, 0
λ_7	−3.14	0	−3.14	0	0, 0
$\lambda_{15,16}$	−0.096	±0.14	−0.2	±0.21	108, 50

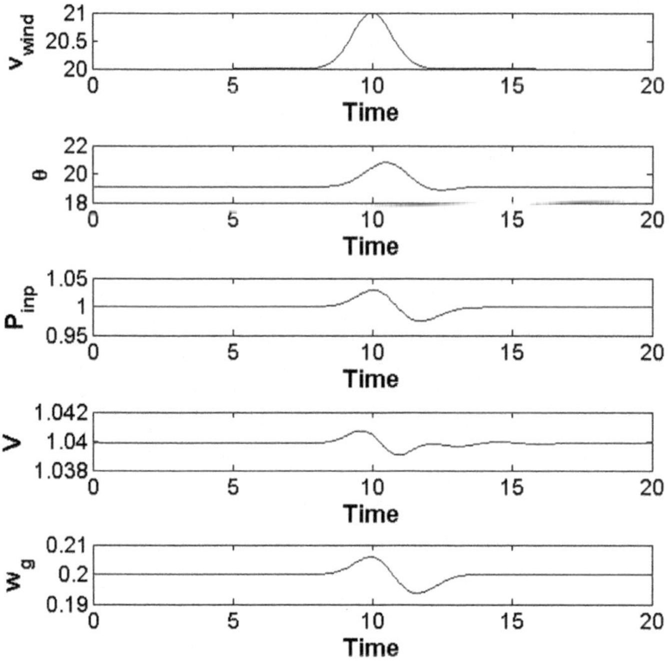

Fig. 18 System response for a given wind profile (upper graph)

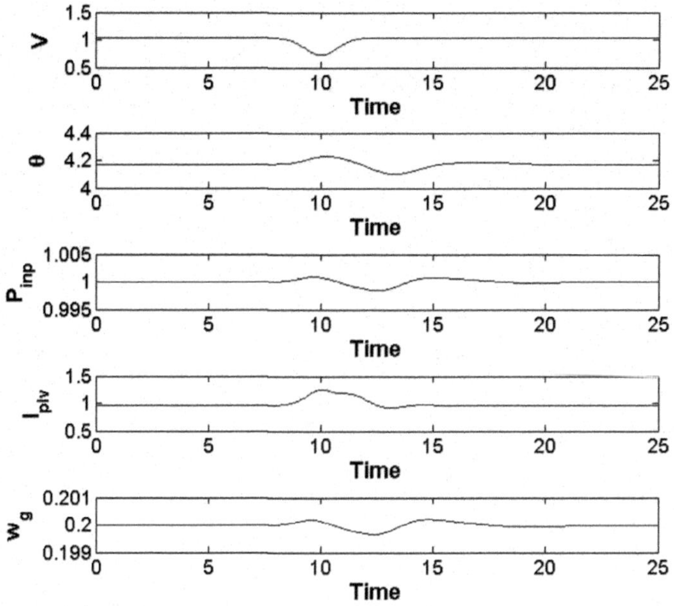

Fig. 19 Drop-clear in the terminal voltage (upper graph) along with the system's response

3.3.2 System Response to a Drop-Clear in Terminal Voltage

Figure 19 shows a drop-clear in terminal voltage and the system's response.

3.3.3 System Response to a Droop-Clear in the Reactance X

Figure 20 shows a droop-clear case of the reactance X. The dropping happens from a stable condition to an unstable one, passing a Hopf bifurcation, as discussed in Sect. 2.3), and then returning to the stable status. The system is capable of stabilizing after clearing the severe disturbance as shown in the simulation.

3.3.4 Basin of Attraction Versus Control Limits

Testing the Attraction Limits Versus Control Limits: As found by strong and detailed study/simulation in [11], the region of attraction around the steady states seems to be larger than the control limits themselves. Let us consider x and x_0 be the vectors of the steady states and initial conditions respectively. Figure 21 represents a simulation for the dynamics having initial condition $x_0 = x + 0.5\,|x|$. The simulation had fixed wind speed. Notice that initial values and the trajectories exceeded the

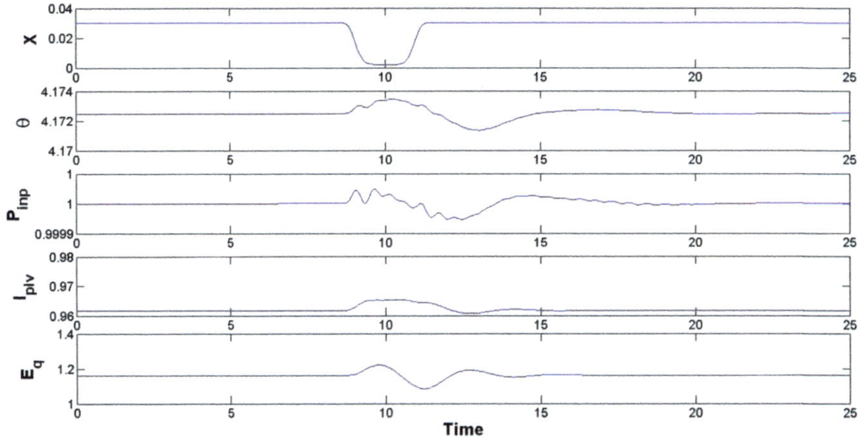

Fig. 20 Droop-clear in the reactance X (upper graph) and the system's response

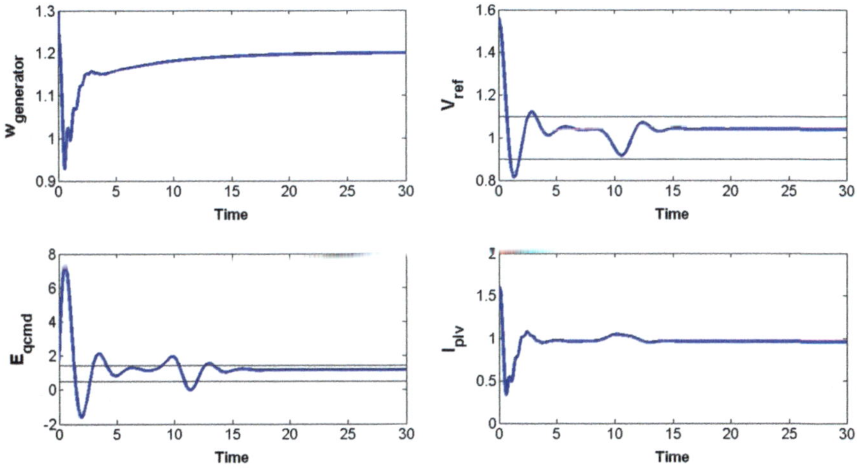

Fig. 21 State trajectories (blue) and limiters (black)

limiters (control limits) in solid black lines and are still attracted to the stable steady state eventually.

Testing the Derivatives of some Mechanical State Variables in Extreme Cases: Since on of the main things of interest for the system is how it responds to sudden disturbances, we tested the system response and both $\frac{dP_{inp}}{dt}$ and $\frac{d\theta}{dt}$ for a sudden but continuous wind disturbance and terminal voltage drop. Figure 22 shows some of the state variables' response to the given wind profile. The derivatives of the power order and the Pitch angle do not exceed their control limits. Figure 23 has the same trajectories response but for a sudden drop in the terminal voltage. Note that one of the states exceeded the control limits. The conducted simulations indicate that the

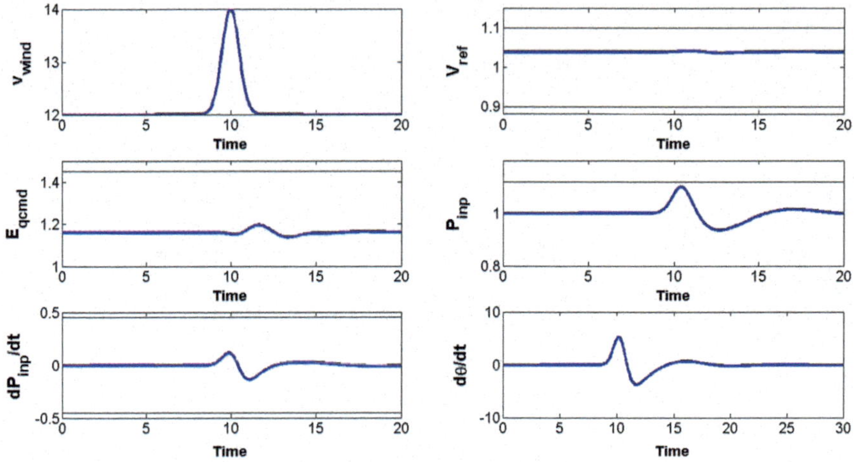

Fig. 22 State trajectories (blue) and limiters (black)

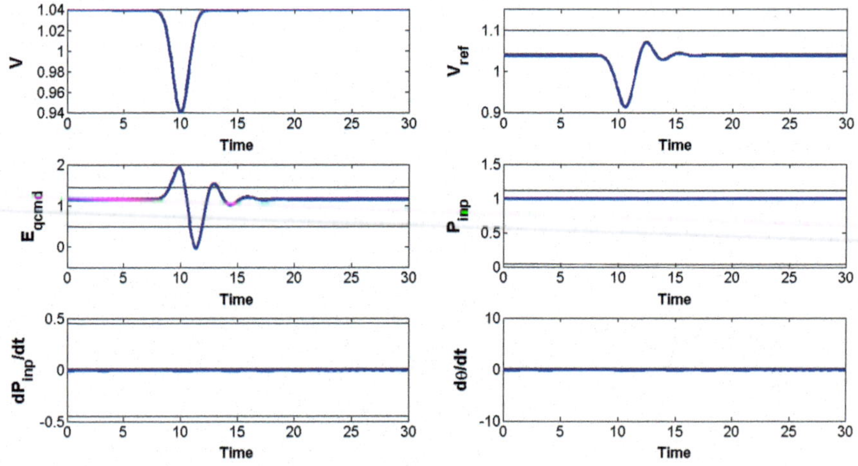

Fig. 23 State trajectories (blue) and limiters (black)

trajectories response and the mechanical derivatives limits respond within the control limits for very disturbing pulse of wind.

3.3.5 Effect of the Q Droop Function

We tested the system by by running a simulation, with and without Q_{droop} function, for a pulse wind speed profile. This should help us emphasize how the integrator variables Q_{wvl} and Q_{wvu} will behave. For the wind profile in Fig. 24, the integrators seem unable to stabilize after the pulse effect. While, In real life the limiters will

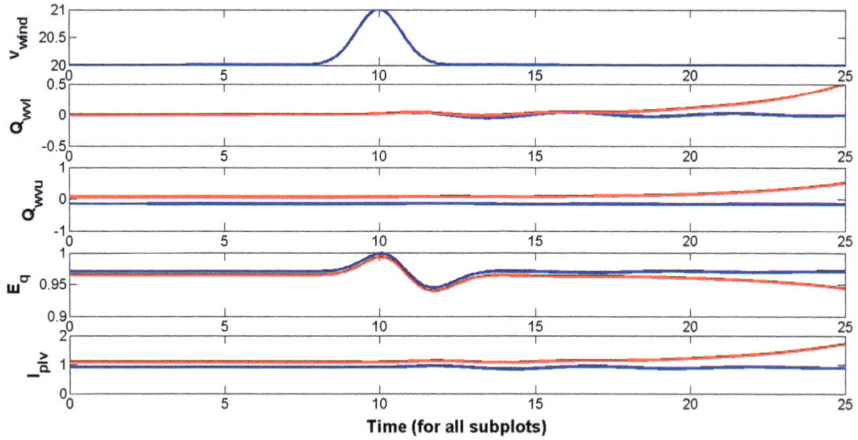

Fig. 24 Trajectories response with (blue) and without (red) Q_{droop}

intervene to prevent such divergences, the same pulse effect could not force the integrators to diverge when the Q_{droop} function is used. This indicates that the Q Droop helps stabilizing the system in extreme cases through improving the performance of the integrator.

3.4 *Verification and Validation of the Model*

3.4.1 Simulink Verification

Simulink projects were built for the transfer functions of the model given by GE [2, 3] and NREL [13]. The purpose is to verify that our differential equations model (Eqs. (29)–(50)) is typical in results when compared to the Simulink simulations. For more information about the verification, the reader is recommended to refer [8]. First project constructed a Simulink project of the system in Fig. 27 and ran the numerical solver ODE15s in Matlab for a fixed wind speed of $v_{wind} = 8.2$ m/s, which makes the model in Region 2 (see Sect. 2.2.2 and Fig. 13). The results of the simulation are typical. Figure 25 represents one of the results for V (the initial condition is the same in both Simulink and ODE15s). The similarity of the results of V is very important as it is the main term to calculate both the active and reactive power. Second project constructed a Simulink project (Fig. 28) with an oscillating wind speed ($v_{wind} = 8 + \sin(10t)$). The dynamics in this case is in Region 1, see Sect. 2.2.2 and Fig. 13. The result was also typical. Figure 26 presents P_{elec} response to the continuous oscillation of v_{wind}.

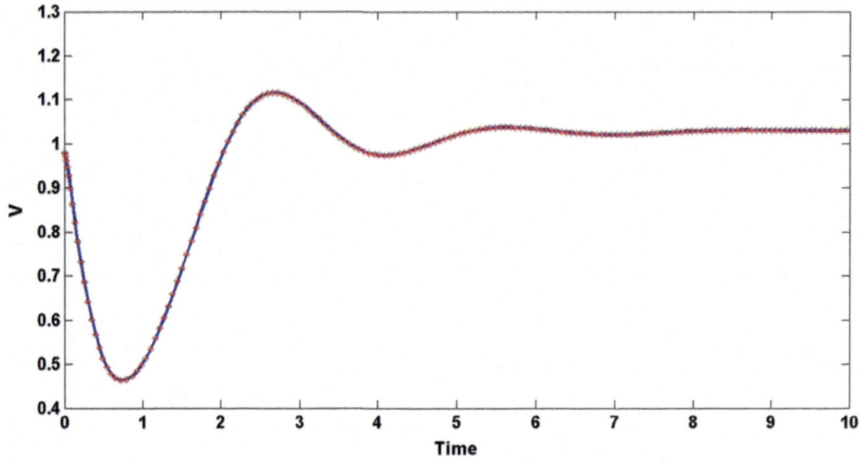

Fig. 25 Response of implementations in Matlab (differential equations) and Simulink to zero initial conditions

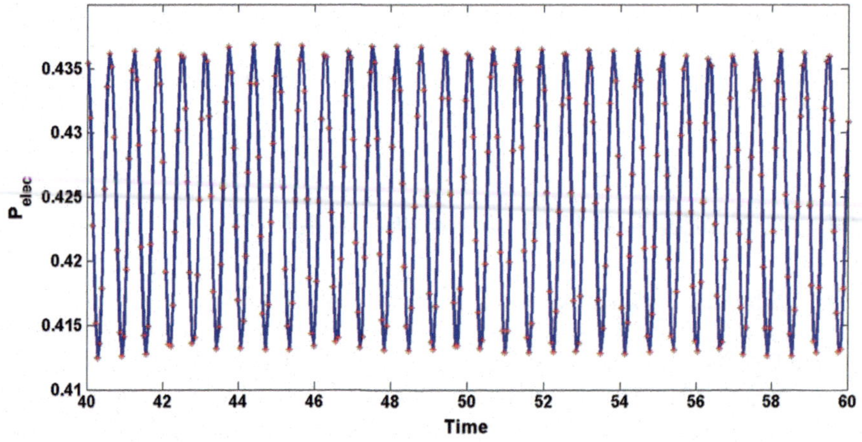

Fig. 26 P_{elec} in the steady response from the model and Simulink

3.4.2 Validation Versus Real Measured Data

In this validation we re-used what we provided in [8, 11]. The models chosen for comparison are [18, 20] as they are a highly cited academic source that also are inclusive in their modeling. We focus on one of the clear differences between our model and theirs. Both the generator and turbine speeds are controlled by the reference speed w_{ref} (see Rotor Model discussion in Sect. 2.1). The rated reference speed is 1.2 pu, however, for lower wind speeds, this is not physically possible. Therefore, the reference speed changes slowly with P_{elec} until it reaches 1.2 pu (see Fig. 10). Our model and the models [18, 20] have different

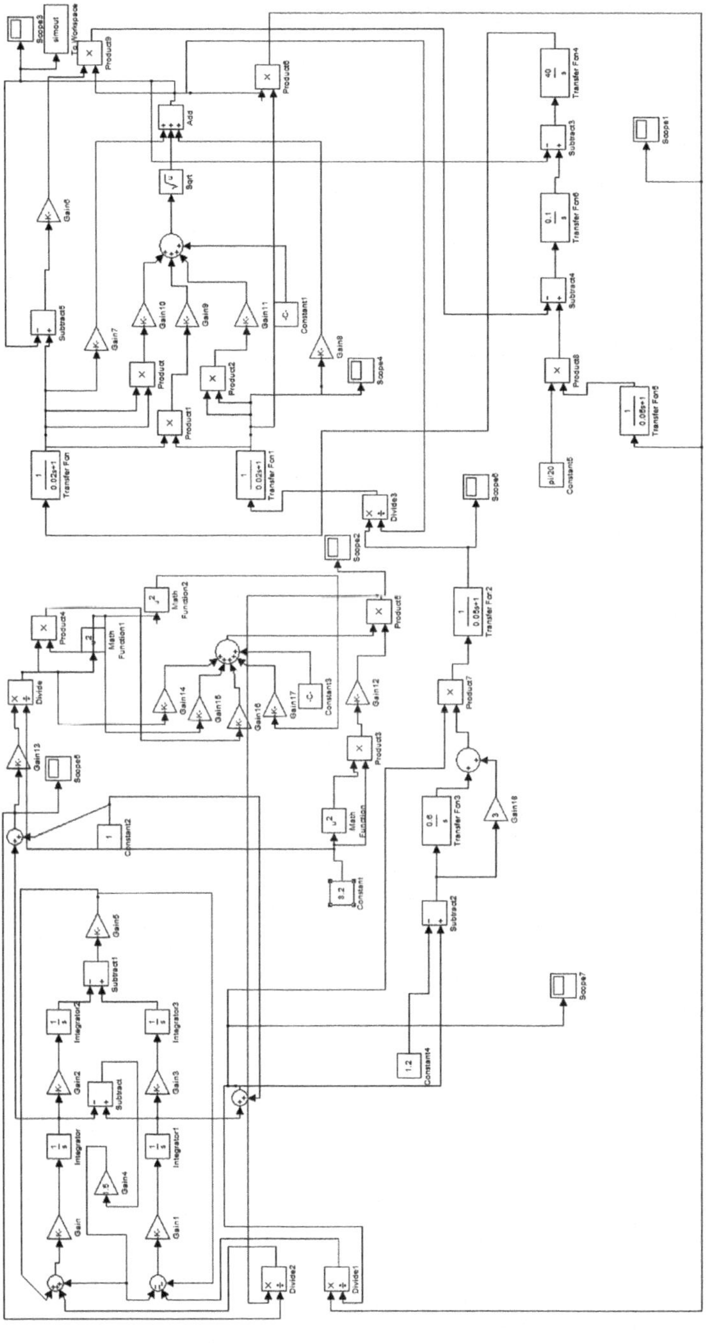

Fig. 27 Simulink project for a fixed wind speed

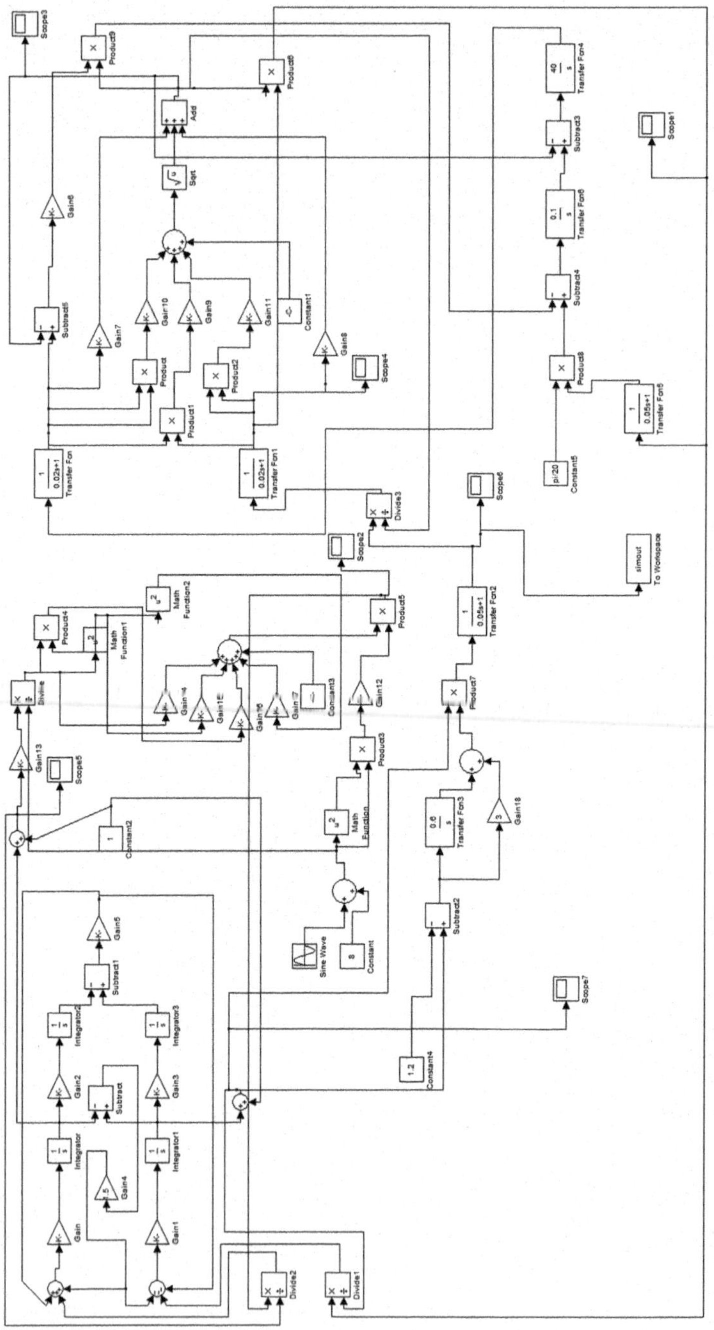

Fig. 28 Simulink project for an oscillating wind speed

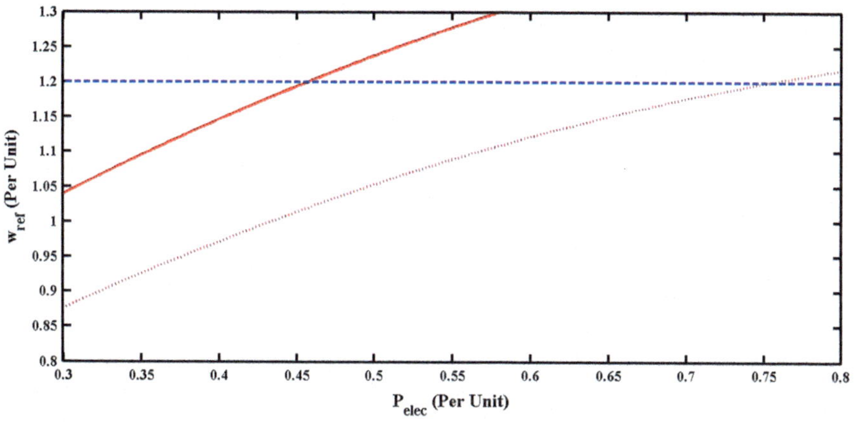

Fig. 29 w_{ref} from our model in solid with [18, 20] doted, and the rated speed dashed

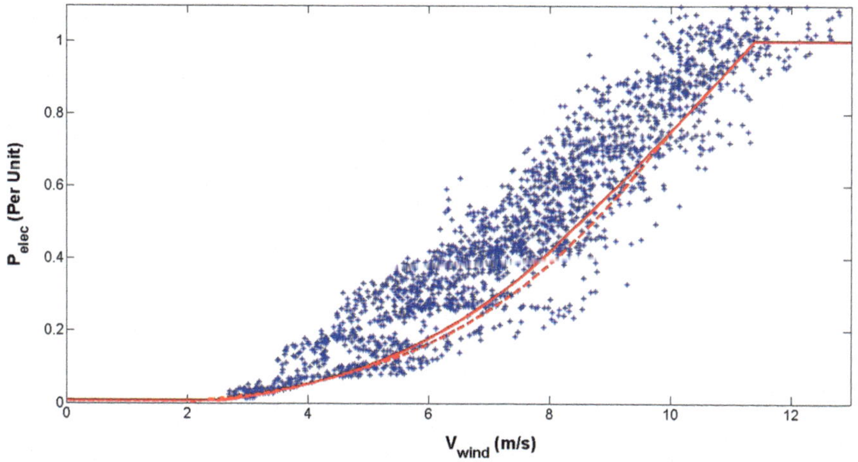

Fig. 30 Real data of a WTG (stars) versus power-wind speed curves for our model in solid and the models [18, 20] dashed. The Fig. is re-used from [8, 11]

curves for lower wind speeds $w_{ref} = -0.67P_{elec}^2 + 1.42P + 0.51$ in our model and $w_{ref} = -0.75P_{elec}^2 + 1.59P + 0.63$ in their models (Fig. 29). Figure 10 shows different w_{ref} curves. To test the effects of this difference between our model and the models considered in the comparison, we generated the power-wind speed profile for all the models (stable steady state of P_{elec} vs. v_{wind}) and plotted them with real measured data (re-use of the same data in [8, 11]). Figure 30 illustrates this comparison and validation with real measure data.

Observations from the comparison and the validation:

1. As seen in Fig. 10 when comparing our model versus [18, 20], we see that w_{ref} reaches the rated value (1.2 pu) at $P_{elec} \approx 0.45$ pu in our model, while it reaches the rated value (1.2 pu) at $P_{elec} \approx 0.75$ pu in their model. The manufacturer (see page 35, [3]) recorded that w_{ref} only has to follow some feed back from P_{elec} below $P_{elec} = 0.46$ pu, as this is the stage where the system can start having $w_{ref} = 1.2$. This indicates that our model is more practical and matches the specifications recorded by the manufacturer.
2. Our model has relatively better power-wind speed profile when compared to [18, 20] and the real time measured data (Fig. 30). The stable steady state of P_{elec} is not an average for the measured data. However, when the size of the measured data is very large, which is the case in our trial, the data may be expected to have some form of normal distribution around the stable steady state. Our model shows better results if we have this explanation in consideration.
3. The power-wind speed profile in Fig. 30 can be generated with either the reactive power control in power factor or supervisory voltage mode (see Fig. 4). The supervisory voltage mode is preferable as it is associated with having the WTG as a member in a group of WTGs as opposed to have the WTG as a separate unit (power factor mode). We remark that stability of P_{elec} and the whole system is not possible without having the Q Droop function in effect (explained in Sect. 3.2). There is almost no explanation about how to implement Q Droop function and/or analyzing its effect and possible cases in other models throughout our search to date in the literature and cited papers that are considered major sources of modeling for WTGs.

4 Conclusion

In this chapter, a state-of-the-art modeling effort, that represents the WTG dynamics and control, is provided. This modeling efforts summarize and collect very recent publications and is validated and verified versus real measured data and Simulink simulations. An advantage from the collective modeling presentation in this chapter, is that one can use numerical solvers such as ODE15s to simulate the WTG system for different scenarios and conditions, without the need for commercial and special simulators. The pitch control, that is essential for the WTG system to keep the power production stable and saturated at the rated requirement, is discussed and analyzed by detail in this chapter. Moreover, the chapter provides the reader with all important technical information regarding this important control as is in industry, such as: stability, sensitivity, resilience in different reactive power modes and a lot of demonstrative simulations.

Copyright

References

1. Department of Energy. Energy Dept. Reports: U.S. Wind Energy Production and Manufacturing Reaches Record Highs 2013. Accessed 25 Sept 2015
2. Miller NW, Price WW, Sanchez-Gasca JJ (2003) Dynamic modeling of GE 1.5 and 3.6 wind turbine-generators. Report, General Electric International, Inc., Oct 2003
3. Clark K, Miller NW, Sanchez-Gasca JJ (2010) Modeling of GE wind turbine-generators for grid studies. Report, General Electric International, Inc., Apr 2010
4. Eisa SA, Wedeward K, Stone W (2016) Sensitivity analysis of a type-3 DFAG wind turbine's dynamics with pitch control. In: 2016 IEEE green energy and systems conference (IGSEC), pp 1–6, Nov 2016
5. Eisa SA, Stone W, Wedeward K (2017) Mathematical modeling, stability, bifurcation analysis, and simulations of a type-3 DFIG wind turbine's dynamics with pitch control. In: 2017 ninth annual IEEE green technologies conference (GreenTech), pp 334–341, Mar 2017
6. Eisa SA, Wedeward K, Stone W (2017) Time domain study of a type-3 DFIG wind turbine's dynamics: Q drop function effect and attraction vs control limits analysis. In: 2017 ninth annual IEEE green technologies conference (GreenTech), pp 350–357, Mar 2017
7. Eisa SA, Stone W, Wedeward K (2018) Mathematical analysis of wind turbines dynamics under control limits: boundedness, existence, uniqueness, and multi time scale simulations. Int J Dyn Control 6(3):929–949
8. Eisa SA, Wedeward K, Stone W (2018) Wind turbines control system: nonlinear modeling, simulation, two and three time scale approximations, and data validation. Int J Dyn Control 1–23
9. Eisa SA (2017) Local study of wind turbines dynamics with pitch activated: trajectories sensitivity. In: 2017 IEEE green energy and smart systems conference (IGESSC), pp 1–6, Nov 2017
10. Eisa SA, Stone W, Wedeward K (2017) Dynamical study of a type-3 DFIG wind turbine while transitioning from rated speed to rated power. In: 2017 IEEE green energy and smart systems conference (IGESSC), pp 1–6, Nov 2017
11. Eisa SA (2019)Modeling dynamics and control of type-3 dfig wind turbines: Stability, Q droop function, control limits and extreme scenarios simulation. Electr Power Syst Res 166:29–42
12. Eisa SA (2018) Investigating periodic attractors of wind turbine dynamics with pitch activated under control limits. Nonlinear Dyn
13. Pourbeik P (2014) Specification of the second generation generic models for wind turbine generators. Report, Electric Power Research Institute
14. WECC Renewable Energy Modeling Task Force (2014) WECC wind power plant dynamic modeling guide. Report, Western Electricity Coordinating Council Modeling and Validation Work Group
15. Tummala A, Velamati RK, Sinha DK, Indraja V, Hari Krishna V (2016) A review on small scale wind turbines. Renew Sustain Energy Rev 56:1351–1371
16. Rahimi M (2014) Dynamic performance assessment of dfig-based wind turbines: a review. Renew Sustain Energy Rev 37:852–866
17. Slootweg JG, Polinder H, Kling WL (2003) General model for representing variable speed wind turbines in power system dynamics simulations. IEEE Trans Power Syst 18(1):144–151

18. Tsourakis G, Nomikos BM, Vournas CD (2009) Effect of wind parks with doubly fed asynchronous generators on small-signal stability. Electr Power Syst Res 79(1):190–200
19. Eisa SA (2017) Mathematical modeling and analysis of wind turbines dynamics. PhD thesis, New Mexico Institute of Mining and Technology
20. Working Group C4.601 (2007) Modeling and dynamic behavior of wind generation as it relates to power system control and dynamic performance. Report, CIGRE TB 328, Aug 2007
21. Miller NW, Sanchez-Gasca JJ, Price WW, Delmerico RW (2003) Dynamic modeling of GE 1.5 and 3.6 MW wind turbine-generators for stability simulations. In: 2003 IEEE power engineering society general meeting, July 2003
22. Heier S (2014) Grid integration of wind energy conversion systems, 2nd edn. Wiley
23. Hiskens Ian A (2012) Dynamics of type-3 wind turbine generator models. IEEE Trans Power Syst 27(1):465–474
24. Rose J, Hiskens IA (2008) Estimating wind turbine parameters and quantifying their effects on dynamic behavior. In: IEEE power and energy society general meeting—conversion and delivery of electrical energy in the 21st century, July 2008
25. Yang L, Xu Z, Ostergaard J, Dong ZY, Wong KP, Ma X (2011) Oscillatory stability and eigenvalue sensitivity analysis of a DFIG wind turbine system. IEEE Trans Energy Convers 26(1)
26. Bowen A, Huskey A, Link H, Sinclair K, Forsyth T, Jager D (2009) Small wind turbine testing results from the national renewable energy lab. In: 2009 conference and exhibition. The American wind energy association WINDPOWER, Illinois, May, Chicago

Distributed Cooperative Control of Wind Farms with On-site Battery Energy Storage Systems

Dinh Hoa Nguyen, Javad Khazaei, Susan W. Stewart and Jennifer Annoni

Abstract This chapter studies the output power regulation in wind farms consisting of doubly-fed induction generator wind turbines, one of the most popular generator configurations in modern multi-MW wind industry, equipped with on-site battery energy storage systems. Traditionally, wind farms were operated as standalone units or were equipped with a central, large-scale, energy storage system. In fact, most of the existing control designs for hybrid wind-storage applications are centralized, where the central unit collects the measured data from all wind turbines and storage systems (if they exist) and compares with the load demand to generate the reference power for each wind turbine controller. Those centralized setups or control structures do not fully exploit the flexibility and locality of the on-site energy storage systems in a wind farm. Moreover, the storage functionality is lost if the central storage system is broken down, while in a distributed architecture, a failure only affects individual on-site storage systems and the others will still operate. Therefore, on-site storage systems and distributed control designs need to be developed for wind farms to provide regulation services for the grid. An approach to tackle that challenge will be presented in this chapter.

D. H. Nguyen (✉)
WPI International Institute for Carbon-Neutral Energy Research (WPI–I2CNER) and Institute of Mathematics for Industry, Kyushu University, 744 Motooka, Nishi-ku, Fukuoka 819-0395, Japan
e-mail: hoa.nd@i2cner.kyushu-u.ac.jp

J. Khazaei
School of Science, Engineering & Technology at Penn State Harrisburg, Harrisburg, PA, USA
e-mail: jxk792@psu.edu

S. W. Stewart
Department of Aerospace Engineering, Pennsylvania State University, University Park, PA, USA
e-mail: sstewart@psu.edu

J. Annoni
National Renewable Energy Laboratory, Golden, CO, USA
e-mail: jennifer.annoni@nrel.gov

R.-E. Precup et al. (eds.), *Advanced Control and Optimization Paradigms for Wind Energy Systems*, Power Systems, https://doi.org/10.1007/978-981-13-5995-8_2

Nomenclature

DFIG	Doubly-fed induction generator.
ESS, BESS	Energy storage system, battery energy storage system.
GSC, RSC	Grid side converter, rotor side converter.
MAS	Multi-agent system.
PMSG	Permanent magnet synchronous generator.
RES	Renewable energy source.
WF, WT	Wind farm, wind turbine.
$t,\ s,\ \tau$	Time, frequency, and time delay variable.
$\mathscr{G},\ \mathscr{A},\ \mathscr{D},\ \mathscr{L}$	Communication graph, its adjacency, degree, and Laplacian matrices.
$\mathbb{R}$	Set of real numbers.
$P_{st,i},\ E_{st,i},\ Q_{st,i}$	Power, energy, and reactive power of the i-th BESS.
$P_{s,i},\ P_{g,i},\ Q_{s,i},\ Q_{g,i}$	Active power of RSC and GSC, reactive power of RSC and GSC, at the i-th WT.
$P_d,\ Q_d$	Active and reactive power demand for the WF.
n	Number of WTs in the WF.
$u_i^{PB},\ u_i^{QB},\ u_i^{Qr},\ u_i^{Qs}$	Control input for active and reactive power of the i-th BESS, and reactive power at the RSC and GSC of the i-th WT.
$K_{st,i}$	Gain to denote the heterogeneity of the i-th WT and BESSs.

1 Introduction

The combustion of fossil fuels for energy production creates problems for the global climate landscape and environment due to green house gas (GHG) and other pollutant emissions. To address these concerns, RESs along with their associated technologies have widely been sought and deployed into electric power grids around the world. Wind generation is now one of the most popular RESs being used worldwide. Large penetration of wind turbines requires planning for long term WF operation with a broad range of ancillary services. These services include but not limited to frequency and output power regulation, voltage control, and low voltage ride through (LVRT) [1, 2], of which, the output power regulation of the WFs was reported as the most critical [3, 4]. This motivates the research to investigate the possibility of incorporating on-site ESSs, in particular BESSs, with individual WTs in WFs along with the needed distributed control strategies to strengthen the flexibility and resiliency of WFs, and providing regulation services for the power grid.

1.1 Literature Review

There are different types of WTs currently available at the utility scale as well as the residential scale. Horizontal-axis wind turbines, as opposed to vertical-axis wind turbines, generally dominate both utility and residential markets [5, 6]. The industry has also moved from fixed speed turbines to nearly entirely all variable speed turbines at the utility scale because of its many advantages such as cost effectiveness, improved performance, and the ability to reduce turbine loads [5, 6]. The current chapter focuses on the DFIG WT since it is one of the most popular types of WTs currently being deployed in wind energy and integration to power grids.

A few research articles have focused on controlling the output power of WFs in smart grids by modifying the existing WFs' controllers [7–11]. For example, three controllers were designed in [7] for the active and reactive power regulation of a WF to predefined set points. As another example, [9] introduced a sliding mode control structure for variable-speed wind systems to ensure stability in WF operating regions despite model uncertainties. It is also worth noting that in practical applications, wake losses are encountered in WFs and this is an issue that could affect the WF output power regulation problem [12, 13]. This effect however has been omitted in this chapter for conciseness and clarity of the control architecture.

The WF control studies mentioned thus far all employed a centralized controller architecture for the output power control, where a centralized unit is dedicated to coordinate the WTs and ESSs. This structure requires significant data to be processed in the central unit and is susceptible to a single point of failure. This centralized approach thus does not guarantee flexibility or reliability and therefore, is not the best option to regulate the generated power in distributed WTs+ESSs [14, 15]. Hence, distributed designs are preferred to synchronize the operation of dispersed WTs+ESSs in smart power grids to maximize the reliability and economic performance of both the grid and the WF.

Several studies have been reported to control wind generations with distributed ESSs in a distributed manner [16–20]. A model predictive control using a semi-distributed algorithm was implemented for BESSs to enhance the output power regulation of WFs [16]. Unfortunately, the designed approach ignored the equal power sharing and did not model the energy balance of BESSs. Furthermore, it was assumed that the BESSs had 100% state of charge at any time, which was not realistic. In another example, a planning strategy was suggested to optimally calculate the power rating and capacity of distributed BESSs using a power flow algorithm [17]. The study did not formulate the output power generation problem of WFs using BESSs and no communication links were considered for energy management studies.

In a recent study, an average consensus method was used for power regulation of distributed WFs with storage devices [19, 20]. The WF controller was taken into account and BESSs were synchronized to control the total power of WTs. However, the energy storage was linked to the DC-link of the WTs and standalone BESSs were not considered. This could greatly limit the capacity of BESSs in output power regulation of WTs. Furthermore, the proposed approach considered all the WTs and

ESSs had the same capacity and reactive power control was not possible due to limited capacity of BESSs. As another limitation, 100% state of charge was assumed for all the ESSs at any moment, which was not realistic. It is worth mentioning that in all the existing research, the communication delays were ignored in the controller designs, which motivated the contribution of this research for the distributed control of ESSs for WTs in presence of communication delays.

In this chapter, a novel control strategy is proposed to synchronize and optimize the operation of WFs using distributed standalone energy storage devices. The proposed controller ensures regulation of active/reactive power demand and solves the capacity limit of BESSs (e.g., in [19]) by a standalone BESS connected to each WT at a point of common coupling (PCC), as shown in Fig. 1. Application of standalone BESS allows reactive power support of WTs and provides more flexibility in control. The standalone application of BESSs for WFs has recently been commercialized, e.g., the Notrees project which was supported by the department of defense (DoD) [21]. The contributions of this work are included in the following.

- A small-signal model of WT and BESS is derived and eigenvalue analysis is conducted to investigate the stability of combined system.
- A distributed control strategy is proposed for standalone BESSs to regulate the output power of WTs in unpredicted wind variations.
- The communication delays were considered in the design that can be any connected undirected graphs. The designed controllers can guarantee the successful regulation of WF output power in case of communication failures, given the fact that a subset of BESSs are still connected.
- A distributed controller is designed for BESSs and WTs to meet the active and reactive power demand based on the capacity of BESSs and WTs.

1.2 Chapter Organization

The rest of the chapter is classified as follows. Section 2 introduces the dynamic modeling of WT and BESSs where detailed mathematical models of the DFIG WT and the BESS together with their electronics parts are presented. Section 3 elaborates the formulation of the power regulation challenge of WTs and BESSs, and discusses their simplified dynamic models for control designs. Next, several case studies are provided in Sect. 4 to demonstrate the proposed theoretical designs. Simulation results show how the proposed consensus controllers can accommodate the on-site BESSs to eliminate the power mismatch between load demand and available wind power, with or without time delays. Finally, Sect. 5 summarizes the chapter.

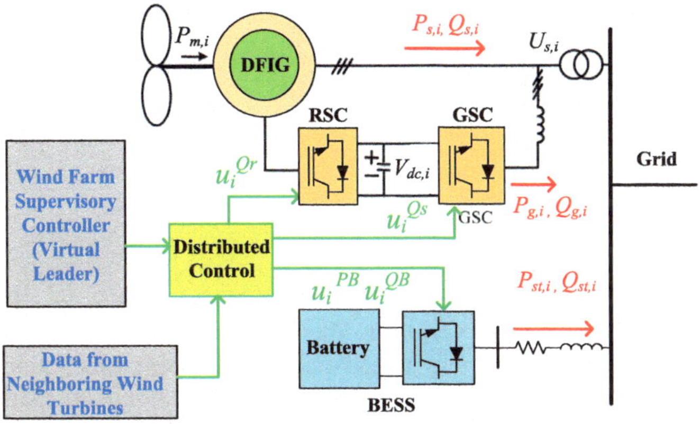

Fig. 1 Distributed control architecture for WTs with on-site BESSs

2 Dynamic Modeling of WT and BESS

An energy storage can help the WF to meet the load demand even in varying wind speed conditions. A typical circuit diagram of a DFIG with a standalone BESS is shown in Fig. 1. Step by step procedure to derive the linearized small signal model of a DFIG wind turbine with a standalone BESS is included in the following.

2.1 Small Signal Model of DFIG Wind Turbine

To find the small signal model of a DFIG WT, the linearized model of the induction generator, drivetrain, pitch control, RSC, GSC, and the DC link capacitor should be derived. Since this chapter focuses on distributed control of WTs with on-site BESSs, the pitch angle control dynamics is ignored. The readers should refer to reference [22] for a detailed small signal model.

The equations of voltage and flux linkage in an induction generator can be used to derive the dynamics of the stator and rotor circuits in a dq reference frame, also known as the generator reference frame [22]. A 6th-order dynamic model is used for the DFIG [23]:

$$\dot{X}_g = AX_g + BU, \quad (1)$$

where, $X_g = [i_{qs}, i_{ds}, i_{os}, i'_{qr}, i'_{dr}, i'_{or}]^T$, $U = [v_{qs}, v_{ds}, v_{os}, v'_{qr}, v'_{dr}, v'_{or}]^T$, and A, B are defined as:

$$B=\begin{bmatrix} \frac{X_{ss}}{\omega_b} & 0 & 0 & \frac{X_M}{\omega_b} & 0 & 0 \\ 0 & \frac{X_{ss}}{\omega_b} & 0 & 0 & \frac{X_{ss}}{\omega_b} & 0 \\ 0 & 0 & \frac{X_{ls}}{\omega_b} & 0 & 0 & 0 \\ \frac{X_{ss}}{\omega_b} & 0 & 0 & \frac{X'_{rr}}{\omega_b} & 0 & 0 \\ 0 & \frac{X_{ss}}{\omega_b} & 0 & 0 & \frac{X'_{rr}}{\omega_b} & 0 \\ 0 & 0 & 0 & 0 & 0 & \frac{X'_{lr}}{\omega_b} \end{bmatrix}, A=-B\begin{bmatrix} r_s & \frac{\omega_e}{\omega_b}X_{ss} & 0 & 0 & \frac{\omega_e}{\omega_b}X_M & 0 \\ -\frac{\omega_e}{\omega_b}X_{ss} & r_s & 0 & -\frac{\omega_e}{\omega_b}X_M & 0 & 0 \\ 0 & 0 & r_s & 0 & 0 & 0 \\ 0 & \frac{\omega_e-\omega_r}{\omega_b}X_M & 0 & r'_r & \frac{\omega_e-\omega_r}{\omega_b}X'_{rr} & 0 \\ -\frac{\omega_e-\omega_r}{\omega_b}X_M & 0 & 0 & \frac{\omega_e-\omega_r}{\omega_b}X'_{rr} & r'_r & 0 \\ 0 & 0 & 0 & 0 & 0 & r'_r \end{bmatrix},$$

where ω_e is the angular velocity of the arbitrary frame. If $\omega_e = \omega_b$, the rotating speed of the reference frame is equal to the nominal electrical angular velocity, 120π rad/s. Such reference frame is named a synchronous reference frame. It is also known that in the analysis of induction machines, the symmetrical rotor variables are transformed to the arbitrary reference frame and transformed variables are shown with a prime "$'$" symbol. In the above representation, ω_r is the rotating speed of the rotor, and $X_{ss}, X_{ls}, X_{rr}, X_M, X_{lr}$ are related to the stator and rotor self/mutual leakage inductances as defined by [24]. The dq frame air gap flux linkage is represented as $\lambda_{qm} = L_M(i_{qs} + i'_{qr})$, $\lambda_{dm} = L_M(i_{ds} + i'_{dr})$, and the torque equation can be represented as $T_e = \lambda_{qm} i'_{dr} - \lambda_{dm} i'_{qr}$.

A two-mass model is considered for the drivetrain as the wind turbine shaft is softer than that of synchronous machines in power plants [25]. The dynamic equations of a two-mass drivetrain are represented as:

$$\begin{bmatrix} \Delta\dot{\omega}_t \\ \Delta\dot{\omega}_r \\ \Delta\dot{T}_g \end{bmatrix} = \begin{bmatrix} \frac{D_t + D_{tg}}{2H_t} & \frac{D_{tg}}{2H_t} & \frac{-1}{2H_t} \\ \frac{D_{tg}}{2H_g} & -\frac{D_g + D_{tg}}{2H_g} & \frac{1}{2H_g} \\ K_{tg}\omega_e & -K_{tg}\omega_e & 0 \end{bmatrix} \begin{bmatrix} \Delta\omega_t \\ \Delta\omega_r \\ \Delta T_g \end{bmatrix} + \begin{bmatrix} \frac{T_m}{2H_t} \\ \frac{-T_e}{2H_g} \\ 0 \end{bmatrix}, \tag{2}$$

where ω_t and ω_r are the turbine and generator angular velocities, respectively, H_g and H_t are the generator and turbine inertias, respectively; T_g is an integral torque of the model; D_t and D_g are the mechanical damping coefficients between two masses, and K_{tg} is the shaft stiffness.

The voltage dynamics of the DC link capacitor, which is located between the GSC and RSC of the DFIG, is represented by a first-order dynamic model:

$$C v_{dc} \dot{v}_{dc} = P_r - P_g, \tag{3}$$

where P_r and P_g are the active powers of the rotor side and grid side converters, respectively, and can be represented by $P_r = 0.5(v'_{qr} i'_{qr} + v'_{dr} i'_{dr})$, $P_g = 0.5(v_{qg} i_{qg} + v_{dg} i_{dg})$. Now, consider that the DFIG WT is connected to the grid through an RL circuit (filter). A synchronous reference frame can be adopted to model the network dynamics. The dynamics of the grid and network model are represented as:

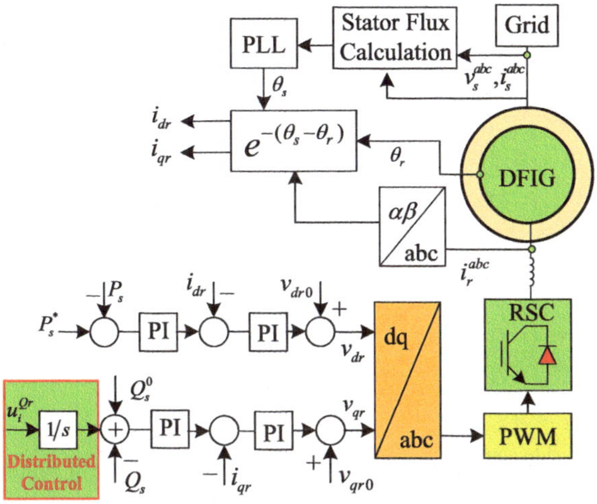

Fig. 2 DFIG's rotor side converter controller

$$\begin{bmatrix} \dot{i}_{wd} \\ \dot{v}_{wq} \end{bmatrix} = \begin{bmatrix} -\frac{R_g}{L_g} & \omega_e \\ -\omega_e & -\frac{R_g}{L_g} \end{bmatrix} \begin{bmatrix} i_{wd} \\ i_{wq} \end{bmatrix} + \begin{bmatrix} \frac{v_{pd} - v_{bd}}{L_g} \\ \frac{v_{pq} - v_{bq}}{L_g} \end{bmatrix}. \tag{4}$$

In the above equation, $i_w = i_{wd} + ji_{wq}$ is the filter current, $i_g = i_{gd} + ji_{gq}$ is the GSC current, $v_b = v_{bd} + jv_{bq}$ is the grid voltage (input), and $v_p = v_{pd} + jv_{pq}$ is the point of common coupling (PCC) voltage, which can be derived as $v_{pd} = v_{gd} - x_{tg}i_{gq}$, $v_{pq} = v_{gq} + x_{tg}i_{gd}$. A basic control structure of the RSC is shown in Fig. 2. To decouple the electromagnetic torque and the reactive power, a stator flux-oriented reference frame rotating at synchronous speed is used, where its d-axis is aligned with the stator flux vector. Variables with superscript s denote the stator flux-oriented reference frame. Since the inner current control loop has a much wider bandwidth than the outer loops (reactive power and speed controllers), fast dynamics of the inner current controller can be ignored [26]. Therefore, simplified dynamics of the rotor side converter can be derived [22], where $K_{iTe} = K_{Te}/T_{Te}$, $K_{iQs} = K_{Qs}/T_{Qs}$,

$$\dot{x}_{Te} = \frac{K_{Te}}{K_{iTe}}(T_e^* - T_e), \tag{5}$$

$$\dot{x}_{Qs} = \frac{K_{Qs}}{K_{iQs}}(Q_s^* - Q_s), \tag{6}$$

and $i_{dr}^s = i_{dr\text{ref}}^s = K_{Qs}(Q_s^* - Q_s) + x_{Qs}$, $i_{qr}^s = i_{qr\text{ref}}^s = -[K_{Te}(T_e^* - T_e) + x_\omega]$. Grid voltage-oriented control, which rotates at synchronous speed with the d-axis aligned with the grid voltage, is used to control the GSC [22]. Two proportional integral (PI) controllers are implemented for regulating the DC link voltage and the

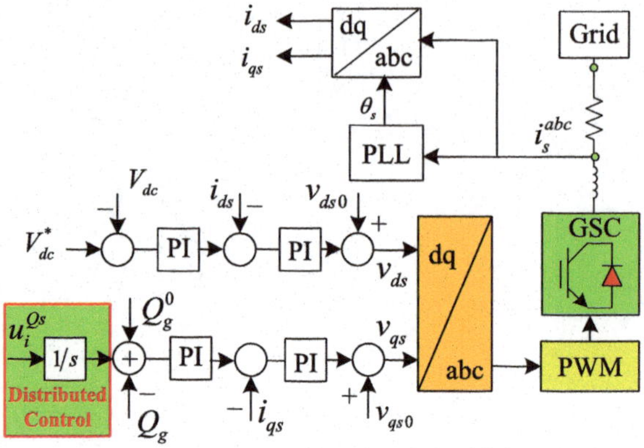

Fig. 3 DFIG's grid side converter controller

reactive power at the grid side, as depicted in Fig. 3. Similar to the RSC derivation, fast dynamics in the inner loop can be omitted, and the dynamics of the GSC can be simplified as follows [22], where $K_{i4} = K_{p4}/T_{p4}$,

$$\dot{x}_V = \frac{K_{PV}}{K_{i4}}(V_{dc}^* - V_{dc}), \tag{7}$$

$$\dot{x}_{Qg} = \frac{K_{p4}}{K_{i4}}(Q_g^* - Q_g), \tag{8}$$

and $i_{gd}^s = i_{gdref}^s = [K_{p4}(V_{dc}^* - V_{dc}) + x_V], i_{qg}^s = i_{qgref}^s = K_{p4}(Q_g^* - Q_g) + x_{Qg}$. Finally, combining (1), (2), (3), (4), (6), and (8), the overall DFIG WT model is derived as a 16th-order differential equation described by:

$$\dot{X} = f(X, U), \tag{9}$$

where $X = [i_{qs}, i_{ds}, i_{os}, i'_{qr}, i'_{dr}, i'_{or}, \Delta\omega_t, \Delta\omega_r, \Delta T_g, v_{dc}, i_{wd}, i_{wq}, x_{Te}x_{Qs}, x_V, x_{Qg}]$. The step by step approach to find the operating-point values of the state variables and parameters of the WT are included in [27, 28].

2.2 Small Signal Model of BESS

The BESS is integrated to the grid via a Voltage Source Converter (VSC) and a filter as shown in Fig. 4. As can be observed, the BESS includes battery packs, a DC/AC converter, a transformer, and an output filter. The BESS is linked to the grid at the point of common coupling (PCC)—point P. Since the dynamics of the battery cell are much slower than the converter dynamics, it is reasonable to ignore the battery cell dynamics for small signal analysis, and therefore, only the fast dynamics of the converter, its controller, the LCL filter, and the grid are considered in this section.

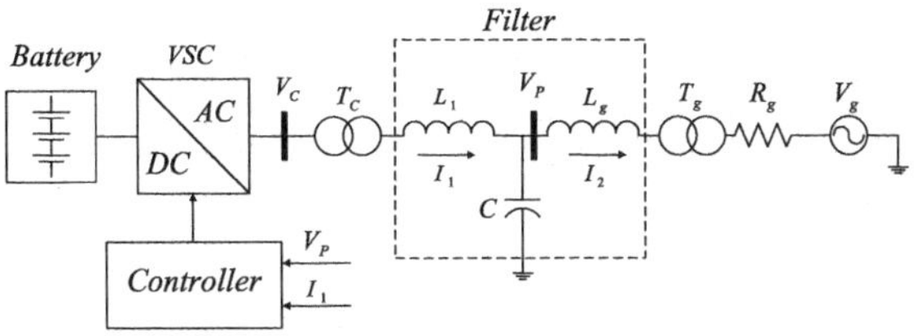

Fig. 4 Top level diagram representation of a BESS connected to the grid

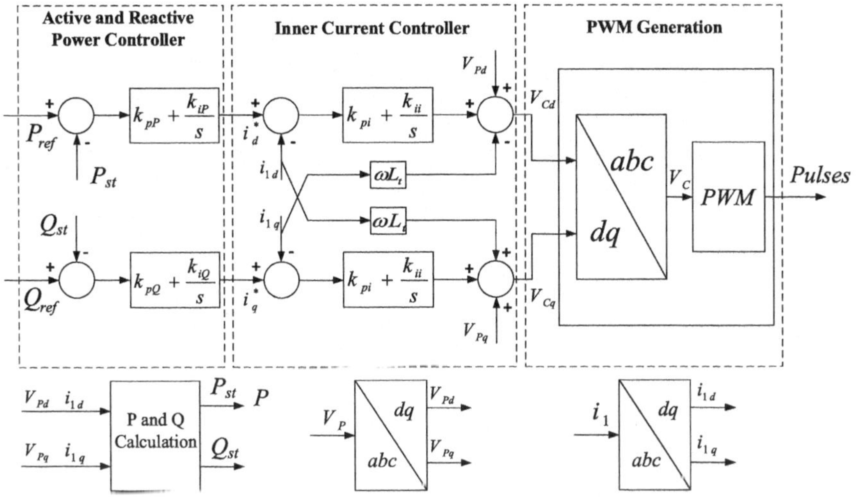

Fig. 5 VSC controller for BESS

When the VSC operates in a grid connected mode, the primary controller is in charge of meeting the active and reactive power demand. The AC voltage and frequency in this mode are controlled by the grid. In the islanded mode, when the connection to the grid is lost, the VSC regulates the AC voltage magnitude and the frequency as well as the active and reactive power demand [29]. A basic structure of the BESS control in the grid supporting mode is illustrated in Fig. 5. It mainly has three control loops; the inner current loop, the active and reactive power loops, and the primary/secondary frequency and voltage loops. Voltages and currents in the *abc*-frame, that are the inputs to the VSC, are measured at the PCC and passed through a first order filter to remove the high frequency harmonics. A phase-locked loop (PLL) is used to measure the real-time frequency of the system. The filter's transfer function is of the form $\frac{1}{\beta s+1}$, where β is the filter's time constant. Since analyzing the small signal stability of the BESS is the primary objective of this section, the effects of the first order filter and computational delays involved in the PCC measurements are ignored for simplicity[30]. In the following sections, the small signal model of the BESS with various control loops are derived.

The dynamics of the AC grid in the system are derived using Fig. 4. Applying the Kirchoff's voltage and current laws (KVL and KCL) around the PCC, the AC dynamics of the system in dq frame are expressed in the following equations:

$$\begin{aligned} L_f \frac{d\bar{i}_1}{dt} &= \bar{v}_c - \bar{v}_p + jL_f\omega_b\bar{i}_1 \\ C\frac{d\bar{v}_p}{dt} &= \bar{i}_1 - \bar{i}_2 + jC\omega_b\bar{v}_p, \\ \hat{L}_g\frac{d\bar{i}_2}{dt} &= \bar{v}_p - \bar{i}_2R_g + jL_g\omega_b\bar{i}_2 - \bar{v}_g, \end{aligned} \tag{10}$$

where ω_b is the rated frequency in rad/s, L_f is the summation of inductances of the converter transformer and converter side filter ($L_f = L_{Tc} + L_1$), and $\hat{L}_g$ is the summation of inductances of the grid side transformer and the grid side filter inductance ($\hat{L}_g = L_g + L_{Tg}$). It is noted that variables with a bar ($^-$) on top represent vectors in dq frame and need to be separated for the small-signal analysis. The inner current controller generates the reference voltages in the dq-frame, which are fed to the pulse width modulation (PWM) generation unit to generate the gate pulses necessary to operate the power electronic switches in the VSC. Dynamics of the inner current control loop in the dq frame can be represented as:

$$\begin{aligned} v_{cd}^* &= \left(k_{pi} + \frac{k_{ii}}{s}\right)(i_d^* - i_{1d}) - \omega L_f i_{1q} + v_{pd}, \\ v_{cq}^* &= \left(k_{pi} + \frac{k_{ii}}{s}\right)(i_q^* - i_{1q}) + \omega L_f i_{1d} + v_{pq}. \end{aligned} \tag{11}$$

If the converter is ideal and dynamics of the PWM generation is ignored, the reference voltages, v_{cd} and v_{cq}, in the dq-frame generated by the inner current controller are the same as the measured voltages in the terminal of the converter. Therefore, $v_{cd}^* \approx v_{cd}$ and $v_{cq}^* \approx v_{cq}$. In this equation, i_d^* and i_q^* are the reference currents generated by the outer control loop, i_{1d} and i_{1q} are the currents measured at the PCC and the feed forward voltage components, v_{pd} and v_{pq}, are measured at the PCC. Furthermore, k_{pi} and k_{ii} are the PI controller gains for the inner current loop controller.

The outer loops of the BESS control the active power P and reactive power Q, delivered by the BESS to the grid. Two PI controllers which track the references P_{ref} and Q_{ref} and generate the reference currents (i_d^* and i_q^*) are used.

$$\begin{aligned} i_d^* &= \left(k_{pP} + \frac{k_{iP}}{s}\right)(P_{ref} - P), \\ i_q^* &= \left(k_{pQ} + \frac{k_{iQ}}{s}\right)(Q_{ref} - Q). \end{aligned} \tag{12}$$

The outer control loop dynamics in the dq-frame are expressed in (12), where P_{ref} and Q_{ref} are active and reactive power demanded by the grid, P and Q are active and reactive power measured at the PCC, which can be calculated using $P = 1.5(v_{pd}i_{1d} + v_{pq}i_{1q})$, $Q = 1.5(-v_{pd}i_{1q} + v_{pq}i_{1d})$ [31].

The PLL block measures the real-time frequency of the system and decouples the dq-frame components so that they can be independently controlled [32]. The PLL block uses a PI controller to force the q component of voltage measured at the PCC, v_{pq}, to zero. The dynamics of the PLL block is represented in (13):

$$\begin{aligned} \omega &= \left(k_p^{PLL} + \frac{k_i^{PLL}}{s}\right) v_{pq} + \omega_0, \\ \dot{x}_{PLL} &= v_{pq}, \\ \omega_{PLL} &= \omega_0 + k_p^{PLL} v_{pq} + k_i^{PLL} x_{PLL}. \end{aligned} \tag{13}$$

2.3 Overall Small Signal Model of the Integrated DFIG WT+BESS

The BESS will be integrated into the DFIG model at the PCC. Therefore, dynamics of the PCC represented in (10) needs to be updated. Referring to Fig. 4, the DFIG will be connected to the point of common coupling (v_p), therefore, a new current component including the DFIG's stator and the grid side converter currents ($i_w = i_s + i_g$) will be added to the PCC dynamics. The new PCC dynamics including the BESS and DFIG model is expressed as:

$$\begin{aligned} C\dot{v}_{pd} &= i_{1d} + i_{sd} + i_{gd} - i_{2d} + C\omega v_{pq}, \\ C\dot{v}_{pq} &= i_{1q} + i_{sq} + i_{gq} - i_{2q} - C\omega v_{pd}. \end{aligned} \tag{14}$$

Therefore, (10) will be updated by (14). The BESS model is then linearized and added to the DFIG's small signal model to form the entire small signal model of the DFIG+BESS system. The MATLAB/LINMOD function is used to linearize the overall system and the eigenvalues of the combined wind turbine and the BESS are shown in Table 1. By comparing the eigenvalues of the DFIG alone and the combined DFIG+BESS system, it is observed that by integrating the BESS with the DFIG, the frequencies of λ_{25}, λ_{26}, λ_{27}, and λ_{28} have decreased from 60 to 6 Hz. However, the overall system remains stable as all the eigenvalues belong to the open left half complex plane.

Table 1 Eigenvalue analysis of a DFIG WT+BESS connected to the grid through an LCL filter

Eigenvalue	f (Hz)	Eigenvalue	f (Hz)	Eigenvalue	f (Hz)
$\lambda_1 = -10.2 + j0.98$	60	$\lambda_{11} = -20$	0	$\lambda_{21} = -39.6$	0
$\lambda_2 = -10.2 - j0.98$	60	$\lambda_{12} = -3.89$	0	$\lambda_{22} = -20$	0
$\lambda_3 = -33.2 + j0.32$	20	$\lambda_{13} = -1$	0	$\lambda_{23} = -9.83 + j2.5$	150
$\lambda_4 = -33.2 - j0.32$	20	$\lambda_{14} = -0.9$	0	$\lambda_{24} = -9.83 - j2.5$	150
$\lambda_5 = -78.7$	0	$\lambda_{15} = -0.001$	0	$\lambda_{25} = -3.6 + j0.1$	6
$\lambda_6 = -108.9 + j0.08$	5	$\lambda_{16} = -1$	0	$\lambda_{26} = -3.6 - j0.1$	6
$\lambda_7 = -108.9 - j0.08$	5	$\lambda_{17} = -20$	0	$\lambda_{27} = -0.1 + j0.1$	6
$\lambda_8 = -20.79$	0	$\lambda_{18} = -48.9$	0	$\lambda_{28} = -0.1 - j0.1$	6
$\lambda_9 = -15.1 + j0.16$	10	$\lambda_{19} = -35.1 + j33$	1980	$\lambda_{29} = -19.2$	0
$\lambda_{10} = -15.1 - j0.16$	10	$\lambda_{20} = -35.1 - j33$	1980		

3 Distributed Cooperative Controller Design for DFIG Wind Turbines with On-site BESSs

The wind farm sizes are explicitly considered and the controller will guarantee that at any moment, the BESSs have enough state of charge to support the load even when the wind speed varies. The main objective is to synchronize the BESSs and WT's GSCs cooperatively for attaining the control purposes below:

- **Supply-demand balance**: The first-priority control objective for a WF is to balance the supply and demand of active and reactive power. The power demands (P_d, Q_d) are obtained from the weather forecast and the optimal power dispatch solution. The time slot for the forecast update is five minutes which is enough for distributed controllers to appropriately regulate the set points for BESSs and WTs such that the power demands (P_d, Q_d) are always met by the WF. In fact, the sizes of WTs might differ in a WF, therefore, each WT is supposed to response to the demand depending on its output power. This means WTs are expected to share the load demand equally with regards to their maximum capacities. The BESSs can play a key role to compensate the wind fluctuations and limited capacity of WTs in demand response. Therefore, the heterogeneity of a WT with its on-site BESS is denoted by a gain, $K_{st,i}$. As the capacity of the on-site BESS is normally the same as the capacity of WT, same gain can be considered for BESS and WT [33]. Therefore, the power balance can be represented by [33]

$$P_d = \sum_{i=1}^{n} K_{st,i}(P_{r,i} + P_{s,i} + P_{st,i}) \Leftrightarrow \sum_{i=1}^{n} K_{st,i} P_{st,i} = P_d - \sum_{i=1}^{n} K_{st,i}(P_{r,i} + P_{s,i}), \tag{15}$$

and $Q_d = \sum_{i=1}^{n} K_{st,i}(Q_{g,i} + Q_{s,i} + Q_{st,i})$. Normally, $Q^*_{s,i}$ is set to zero, since the DFIG operates at unity power factor, hence $Q_{s,i}$ is also zero. Thus,

$$Q_d = \sum_{i=1}^{n} K_{st,i} Q_{st,i} + \sum_{i=1}^{n} K_{st,i} Q_{g,i}. \tag{16}$$

- **Equal sharing of power demands**: Since the capacities of BESSs are different, it is expected that they equally share active and reactive power demands [33], i.e.,

$$\begin{aligned} K_{st,i} P_{st,i} &= K_{st,j} P_{st,j}, \\ K_{st,i}(Q_{g,i} + Q_{st,i}) &= K_{st,j}(Q_{g,j} + Q_{st,j}). \end{aligned} \tag{17}$$

Furthermore, since the capacity of GSC is normally up to 30% of the rated DFIG power, only 30% of reactive load demand can be supplied by the DFIG and the other 70% should be supported by BESSs

$$\sum_{i=1}^{n} K_{st,i} Q_{st,i} = 0.7 Q_d, \quad \sum_{i=1}^{n} K_{st,i} Q_{g,i} = 0.3 Q_d. \tag{18}$$

- **BESS Energy Synchronization**: It is also expected that the state of charges of BESSs are enough at any moment so that no BESS is out of energy while the others are still full. Therefore, another control objective in this chapter is to guarantee:

$$E_{st,i} = E_{st,j}. \tag{19}$$

In order to simplify the control designs for achieving the given control objectives in (15)–(19), the simplified models of the DFIG WT and the BESS are developed in the following subsections.

3.1 Simplified Model of DFIG Wind Turbine

The connection between a DFIG WT and the grid is at the terminals of both the rotor and the stator. The objective of such connection is to supply the active and reactive powers from the WF to the grid. More specifically, the DFIG's stator will supply the active and reactive powers $P_{s,i}$ and $Q_{s,i}$. Next, two back-to-back converters called the GSC and the RSC are utilized to connect the DFIG's rotor to the grid. For a DFIG WT, the GSC usually accounts for 25–30% of its nominal power, and the GSC's active power and reactive power $P_{g,i}$ and $Q_{g,i}$ could be individually regulated. On the other hand, the RSC independently regulates the stator-side active power and reactive power $P_{s,i}$ and $Q_{s,i}$ by regulating the rotor current in a synchronous reference frame through two control loops. The external loop individually controls $P_{s,i}$ and

$Q_{s,i}$, whilst the internal loop controls the rotor current (see Fig. 2 for details). Then the distributed controller for reactive power is added to the external loop. Suppose that the inner and the outer loops can successfully track the reference values, i.e., $P_s^* \approx P_s$, $Q_s^* \approx Q_s$, the RSC dynamics is simplified by:

$$K_{st,i}\dot{Q}_{s,i} = u_i^{Qr}, \tag{20}$$

where u_i^{Qr} is a control input synthesized for the equal sharing of reactive power of the DFIG i with an on-site BESS using RSC. The GSC's control regulates the voltage $V_{dc,i}$ of the DC-link and the exchanged reactive power $Q_{g,i}$ by employing a synchronous frame based vector control method. Similar to RSC, two control loops exist for the GSC (see Fig. 3 for details), in which a control input is incorporated to the external loop for regulating the reactive power of DFIG WT+BESS in a distributed manner. Assuming that the controllers for the DC-link and reactive power can track the reference values fast enough, i.e., $V_{dc}^* \approx V_{dc}$ and $Q_g^* \approx Q_g$, the GSC controller dynamics is simplified by:

$$K_{st,i}\dot{Q}_{g,i} = u_i^{Qs}, \tag{21}$$

where u_i^{Qs} is the incorporated distributed input to be designed for the equal sharing of the DFIG WT and an on-site BESS reactive power.

3.2 BESS Simplified Model

Vector current control strategy is used for the BESS, as shown in Fig. 6, to independently regulate the active power $P_{st,i}$ and reactive power $Q_{st,i}$ transferred to the grid. More precisely, the active power of BESSs is utilized to compensate the variations on the WT' output active power in presence of fluctuating wind velocity. Furthermore, BESSs' reactive powers are controlled to accommodate the sharing of reactive power from WTs+BESSs in a WF. Because BESSs' active power and reactive power could be individually controlled, two distributed control inputs will be designed accordingly. Recently, the research in [34] has shown that a simplified model, which were validated through real-time simulations, can be employed for BESSs to develop distributed control algorithms. Hence, a similar simplified BESS dynamics is used here for synthesizing distributed control algorithms for WTs with on-site BESSs. Note that the on-site BESSs may have different sizes because of the difference on WTs' sizes. To take into account the heterogeneity of on-site BESSs, the following model is employed,

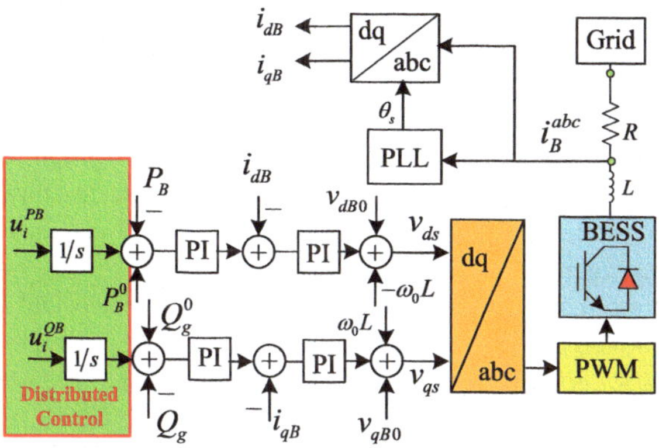

Fig. 6 Distributed control structure for the BESS

$$\dot{E}_{st,i} = \frac{-K_{st,i}}{3600} P_{st,i}, \tag{22a}$$

$$K_{st,i} \dot{P}_{st,i} = u_i^{PB}, \tag{22b}$$

$$K_{st,i} \dot{Q}_{st,i} = u_i^{QB}. \tag{22c}$$

3.3 Problem Formulation

Based on (21), (22), and the given control objectives, an overall simplified model for DFIG WTs with on-site BESSs is described below,

$$\dot{x}_i = Ax_i + Bu_i, \tag{23}$$

$$x_i \triangleq \begin{bmatrix} E_{st,i} \\ K_{st,i} P_{st,i} \\ K_{st,i} Q_{st,i} \\ K_{st,i} Q_{g,i} \end{bmatrix}, u \triangleq \begin{bmatrix} u_i^{PB} \\ u_i^{QB} \\ u_i^{Qs} \end{bmatrix}, A \triangleq \begin{bmatrix} 0 & \frac{-1}{3600} & 0 & 0 \\ 0 & 0 & 0 & 0 \\ 0 & 0 & 0 & 0 \\ 0 & 0 & 0 & 0 \end{bmatrix}, B \triangleq \begin{bmatrix} 0 & 0 & 0 \\ 1 & 0 & 0 \\ 0 & 1 & 0 \\ 0 & 0 & 1 \end{bmatrix}.$$

Remark 1 It is worth noting that $Q_{s,i} \neq 0$ when WTs do not work at unity power factor, hence $Q_{s,i}$ will appear in the equations (16) and (17). Hence, a state variable $K_{st,i} Q_{s,i}$ and a control input u_i^{Qr} must be added to the overall system state-space model in (23), and matrices A, B should be modified correspondingly. Therefore, the distributed designs to be proposed in Sect. 3.4 need to be modified accordingly, however, the design principle is still the same.

3.4 Control Design

To synthesize distributed controllers for the WF, MAS theory is utilized, in which each WT+BESS is considered as an agent, and an undirected graph $\mathcal{G}$, with its vertex set $\mathcal{V}$, and edge set $\mathcal{E}$ is considered for the communication structure among agents. Each vertex constitutes an agent and each edge $(k, j) \in \mathcal{E}$ matches to the communication between agent k and agent j. Then $\mathcal{N}_k \triangleq \{j \in \mathcal{V} : (k, j) \in \mathcal{E}\}$ stands for agent k's neighboring set. Denote $\mathcal{A}$ the graph adjacency matrix and a_{kj} its elements, where $a_{kj} = 1$ if $(k, j) \in \mathcal{E}$, $a_{kj} = 0$ otherwise. Next, let $\mathcal{D} = \text{diag}\{d_k\}_{k=1,\ldots,n}$ be the degree matrix of $\mathcal{G}$, in which $d_k \triangleq \sum_{j\in\mathcal{N}_k} a_{kj}$. Subsequently, $\mathcal{L} \triangleq \mathcal{D} - \mathcal{A}$ is called the Laplacian matrix.

To account for the difference on the demand from the main grid and the WF output power, the WF supervisory controller is equipped with a virtual leader which sends an information on the power imbalance to at least one agent in the MAS. As a result, all WTs+BESSs can equally share that power imbalance throughout the communications among agents. The leader is denoted through the sub-index 0 and $\mathcal{N}_0$ denotes its neighboring set. Consequently, define $a_{0i} = 1$ if $i \in \mathcal{N}_0$, i.e., there is a connection between the leader and agent i while $a_{0i} = 0$ otherwise. Finally, let $\Gamma \triangleq \text{diag}\{a_{0i}\}_{i=1,\ldots,N}$.

3.4.1 Control Design without Communication Delays

To track a power demand from the grid, the WF supervisory controller contains a virtual leader with the following dynamics

$$\dot{x}_0 = Ax_0, \tag{24}$$

where $x_0 \in \mathbb{R}^4$ is its state vector capturing the power imbalance information. Then the design of distributed controllers for WTs with on-site BESSs is proposed as follows.

Theorem 1 *Having $\mathcal{G}$ connected, with at least once connection from the virtual leader to an agent, the virtual leader initial states are set to be*

$$x_0(0) = \left[0, \frac{P_d - \sum K_{st,i}(P_{r,i} + P_{s,i})}{n}, \frac{0.7Q_d}{n}, \frac{0.3Q_d}{n}\right]^T. \tag{25}$$

The distributed controller for WTs+BESSs is designed as follows,

$$u_i = -cK \sum_{j\in\mathcal{N}_i} a_{ij}(x_i - x_j) - c_0 a_{i0} K(x_i - x_0), \tag{26}$$

with any $c > 0, c_0 > 0$; $K = \begin{bmatrix} k_1 & k_2 & 0 & 0 \\ 0 & 0 & k_3 & 0 \\ 0 & 0 & 0 & k_4 \end{bmatrix}$, $k_1 < 0, k_2 > 0, k_3 > 0, k_4 > 0$, *and* $i = 1, 2, \dots, n$. *Then, the following quantities reach consensus: energy levels, proportional active powers, and proportional reactive powers of all BESSs, proportional reactive powers of GSCs. Moreover, the demand power tracking is achieved.*

Proof Denote $\zeta = x - \mathbf{1}_n \otimes x_0$. We then obtain

$$u = -[c_0 \Gamma \otimes K]\zeta - c[\mathcal{L} \otimes K]x. \tag{27}$$

Therefore,

$$\begin{aligned} \dot{\zeta} &= [I_n \otimes A]\zeta + [I_n \otimes B]u \\ &= [I_n \otimes A]\zeta - c[I_n \otimes B][\mathcal{L} \otimes K]x - [I_N \otimes B][c_0 \Gamma \otimes K]\zeta \\ &= [I_n \otimes A - c\mathcal{L} \otimes BK - c_0 \Gamma \otimes BK]\zeta - c[\mathcal{L} \otimes BK](\mathbf{1}_n \otimes x_0), \\ &= [I_n \otimes A - (c\mathcal{L} + c_0 \Gamma) \otimes BK]\zeta, \end{aligned} \tag{28}$$

because $\mathcal{L}\mathbf{1}_n = 0$. Let us denote $\mathbf{A} \triangleq I_n \otimes A - (c_0 \Gamma + c\mathcal{L}) \otimes (BK)$, then the dynamics of ζ in (28) is rewritten as

$$\dot{\zeta} = \mathbf{A}\zeta. \tag{29}$$

Note that $c_0 \Gamma + c\mathcal{L}$ is positive definite because of the connectedness of $\mathcal{G}$ and the connection from the virtual leader to an agent [35]. Denote the eigenvalues of $c_0 \Gamma + c\mathcal{L}$ by λ_i, then $\lambda_i > 0 \; \forall \, i = 1, \dots, n$. Moreover, $c_0 \Gamma + c\mathcal{L}$ is diagonalizable by an orthogonal matrix $U \in \mathbb{R}^{n \times n}$, which leads to

$$\mathbf{A} = (U \otimes I_4)\text{diag}\{A - \lambda_i BK\}_{i=1,\dots,N}(U^T \otimes I_4), \tag{30}$$

Therefore, all eigenvalues of matrix $\mathbf{A}$ are actually indicated by the eigenvalues of matrices $A - \lambda_i BK$ for $i = 1, \dots, n$. It can be easily shown that the characteristic equation of the system matrix $A - \lambda_i BK$ is

$$(s + \lambda k_4)(s + \lambda k_3)\left(s^2 + \lambda_i k_2 s - \frac{1}{3600}\lambda_i k_1\right) = 0. \tag{31}$$

Thus, $A - \lambda_i BK$ is stable, and hence, $\mathbf{A}$ is stable, if and only if $\lambda_i k_4 > 0$, $\lambda_i k_3 > 0$, $\lambda_i k_2 > 0$ and $\frac{1}{3600}\lambda_i k_1 < 0$, which are equivalent to $k_2 > 0, k_3 > 0, k_4 > 0$, and $k_1 < 0$. This means the consensus is achieved for the energy levels, proportional active powers, and proportional reactive powers of all BESSs. Further, the demand power tracking is achieved because of the initial conditions setting (25). □

3.4.2 Control Design in Presence of Communication Delays

Suppose that all communication channels have a uniform time delay which is equal to $\tau > 0$. Then the distributed controller design in Section 3.4.1 needs to be revised to account for time delays, as follows.

Theorem 2 *Having $\mathcal{G}$ connected, with at least one connection from the virtual leader to an agent, the virtual leader initial states are set by (25). The delayed distributed controller is designed as follows,*

$$u_i(t) = cK \sum_{j\in\mathcal{N}_i} a_{ij}[x_i(t-\tau) - x_j(t-\tau)] - c_0 a_{i0} K(x_i(t-\tau) - x_0(t-\tau)), \tag{32}$$

where K is defined in Theorem 1, $c > 0,\ c_0 > 0$, and

$$\tau < \frac{\arccos \frac{1}{\frac{k_2^2\lambda_n}{2\alpha k_1} + \sqrt{(\frac{k_2^2\lambda_n}{2\alpha k_1})^2+1}}}{\sqrt{\frac{k_2^2\lambda_n^2}{2} + \sqrt{(\frac{k_2^2\lambda_n^2}{2})^2 + \alpha^2 k_1^2\lambda_n^2}}}, \tag{33}$$

with λ_n denotes the maximum eigenvalue of $c_0\Gamma + c\mathcal{L}$ and $\alpha \triangleq -\frac{1}{3600}$. Consequently, the consensus is achieved for all BESSs for their energy levels, proportional active powers, and proportional reactive powers. The consensus is also achieved for proportional reactive powers of GSCs. Moreover, the demand power tracking is achieved.

Proof The closed-loop system dynamics in presence of the controller (32) is

$$\dot{\zeta}(t) = (I_n \otimes A)\zeta(t) - (c_0\Gamma + c\mathcal{L}) \otimes (BK)\zeta(t-\tau). \tag{34}$$

Subsequently, utilizing the similar approach as Theorem 1's proof leads to the result that $\zeta(t)$ becomes 0 if and only if the closed-loop transfer function $sI_4 - (A - \lambda_i BKe^{-\tau s})$ is stable. This is achieved if all roots of the following characteristic equation lie on the open left hand side of the complex plane,

$$(s + \lambda k_4 e^{-\tau s})(s + \lambda k_3 e^{-\tau s}) \left(s^2 + \lambda_i k_2 e^{-\tau s} s - \frac{1}{3600}\lambda_i k_1 e^{-\tau s}\right) = 0. \tag{35}$$

The equations $s + \lambda k_4 e^{-\tau s} = 0$ and $s + \lambda k_4 e^{-\tau s} = 0$ have all roots on the open left hand side of the complex plane if and only if $k_4 > 0$ and $k_3 > 0$. Next, we can prove that all roots of the following equation also lie on the open left hand side of the complex plane,

$$s^2 + \lambda_i k_2 e^{-\tau s} s - \frac{1}{3600}\lambda_i k_1 e^{-\tau s} = 0, \tag{36}$$

if (33) is satisfied, by employing the same technique in [36]. □

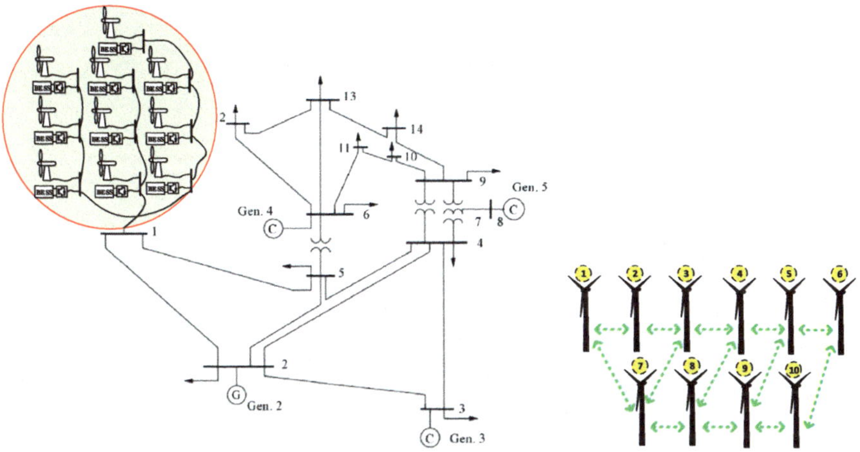

Fig. 7 The tested IEEE 14-bus system with 10 WTs and BESSs, and their communication graph

4 Case Studies

The system under consideration, which is depicted in Fig. 7, is the modified version of the classical IEEE 14-bus system whose parameters can be found in [37]. Ten WT+BESS modules are added to the system at Bus 1 as a replacement of the generator. Distributed control inputs are then developed and supplemented to the system based on (26) and (32). For the control design, the MatDyn toolbox is utilized [38]. The undirected communication structure of the system under consideration shown in Fig. 7 means that two-way communication channels are available for neighboring WTs+BESSs. To validate the proposed approach's performance in the regulation of the WF output power, a few case studies are considered. Parameters of the design can be found in [33]. BESS parameters can be found in [34].

4.1 Case 1: Variation of Wind Speed and Demand Response

In this scenario, the proportional power sharing and energy synchronization are validated for BESSs in terms of fluctuating wind velocity and a sudden demand change. To account for unpredicted wind speed variations due to weather conditions, the wind speed is randomly changed, which results in WF output power change accordingly. In addition, a step increase on the active power demand P_d is applied after 4000 s. Simulation results are displayed in Fig. 8. It can be seen that the energy levels of BESSs are synchronized and the power is equally shared between BESSs.

As also shown in the last subplot, the mismatch between the total load demand and generated WT powers at anytime (red line) is compensated by energy storage devices (black line).

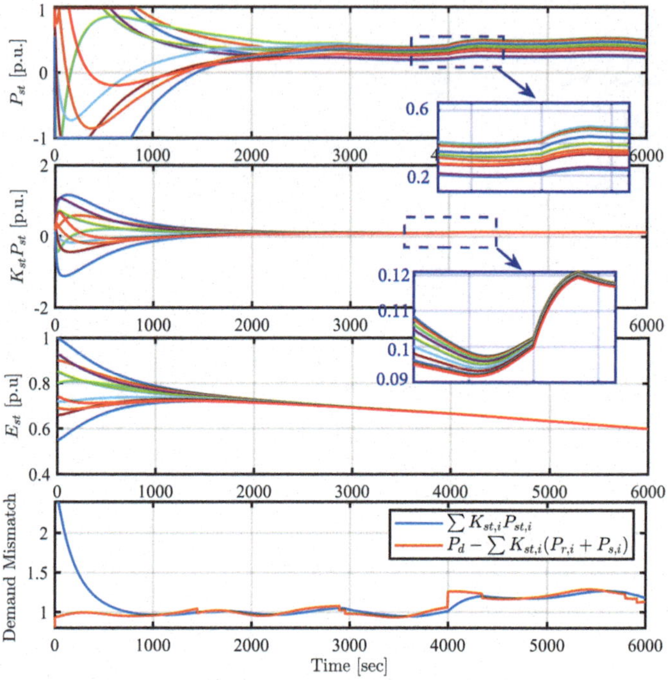

Fig. 8 Simulation results for case 1

In order to have a faster synchronization, the consensus gains can be adjusted, but the BESS output power will experience higher transients. The information on how to adjust the consensus gains to achieve limited transients in the output power was further investigated in [34], the interested readers can check this reference for details.

4.2 Case 2: Invariant Wind Velocity and Communication Delays

The requirements on allowed communication delays in smart grids were developed by the IEEE (P2030—Smart Grid Interoperability Standard and 1527—Interconnection Standard) and the International Electrotechnical Commission (IEC 61850—Power Utility Automation) to ensure the grid reliability and to avoid possible transient problems. Utilizing those standards, the maximum time delay for communication in the monitoring and control is defined as 16 milliseconds (ms) [39], which is adopted in the current analysis. To verify the effect of time delays, the delayed control inputs in (32) are supplemented to the WTs and BESSs. Two scenarios are investigated: (i)

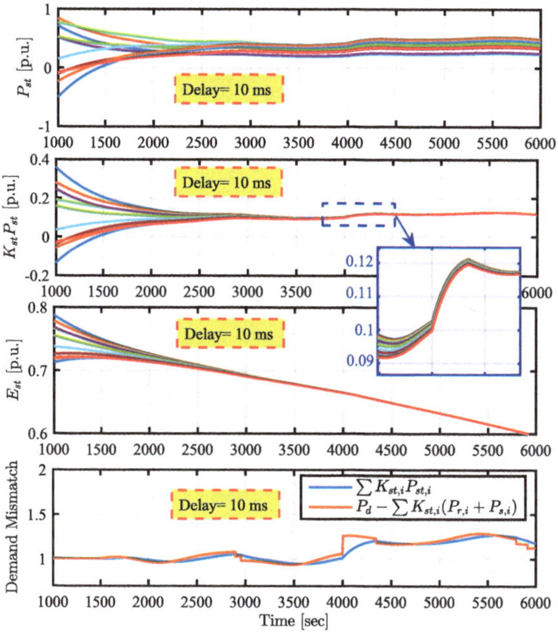

Fig. 9 Simulation results for communication delays in case 2

10 ms time delays; and (ii) time delays start at 20 ms then increase to 30 ms after 4050 s.

Figure 9 depicts the simulation results for 10 ms communication delays. It can be observed that the synthesized controller successfully shares the active power between BESSs, synchronizes the state of charge levels of BESSs, and compensates the demand mismatch (as seen in the last subplot). It is also observed that the synchronization takes more time, around 3000 s, to obtain the consensus of active power sharing (first subplot). However, it takes less than 2000 s to achieve consensus when there is no communication delays. As a result, the synchronization is slower when time delays exist.

Figure 10 shows the simulation results when there exist 20 ms time delays for the first 4050 s and when there are 30 ms communication delay after 4050 s. The obtained results show that the system synchronizes the WTs+BESSs successfully when the communication delays are 20 ms. In contrast, by increasing the communication delays to 30 ms, the designed controller cannot coordinate the power sharing and the system loses its stability. Based on the IEEE standard, the maximum allowable communication delay for a control system is 16 ms [39]. Therefore, the designed approach can successfully operate in standard communication systems.

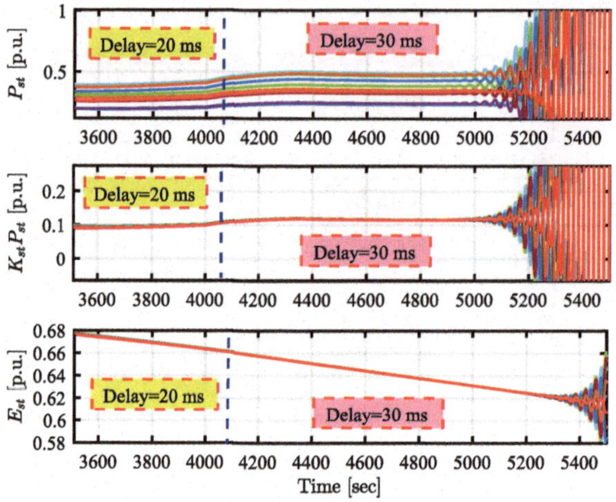

Fig. 10 Simulation results for 20 and 30 ms time delays

4.3 *Case 3: Failure in a Communication Link*

This case study investigates the capability of the developed design in case of communication failures. It should be noted that all available consensus designs (including the developed approach) are valid in connected communication graphs. This means, if the communication structure in the system becomes unconnected due to communication failures, the synchronization of the entire system is impossible. However, the synchronization can be attained in each sub-network associated with each sub-graph which is connected.

Next, assume that a communication failure happens for WT+BESS number 1 where all communication links from this WT+BESS to its neighbors are failed. Consequently, WT+BESS number 1 is separated from the rest of the system. Based on the above clarification, the synchronization in this case should be achieved for WT+BESS number 2–10, as shown in Fig. 11. It is observed that WT+BESS number 2–10 could still successfully support the demand mismatch without help from BESS number 1. In addition, the consensus of active power sharing and energy levels is still attained for the connected WTs+BESSs (WT+BESS numbers 2–10). The results validate that the designed approach operates even when communication failures occur.

4.4 *Case 4: Verification of Reactive Power Demand*

This section investigates the ability of the designed distributed controllers in reactive power sharing. The purpose is to equally share the demand on reactive power

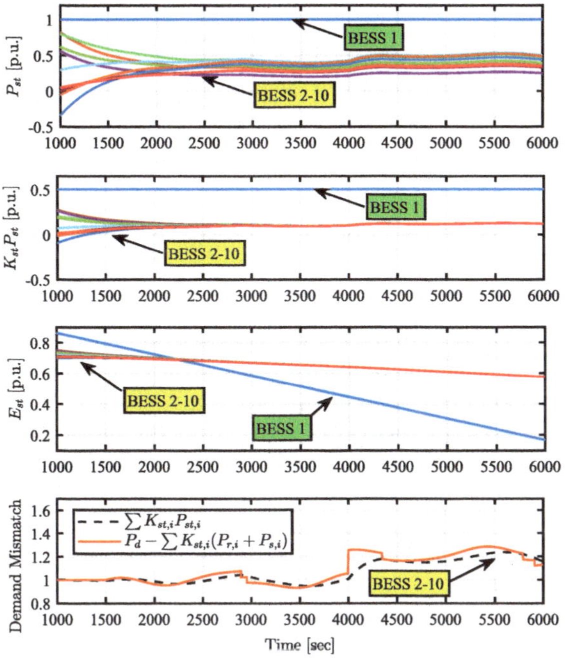

Fig. 11 Simulation results for communication failure

among all GSCs of WTs and their on-site BESSs based on their respective maximum capacities, as shown in (16).

The reactive power demand is initially set to 1 p.u., where after 1500 s the reactive power demand is suddenly increased by 50%. Subsequently, after 4000 s, the demand is decreased to 0.5 p.u. Simulation results are depicted in Fig. 12. It is seen that the WT+BESS units can successfully share the demand on reactive power even with sudden demand changes. In addition, the proposed controller operates successfully in presence of maximum 20 ms time delays. If the time delay is increased from 20 to 30 ms, the system becomes unstable. Hence, as discussed in case 2, the proposed approach meets the demand on reactive power for standard communication systems. Also, the fast synchronization of reactive power sharing can be achieved as the reactive power does not depend on BESS energy. By adjusting the consensus input gains, k_3 and k_4, different synchronization speeds are achieved.

5 Conclusion and Remarks

The problem of WF output power regulation using standalone BESS is considered in this chapter. First, the small signal stability analysis is conducted on full models of

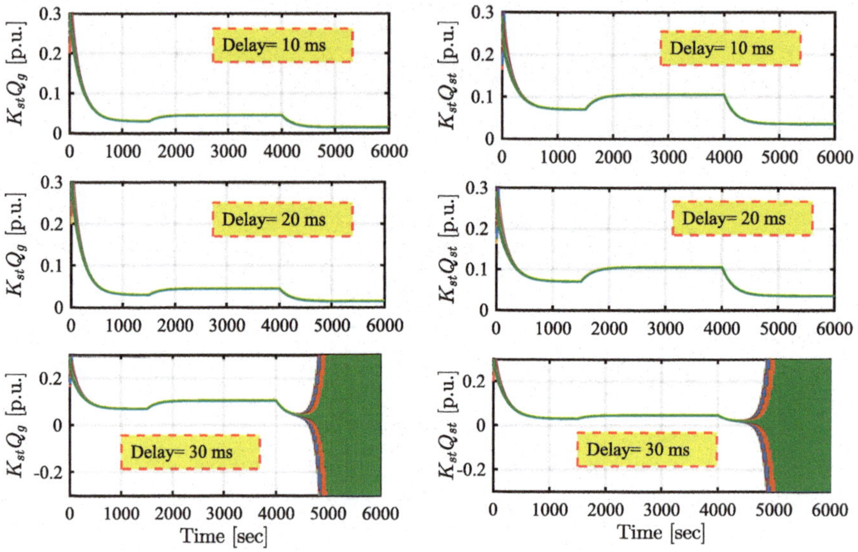

Fig. 12 Simulation results for reactive power control

the DFIG WT and on-site BESS system when they are connected to the grid. Next, a novel control architecture based on the leader-follower consensus theory is proposed for DFIG WTs with on-site BESSs. The proposed approach could synchronize the energy levels, the active power sharing, and the reactive power sharing of individual WTs with on-site BESSs. Two distributed consensus design approaches are derived for the scenarios of having no time delay and delayed communication links. In case of no communication delays, the delay-free consensus design can successfully meet the power demand mismatch and compensate for the variations in wind velocity or power demands. When there are time delays in the communication links, the proposed distributed delayed design can successfully ensure the cooperation of WTs with on-site BESSs. In case of failures on communication links, the proposed approach can also synchronize the operation of WTs with on-site BESSs if the rest of the system is still connected.

References

1. Mauricio JM, Marano A, Gómez-Expósito A, Ramos JLM (2009) Frequency regulation contribution through variable-speed wind energy conversion systems. IEEE Trans Power Syst 4(1):173–180
2. Howlader AM, Urasaki N, Yona A, Senjyu T, Saber AY (2013) A review of output power smoothing methods for wind energy conversion systems. Renew Sustain Energy Rev 26:135–146

3. Muyeen SM, Ali MH, Takahashi R, Murata T, Tamura J (2007) Wind generator output power smoothing and terminal voltage regulation by using STATCOM/ESS. In: Proceedings of power tech, 2007 IEEE Lausanne, pp 1232–1237. IEEE
4. Østergaard PA (2006) Ancillary services and the integration of substantial quantities of wind power. Appl Energy 83(5):451–463
5. Pao LY, Johnson KE (2011) Control of wind turbines: approaches, challenges, and recent developments. IEEE Control Syst Mag 31(2):44–62
6. Pao LY, Johnson KE (2009) A tutorial on the dynamics and control of wind turbines and wind farms. In: Proceedings of 2009 American control conference, pp 2076–2089
7. Fernandez LM, Garcia CA, Jurado F (2008) Comparative study on the performance of control systems for doubly fed induction generator (DFIG) wind turbines operating with power regulation. Energy 33(9):1438–1452
8. Boukhezzar B, Siguerdidjane H (2005) Nonlinear control of variable speed wind turbines for power regulation. In: Proceedings of 2005 IEEE conference on control applications, CCA 2005, pp 114–119, Aug 2005
9. Beltran B, Ahmed-Ali T, Benbouzid MEH (2008) Sliding mode power control of variable-speed wind energy conversion systems. IEEE Trans Energy Convers 23(2):551–558
10. Wang Q, Chang L (2004) An intelligent maximum power extraction algorithm for inverter-based variable speed wind turbine systems. IEEE Trans Power Electron 19(5):1242–1249
11. Chinchilla M, Arnaltes S, Burgos JC (2006) Control of permanent-magnet generators applied to variable-speed wind-energy systems connected to the grid. IEEE Trans Energy Convers 21(1):130–135
12. Fleming P, Aho J, Gebraad P, Pao L, Zhang Y (2016) Computational fluid dynamics simulation study of active power control in wind plants. In: Proceedings of 2016 American control conference, pp 1413–1420
13. van Wingerden J-W, Pao L, Aho J, Fleming P (2017) Active power control of waked wind farms. IFAC PapersOnLine 50(1):4484–4491
14. Dragičević T, Guerrero JM, Vasquez JC, Škrlec D (2014) Supervisory control of an adaptive-droop regulated DC microgrid with battery management capability. IEEE Trans Power Electron 29(2):695–706
15. Xu Y, Zhang W, Hug G, Kar S, Li Z (2015) Cooperative control of distributed energy storage systems in a microgrid. IEEE Trans Smart Grid 6(1):238–248
16. Khalid M, Savkin AV (2014) Minimization and control of battery energy storage for wind power smoothing: aggregated, distributed and semi-distributed storage. Renew Energy 64:105–112
17. Sedghi M, Ahmadian A, Aliakbar-Golkar M (2016) Optimal storage planning in active distribution network considering uncertainty of wind power distributed generation. IEEE Trans Power Syst 31(1):304–316
18. Cao Q, Song Y-D, Guerrero JM, Tian S (2016) Coordinated control for flywheel energy storage matrix systems for wind farm based on charging/discharging ratio consensus algorithms. IEEE Trans Smart Grid 7(3):1259–1267
19. Baros S, Ilić MD (2018) A consensus approach to real-time distributed control of energy storage systems in wind farms. IEEE Trans Smart Grid. https://doi.org/10.1109/TSG.2017.2749379
20. Baros S, Ilić MD (2017) Distributed torque control of deloaded wind DFIGs for wind farm power output regulation. IEEE Trans Power Syst 32(6):4590–4599
21. Wehner J, Mohler D, Gibson S, Clanin J, Faris D, Hooker K, Rowand M (2015) Technology performance report: Duke energy Notrees wind storage demonstration project. Technical report, Duke Energy Renewables, Charlotte, NC (United States)
22. Yang L, Xu Z, Østergaard J, Dong ZY, Wong KP, Ma X (2011) Oscillatory stability and eigenvalue sensitivity analysis of a DFIG wind turbine system. IEEE Trans Energy Convers 26(1):328–339
23. Miao Z, Fan L (2008) The art of modeling and simulation of induction generator in wind generation applications using high-order model. Simul Model Pract Theory 16(9):1239–1253
24. Krause P, Wasynczuk O, Sudhoff SD, Pekarek S (2013) Analysis of electric machinery and drive systems, vol 75. Wiley

25. Akhmatov V (2005) Induction generators for wind power. Multi-Science Pub
26. Pena R, Clare JC, Asher GM (1996) Doubly fed induction generator using back-to-back PWM converters and its application to variable-speed wind-energy generation. IEE Proc Electr Power Appl 143(3):231–241
27. Mohammadpour HA, Santi E (2015) Analysis of sub-synchronous resonance (SSR) in doubly-fed induction generator (DFIG)-based wind farms. Synth Lect Power Electron 5(3):1–64
28. Fan L, Kavasseri R, Miao ZL, Zhu C (2010) Modeling of DFIG-based wind farms for SSR analysis. IEEE Trans Power Deliv 25(4):2073–2082
29. Rocabert J, Luna A, Blaabjerg F, Rodriguez P (2012) Control of power converters in AC microgrids. IEEE Trans Power Electron 27(11):4734–4749
30. Harnefors L, Bongiorno M, Lundberg S (2007) Input-admittance calculation and shaping for controlled voltage-source converters. IEEE Trans Ind Electron 54(6):3323–3334
31. Lorduy A, Lzaro A, Barrado A, Fernndez C, Quesada I, Lucena C (2010) Simplified synchronous reference frame control of the three phase grid connected inverter. In: 2010 twenty-fifth annual IEEE applied power electronics conference and exposition (APEC), pp 1026–1033, Feb 2010
32. Du W, Chen X, Wang H (2018) Parameter tuning of the PLL to consider the effect on power system small-signal angular stability. IET Renew Power Gener 12(1):1–8
33. Khazaei J, Nguyen DH (2018) Distributed consensus for output power regulation of DFIGs with on-site energy storage. IEEE Trans Energy Convers 1–1. Early access. https://doi.org/10.1109/TEC.2018.2871575
34. Khazaei J, Miao Z (2018) Consensus control for energy storage systems. IEEE Trans Smart Grid. https://doi.org/10.1109/TSG.2016.2624144
35. Nguyen DH (2015) A sub-optimal consensus design for multi-agent systems based on hierarchical LQR. Automatica 55:88–94
36. Peng L, Yingmin J, Junping D, Shiying Y (2007) Distributed consensus control for second-order agents with fixed topology and time-delay. In: Proceedings of 26th Chinese control conference, pp 577–581
37. Milano F (2005) An open source power system analysis toolbox. IEEE Trans Power Syst 20(3):1199–1206
38. Cole S, Belmans R (2011) Matdyn, a new matlab-based toolbox for power system dynamic simulation. IEEE Trans Power Syst 26(3):1129–1136
39. Gungor VC, Sahin D, Kocak T, Ergut S, Buccella C, Cecati C, Hancke GP (2013) A survey on smart grid potential applications and communication requirements. IEEE Trans Ind Inform 9(1):28–42

Sensitivity Analysis of Frequency Regulation Parameters in Power Systems with Wind Generation

Julian Patiño, José David López and Jairo Espinosa

Abstract In power systems, frequency constitutes a parameter indicating the equilibrium between power demanded by load and energy produced by generation systems. This chapter studies the effects of varying different system parameters on the overall performance of the traditional frequency regulation system when including contributions of renewable energy sources. A model for the inclusion of variable speed wind turbines in the frequency control loops is analyzed, and parametric sensitivity functions are established using linearized models and transfer function representations for the system components. Through both theoretical analysis and performance simulations, the impact of an inaccurate representation of system inertia in frequency performance is established. Stability analysis for inertia sensitivity of frequency regulation involving wind generation is also provided. Results indicate more robustness to parameter variations for systems including wind turbine participation. However, the frequency deviation rate increases when the uncertainty in system parameters grows. This behavior might lead to instability scenarios under frequency disturbances for the power system.

1 Introduction

In power systems, frequency constitutes a parameter indicating the equilibrium between power demanded by load and energy produced by generation systems [1]. When this relationship is unbalanced, control structures are in place to return the

J. Patiño (✉) · J. Espinosa
Universidad Nacional de Colombia, Manizales, Colombia
e-mail: japatin0@unal.edu.co

J. Espinosa
e-mail: jespinov@unal.edu.co

J. Patiño
I. U. Pascual Bravo, Calle 73 No. 73A-226, Medellín, Colombia

J. D. López
Universidad de Antioquia UDEA, Calle 70 No. 52-21, Medellín, Colombia
e-mail: josedavid@udea.edu.co

R.-E. Precup et al. (eds.), *Advanced Control and Optimization Paradigms for Wind Energy Systems*, Power Systems, https://doi.org/10.1007/978-981-13-5995-8_3

system frequency to operational values in the so-called Load Frequency Control (LFC) of power systems. However, traditional frequency control strategies have been developed for power systems with almost complete reliance on conventional energy sources, but penetration of Renewable Energy Sources (RES) may require the participation of these new units in the control loops [2].

Wind is the fastest growing and most widely deployed renewable energy source worldwide [3]. Variable Speed Wind Turbine Generators (VSWTG) are the most popular devices for extracting electrical power from wind. VSWTG operation requires action of electronic power converters, which effectively decouple the rotor angular speed of the wind generating unit from the electrical frequency of the grid. Consequently, VSWTG does not contribute directly to the natural inertial response of the system under frequency disturbances and other operational events [4].

However, the increase of wind generators and the unpredictability and variability associated with the wind resource also raise the difficulty level of the frequency regulation tasks in power systems. In future grid operating normative, the participation of wind turbine generators in system frequency regulation might become mandatory. Hence, several studies have been proposed about control strategies for the active inclusion of VSWTG in LFC loops. Complete reviews can be found in [4–6]. Among many techniques, the required primary reserves for frequency regulation contributions from Wind Turbine generators (WT) can be supplied through de-loaded operation (under the point of maximum power extraction) of the WT and the addition of control loops, emulating the response of conventional units (inertia and droop controllers) [4, 7, 8].

This chapter focuses on analyzing the effects of parameter variations in the frequency regulation structure of power systems involving wind generation. Sensitivity functions are established using linearized models and transfer function representations for the system components. Sensitivity analysis has been previously employed in the assessment of the dynamic performance of power systems. Nanda and Kaul [9] explored the optimal tuning of automatic generation controllers in a multi-area power system with conventional steam generation units. The role of inertia was studied through sensitivity tests in the development of a composite load model with conventional machines [10]. For scenarios considering renewable energy sources, the impacts of damping and inertia in the dynamic performance of grid frequency were studied in [11], where authors analyzed different locations to provide emulated inertia. Additionally, the behavior of Doubly-Fed Induction Generator (DFIG) wind units in power systems was studied using eigenvalue sensitivities about inertia variations [12]. However, these works did not consider an explicit function for system sensitivities. In this regard, transfer functions for power systems frequency regulation elements are developed in [13–16] for conventional-only scenarios, and extended in [7] for assessing the effects of load damping including wind farms.

Amidst this context, this chapter studies the effects of varying different system parameters on the overall performance of the traditional LFC system when including contributions of renewable energy systems such as VSWTG. Through both theoretical analysis and performance simulations, the impact of an inaccurate representation of system inertia is established. Results indicate more robustness to parameter variations for systems that include wind turbine participation. However, the frequency deviation rate increases when the uncertainty in system parameters grows. This behavior would lead to instability scenarios under frequency disturbances for the power system.

2 Modeling of the Variable Speed Wind Turbine Generator Control Loops

The relationship between rotor speed and power for a wind turbine (WT) is inherently non-linear. The maximum power point (MPP) is the operational rotor speed value that causes the maximum power extraction from the WT. Given the non-linear dynamic of the system, the MPP needs to be tracked continually in order to guarantee the efficiency of the WT. The MPP tracking (MPPT) process starts with Eq. (1), which represents the mechanical power of the WT:

$$P_m = \frac{1}{2}\rho A C_p\left(\lambda_w, \beta_w\right) v_w^3. \quad (1)$$

where A denotes the turbine sweeping area, ρ represents the air density, β_w is the pitch-angle, λ_w is Tip Speed Ratio (TSR), $C_p\left(\lambda_w, \beta_w\right)$ is the performance coefficient for the WT, and v_w represents the per-unitized values of measured wind speed.

By definition, the TSR is a relation between the tangential speed at the tip of a WT blade and the actual wind speed. Therefore, in per-unit system, $\lambda_{w,p.u.} = \frac{\omega}{v}$, with v and ω denoting the rotor speed and the wind speed per-unit for a reference speed of 12 m/s. Additionally, Eq. (2) presents the mechanical torque $T_{m,p.u.}$ of the WT in per-unit system, where ω_m represents the per-unitized values of measured rotor speed, $C_{p,\max}$ is the maximum value of $C_p\left(\lambda_w, \beta_w\right)$, and K_w is a constant defined in Eq. (3).

$$T_{m,p.u.} = \frac{0.5\,\rho\,A\,C_p(\lambda_w, \beta_w)v_w^3}{\omega_m P_{\text{base}}} = \frac{K_w C_p v^3}{\omega}, \quad (2)$$

$$\text{with}\quad K_w = \frac{0.5\,\rho\,A\,C_{p,\max}}{P_{\text{base}}} \quad (3)$$

If wind turbines are contributing to frequency regulation, the linearized representation must include changes of the area frequency Δf and wind speed variations v. In the same way as the LFC, frequency regulation contributions can come from the variation of mechanical or electrical power in response to grid frequency changes.

Frequency contribution from wind turbines is possible due to the so called de-loaded operation [4]. In this mode, the wind turbine operates under the MPP to generate the active power reserve employed to contribute to frequency control tasks. This active power reserve P_{cont} (p.u.) is:

$$P_{cont} = P_{MPP} - P_{und} = (1 - X_u)\, K_{wind}\, \omega^3 \tag{4}$$

where ω denotes the rotor speed, $X_u \in (0, 1)$ is a weighting factor expressing the fraction of maximum power at de-loaded operating point [7], P_{und} is the de-loaded power, P_{MPP} is the maximum power extracted at the operating speed, and K_{wind} is a constant parameter depending on the characteristics of the wind turbine [17]. In operation, ω_{MPP} and ω_{und} are the rotor angular speeds at P_{MPP} and P_{und}, respectively. According to [18], the power P_{uref} and torque τ_{uref} for operating at determined wind speed are:

$$P_{uref} = P_{und} + (P_{MPP} - P_{und})\frac{(\omega_{und} - \omega)}{(\omega_{und} - \omega_{MPP})} \tag{5}$$

$$= X_u\, K_{wind}\, \omega^3 + K_{wind}\, \omega^3 \frac{(1 - X_u)\,(\omega_{und} - \omega)}{(\omega_{und} - \omega_{MPP})} \tag{6}$$

$$\tau_{uref} = X_u\, K_{wind}\, \omega^2 + K_{wind}\, \omega^2 \, \frac{(1 - X_u)\,(\omega_{und} - \omega)}{(\omega_{und} - \omega_{MPP})} \tag{7}$$

Figure 1 illustrates the model of WT with control loops for frequency contribution from wind turbines. According to this figure, variations in electrical torque $\Delta\tau_{e,cont}$ can be expressed as:

$$\Delta\tau_{e,cont}(s) = \frac{\Delta P_{e,cont}(s)}{\omega_0} = (-1)\left[\frac{1/R_w + K_n s}{\omega_0}\right]\Delta f(s) \tag{8}$$

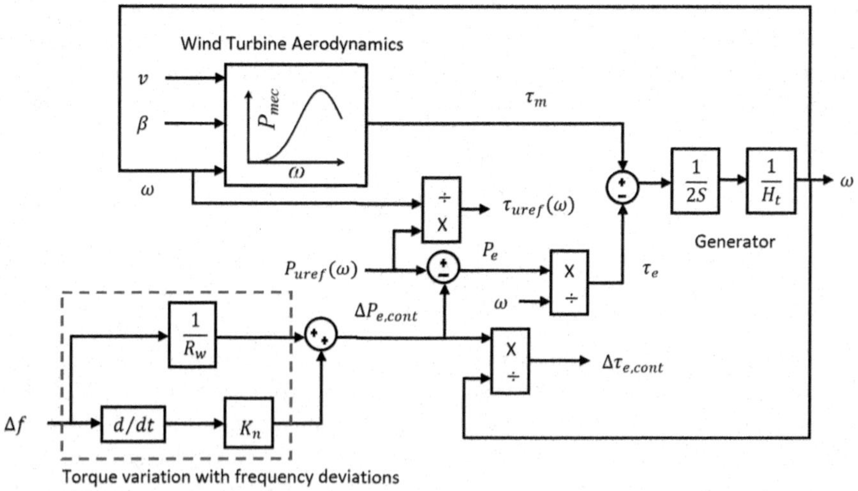

Fig. 1 Model of variable speed WT for LFC support

In Eq. (8), $\Delta P_{e,cont}$ denotes electrical power variations related to changes in frequency, ω_0 is the current angular speed of the rotor, K_n is a parameter weighting the inertia control loop, and R_w is a parameter similar to the speed droop R of the LFC regulation. The latter parameters are responsible for inertial emulation from WT, with dynamic performance improving with both low values of R_w and high values of K_n [18]. Let the linearized dynamic operating electrical torque be:

$$\Delta\tau_{uref}(s) = \frac{\partial\tau_{uref}}{\partial\omega}\Delta\omega(s) = \tag{9}$$

$$\left[2K_{wind}\,\omega_0\,X_u + 2K_{wind}\,\omega_0(1-X_u)\cdot\frac{(\omega_{und}-\omega_0)}{(\omega_{und}-\omega_{MPP})} - \frac{K_{wind}\,\omega_0^2\,(1-X_u)}{(\omega_{und}-\omega_{MPP})}\right]\Delta\omega(s).$$

The total variation in electrical torque ($\Delta\tau_e$) can be presented in terms of $\Delta f(s)$ and $\Delta\omega(s)$ as follows:

$$\Delta\tau_e(s) = \Delta\tau_{e,cont}(s) + \Delta\tau_{uref}(s) = (-1)\left[\frac{1/R_w + K_n s}{\omega_0}\right]\Delta f(s) + \frac{\partial\tau_{uref}}{\partial\omega}\Delta\omega(s). \tag{10}$$

Denote v_0 as the wind speed (p.u.) and ω_0 as the initial angular speed of the rotor at de-loading operating point, and let, λ_{ref} and $C_{p,ref}$ be the operational values of the tip speed ratio λ and the performance coefficient $C_p(\lambda, \beta)$ of WT for a reference pitch-angle β_{ref}. The mechanical torque τ_m of the turbine is:

$$\tau_m = \frac{K_{wind} C_p v^3}{\omega} \tag{11}$$

From Eq. (11), the linearized mechanical torque $\Delta\tau_m(s)$ can be expressed as:

$$\Delta\tau_m(s) = \frac{\partial\tau_m}{\partial\omega}\cdot\Delta\omega(s) + \frac{\partial\tau_m}{\partial v}\cdot\Delta v(s) + \frac{\partial\tau_m}{\partial f}\cdot\Delta f(s);\quad \text{with} \tag{12}$$

$$\frac{\partial\tau_m}{\partial\omega} = \left(\frac{K_{wind}\,\varsigma\,{v_0}^2}{\omega_0} - \frac{K_{wind}\,C_{p,\text{ref}}\,{v_0}^3}{{\omega_0}^2}\right)$$

$$\frac{\partial\tau_m}{\partial v} = \left(\frac{3K_{wind}\,C_{p,\text{ref}}\,{v_0}^2}{\omega_0} + \frac{K_{wind}\,\varsigma\,\lambda_{\text{ref}}\,{v_0}^2}{\omega_0}\right)$$

$$\frac{\partial\tau_m}{\partial f} = \left(\frac{\varepsilon\,K_{wind}\,K_b\,{v_0}^3}{\omega_0}\right).$$

The expressions for $\varepsilon = \frac{\partial C_p}{\partial\beta}|_{op}$ and $\varsigma = \frac{\partial C_p}{\partial\lambda}|_{op}$ should be calculated depending on the operational conditions λ_{ref} and $C_{p,ref}$. Also, $K_b = \frac{\Delta\beta}{\Delta f} = \frac{\partial\beta}{\partial f}$.

For a wind turbine with a inertia H_w, the swing equation can be employed to obtain and expression for the variations in rotor angular speed $\Delta\omega$ in terms of the grid frequency variations Δf and wind speed changes Δv. Using the functions for $\Delta\tau_e(s)$ and $\Delta\tau_m(s)$, the power swing equation becomes:

$$\Delta\omega(s) = \frac{\Delta\tau_m(s) - \Delta\tau_e(s)}{2H_w\, s}. \tag{13}$$

As illustrated in [19], the current parameter values related with active power production in WT can be obtained with small-signal analysis. Denote by ΔP_e, $\Delta\tau$ and $\Delta\omega$ the small deviations in electrical power P_e, electrical torque τ_e and the angular rotor speed ω respectively. Then,

$$\omega = \omega_0 + \Delta\omega; \tag{14}$$

$$\tau_e = \tau_{e0} + \Delta\tau_e; \tag{15}$$

$$P_e = P_{e0} + \Delta P_e = \tau_e\,\omega = (\tau_{e0} + \Delta\tau_e)\,(\omega_0 + \Delta\omega)\,, \tag{16}$$

where P_{e0}, τ_{e0}, and ω_0 are the initial values for the corresponding parameters previously mentioned. After expanding Eq. (16) and neglecting the terms of superior order, ΔP_e can be expressed in terms of the angular speed and electrical torque deviations as:

$$\Delta P_e = \omega_0\,\Delta\tau_e + \tau_{e0}\,\Delta\omega. \tag{17}$$

In this context, the initial operating electrical torque τ_{e0} equates the de-loaded torque τ_{uref} presented in Eq. (7). Using the previously developed expressions for $\Delta\omega$, $\Delta\tau_e$, and τ_{e0}, the linearized electrical power ΔP_e for a wind turbine in terms of the grid frequency variations Δf and wind speed changes Δv is:

$$\Delta P_e(s) = \left(\frac{-1}{s+r}\right)\left(\frac{as^2+bs+c}{qs+1}\right)\Delta f(s) + \left(\frac{g}{qs+1}\right)\Delta v(s),\ with \tag{18}$$

$$G_{wf}(s) = \left(\frac{-1}{s+r}\right)\left(\frac{as^2+bs+c}{qs+1}\right) \tag{19}$$

$$G_{wv}(s) = \left(\frac{g}{qs+1}\right). \tag{20}$$

Parameters g, q, r, a, b, and c are constant terms left omitted because of their extension. They are the result of long algebraic operations among system parameters.

3 Extraction of Sensitivity Functions

Frequency regulation can be classified in three main stages according to the nature and timescales of the control efforts: *primary* actions proportional to the frequency deviations, *secondary* actions allowing correction of steady-state errors, and *tertiary* actions related with predefined dispatches and some emergency conditions. These three stages constitute the Load Frequency Control (LFC) system [1, 20]. Grid elements must be modeled for the design of LFC controllers. First order models are

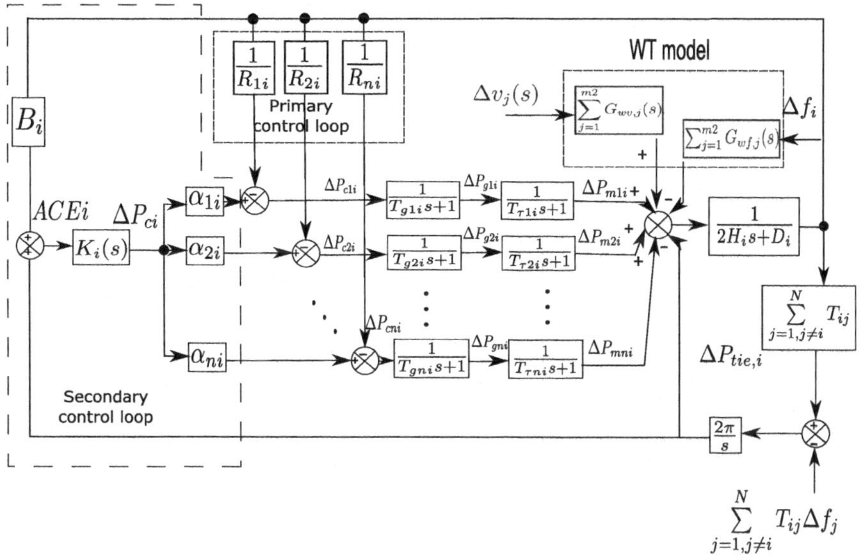

Fig. 2 Load frequency regulation system with linearized WT model for RES power contributions

assumed for the governor and turbine of conventional units, and for the representation of the frequency response characteristic of any control area in the power system.

Figure 2 illustrates the complete system representation for m_1 conventional units with contribution of m_2 wind generation systems. Note that wind speed variations Δv_j are acting as input parameters, just as the load variations (ΔP_L). From Fig. 2 parameters for the i-th area are: ΔP_{mki} the change in mechanical power of the generator k, ΔP_{gki} the change in the active power output of generator k, ΔP_L the load perturbation, Δf_i the frequency change, D_i the damping coefficient. H_i the equivalent inertia, ΔP_{cki} the control action of the LFC for the k-th generator, T_{ij} the power exchange coefficient between area i and area j, ΔP_{tiei} the total change in the power exchanged between area i and other areas and Δf_j the change in the frequency of area j connected to area i. Also, B_i denotes the bias factor for modulation of the error signal in secondary regulation, $K_i(s)$ is the transfer function of the secondary controller and α_i the participation factor of each generator in secondary control. The total inertia H_i represents the sum of the aggregated inertias of conventional units and wind turbines. Usually, this parameter is calculated empirically, but the analysis of previous works has shown the impacts of inertia in frequency regulation [21, 22]. The sensitivity analysis proposed in this chapter shows the impacts on frequency regulation when the measured value of H_i is different from the expected. This is also justified as the inertial emulation from wind turbines could lead to inertia variations [13, 14]. From Fig. 2, the total variations on system frequency Δf can be obtained as the linear composition of the individual responses to each input signal. Denote with Δf_L the variations in system frequency with respect to load changes ΔP_L. Making $\Delta v_j = 0$, Δf_L is given by:

$$\Delta f_L(s) = \frac{\frac{1}{2H_i s+D}\{-\Delta P_L(s)\}}{1+\frac{1}{2H_i s+D}\left\{K(s)+\sum_{i=1}^{m_1} M_i(s)+\sum_{j=1}^{m_2} G_{wf,j}(s)\right\}} \quad (21)$$

In the same way, denoting with Δf_v the variations in system frequency with respect to wind speed changes Δv_j and making $\Delta P_L = 0$, the following expression is obtained:

$$\Delta f_v(s) = \frac{\frac{1}{2H_i s+D}\left\{\sum_{j=1}^{m_2} G_{wv,j}(s)\Delta v_j(s)\right\}}{1+\frac{1}{2H_i s+D}\left\{K(s)+\sum_{i=1}^{m_1} M_i(s)+\sum_{j=1}^{m_2} G_{wf,j}(s)\right\}} \quad (22)$$

Finally, the total variations on system frequency $\Delta f(s)$ can be expressed from Eqs. (21) and (22), as:

$$\begin{aligned}\Delta f(s) &= \Delta f_L(s)+\Delta f_v(s) \quad (23)\\ &= \frac{\left\{-\Delta P_L(s)+\sum_{j=1}^{m_2} G_{wv,j}(s)\Delta v_j(s)\right\}}{2H_i s+D+\left\{K(s)+\sum_{i=1}^{m_1} M_i(s)+\sum_{j=1}^{m_2} G_{wf,j}(s)\right\}}\end{aligned}$$

3.1 *Sensitivity to Inertia H_i*

To represent the effects of inertia variations in frequency regulation, the calculation of $\frac{\partial \Delta f(s)}{\partial H_i}$ is required. Extracting the partial derivative of Δf with respect to H_i from Eq. (23), the following expression is obtained:

$$\begin{aligned}\frac{\partial \Delta f(s)}{\partial H_i} &= \frac{2s\left[\Delta P_L(s)-\sum_{j=1}^{m_2} G_{wv,j}(s)\Delta v_j(s)\right]}{\left[2H_i s+D+\left\{K(s)+\sum_{i=1}^{m_1} M_i(s)+\sum_{j=1}^{m_2} G_{wf,j}(s)\right\}\right]^2} \quad (24)\\ &= \{2s\}\left[\frac{\Delta f(s)}{\left\{\Delta P_L(s)-\sum_{j=1}^{m_2} G_{wv,j}(s)\Delta v_j(s)\right\}}\right]^2\left\{\Delta P_L(s)-\sum_{j=1}^{m_2} G_{wv,j}(s)\Delta v_j(s)\right\}\end{aligned}$$

The unit-less sensitivity function S_H with respect to total inertia H_i can be also calculated by definition as:

$$\begin{aligned}S_H &= \frac{d\Delta f(s)}{\Delta f(s)}\Big/\frac{dH_i}{H_i} = \frac{\partial \Delta f(s)}{\partial H_i}\cdot\frac{H_i}{\Delta f(s)}\\ &= \frac{\Delta f(s)}{\left\{\Delta P_L(s)-\sum_{j=1}^{m_2} G_{wv,j}(s)\Delta v_j(s)\right\}}\cdot\{2sH_i\} \quad (25)\end{aligned}$$

In [14], a similar sensitivity analysis was performed for a power system but only with conventional hydraulic machines. From those results, frequency variations Δf_{base} for a purely conventional system (no RES penetration) can be obtained. In this sense, the sensitivity expression Δf_{base} with respect to to inertia for conventional power systems is

$$\frac{\partial \Delta f_{base}(s)}{\partial H} = \left[\frac{\Delta f_{base}(s)}{\Delta P_L(s)}\right]^2 \cdot \Delta P_L(s) \cdot 2s. \tag{26}$$

Similarly, the aforementioned unit-less sensitivity expression $S_{H,base}$ of Δf_{base} with respect to H_{base} for a completely conventional power system is [13]:

$$S_{H,base} = \frac{d\Delta f(s)}{\Delta f(s)} \Big/ \frac{dH_{base}}{H_{base}} = \frac{\partial \Delta f(s)}{\partial H_{base}} \cdot \frac{H_{base}}{\Delta f(s)} = \frac{\Delta f_{base}(s)}{\Delta P_L(s)} 2H_{base}s. \tag{27}$$

The comparison among sensitivities with respect to inertia for power systems with and without wind turbine contributions to frequency variations can be established from Eqs. (24) to (27). As expected, the inclusion of wind turbines means including wind speed in frequency sensitivity through the term $\sum_{j=1}^{m_2} G_{wv,j}(s)\Delta v_j(s)$ (see Eqs. (24) and (25)). It is expected that the intrinsic unpredictable and variable nature of the wind resource will impact the power generated from renewable units. In consequence, the dynamic characteristics of the frequency regulation are being modified according to the wind profile for a determined inertia value. This will be illustrated through simulation in subsequent sections.

4 Stability Analysis of Inertia Sensitivity of LFC with WT

Traditionally, grid frequency variations $\Delta f(s)$ have been exclusively analyzed under the influence of load disturbances $\Delta P_L(s)$ [1, 19, 20]. This approximation was developed for systems involving only conventional generation, resulting in the following expression:

$$d\Delta f(s) = \frac{\partial \Delta f(s)}{\partial \Delta P_L(s)} \cdot d\Delta P_L(s). \tag{28}$$

For systems with contribution of wind turbines to frequency regulation, wind speed variations Δv_j need to be considered in the analysis of frequency deviation. Therefore, for a system with frequency regulation contributions from m_2 wind turbines, Eq. (28) becomes:

$$d\Delta f(s) = \frac{\partial \Delta f(s)}{\partial \Delta P_L(s)} \cdot d\Delta P_L(s) + \sum_{j=1}^{m_2} \frac{\partial \Delta f(s)}{\partial \Delta v_j(s)} \cdot d\Delta v_j(s). \tag{29}$$

Nevertheless, the impacts of the system inertia coefficient are not considered neither in Eq. (28) nor Eq. (29). This omission could offer incomplete information, because a loss of generation or an interruption event may result in a variation of the inertia coefficient for a given machine, affecting the aggregated system inertia [23]. These events have an high probability of occurrence in an environment with variable and unpredictable renewable energy sources, where resource intermittence or generation drops may result in changes from the initial inertia calculation [22]. Additionally, inertia could have been estimated form an outdated generation profile. All these phenomena suggest that frequency deviation Δf should consider the effects of the generator inertia coefficient rather than being function of the external disturbances ΔP_L and Δv_j exclusively. In consequence, the impact of the inertia coefficient in the grid frequency regulation performance must be determined.

4.1 Extraction of Differential Equation for Frequency Deviation

Assuming mutual independence among ΔP_L, Δv_j, and inertia coefficient H_i, Eq. 29 is modified by adding variations with respect to H_i, as follows:

$$d\Delta f(s) = \frac{\partial \Delta f(s)}{\partial \Delta P_L(s)} \cdot d\Delta P_L(s) + \sum_{j=1}^{m_2} \frac{\partial \Delta f(s)}{\partial \Delta v_j(s)} \cdot d\Delta v_j(s) + \frac{\partial \Delta f(s)}{\partial \Delta H_i} d\Delta H_i. \tag{30}$$

From Eq. (24),

$$\frac{\partial \Delta f(s)}{\partial H_i} dH_i = 2s(dH_i) \left[\frac{\Delta f(s)}{\Delta P_L(s) - \sum_{j=1}^{m2} G_{wv,j}(s)\Delta v_j(s)} \right]^2 \left\{ \Delta P_L(s) - \sum_{j=1}^{m2} G_{wv,j}(s)\Delta v_j(s) \right\} \tag{31}$$

Additionally, taking partial derivatives with respect to ΔP_L and Δv_j from Eq. 23, we can show that:

$$\frac{\partial \Delta f(s)}{\partial \Delta P_L(s)} = \frac{\Delta f(s)}{\Delta P_L(s)}; \qquad \frac{\partial \Delta f(s)}{\partial \Delta v_j(s)} = \frac{\Delta f(s)}{\Delta v_j(s)} \tag{32}$$

Replacing Eqs. (31) and (32) in Eq. (30), we get:

$$d\Delta f(s) = \frac{\Delta f(s)}{\Delta P_L(s)} \cdot d\Delta P_L(s) + \sum_{j=1}^{m2} \frac{\Delta f(s)}{\Delta v_j(s)} \cdot d\Delta v_j(s) \tag{33}$$

$$+ 2s(dH_i) \left[\frac{\Delta f(s)}{\Delta P_L(s) - \sum_{j=1}^{m2} G_{wv,j}(s)\Delta v_j(s)} \right]^2 \left\{ \Delta P_L(s) - \sum_{j=1}^{m2} G_{wv,j}(s)\Delta v_j(s) \right\}$$

Laplace inverse transformation is employed to get the time domain representation of Eq. (33), resulting in:

$$d\Delta f(t) = L^{-1}\left[\frac{\Delta f(s)}{\Delta P_L(s)}\cdot d\Delta P_L(s)\right] + L^{-1}\left[\sum_{j=1}^{N}\frac{\partial \Delta f(s)}{\partial \Delta v_j(s)}\cdot d\Delta v_j(s)\right] + \ldots$$
$$\ldots L^{-1}\left\{\left[\frac{\Delta f(s)}{\Delta P_L(s) - \sum_{j=1}^{m2} G_{wv,j}(s)\Delta v_j(s)}\right]^2\left[\Delta P_L(s) - \sum_{j=1}^{m2} G_{wv,j}(s)\Delta v_j(s)\right]\cdot 2sdH_i\right\}. \tag{34}$$

Further, taking integration of Eq. (34), we get $\Delta f(t)$ as

$$\int d\Delta f(t) = \Delta f(t) = \int L^{-1}\left[\frac{\Delta f(s)}{\Delta P_L(s)}\cdot d\Delta P_L(s)\right] + \int L^{-1}\left[\sum_{j=1}^{N}\frac{\partial \Delta f(s)}{\partial \Delta v_j(s)}\cdot d\Delta v_j(s)\right] + \cdots$$
$$\cdots L^{-1}\left\{\left[\frac{\Delta f(s)}{\Delta P_L(s) - \sum_{j=1}^{m2} G_{wv,j}(s)\Delta v_j(s)}\right]^2\left[\Delta P_L(s) - \sum_{j=1}^{m2} G_{wv,j}(s)\Delta v_j(s)\right]\cdot 2sdH_i\right\}. \tag{35}$$

Equation (35) presents the total differential equation of frequency deviation $\Delta f(t)$ in time domain considering inertia effects. In order to determine the impacts of the inertia coefficient, a stability analysis of $\Delta f(t)$ is presented in the following subsection.

4.2 Stability Analysis

After a disturbance, all characteristic poles of the transfer function of a power system are located on the left half-plane in the s-domain if the system is stable. In this case, the finite time-domain input signals $\Delta v_j(t)$ and $\Delta P_L(t)$ would not be producing infinite time-domain responses on system output $\Delta f(t)$. From control theory, that is equivalent to show that the norm of the transfer function of the system is bounded. From Eq. (35), this would represent that the transfer functions listed in Eq. (36) are already bounded:

$$\left\{\begin{array}{l} \left\|\Delta f(s)\Big/\left[\Delta P_L(s) - \sum_{j=1}^{m_2} G_{wv,j}(s)\Delta v_j(s)\right]\right\| \\ \left\|\Delta f(s)\Big/\Delta P_L(s)\right\| \\ \left\|\sum_{j=1}^{N}\Delta f(s)\Big/\Delta v_j(s)\right\| \end{array}\right. < \infty \text{ for } \forall_t \in (0,\infty) \tag{36}$$

Moreover, frequency regulation in power systems is designed to keep $\Delta f(t)$ inside a determined finite band despite variations in Δv_j and ΔP_L. Assuming an stable system, the bounds of frequency variations should be determined. Considering both Δv_j and ΔP_L as step functions, and using triangle inequality properties in the expression of Eq. (35) we get:

$$
\begin{aligned}
\|\Delta f(t)\| \leq & \left\| \int L^{-1}\left[\frac{\Delta f(s)}{\Delta P_L(s)} \cdot d\Delta P_L(s)\right] + \int L^{-1}\left[\sum_{j=1}^{m_2} \frac{\Delta f(s)}{\Delta v_j(s)} \cdot d\Delta v_j(s)\right] + \cdots \right.\\
& \left. \cdots \int L^{-1}\left\{\left[\frac{\Delta f(s)}{\Delta P_L(s) - \sum_{j=1}^{m_2} G_{wv,j}(s)\Delta v_j(s)}\right]^2 \cdot \left[\Delta P_L(s) - \sum_{j=1}^{m_2} G_{wv,j}\Delta v_j(s)\right] \cdot 2sdH_i\right\} \right\| \\
\leq & \left\| \int L^{-1}\left\{\left\|\frac{\Delta f(s)}{\Delta P_L(s)}\right\| d\Delta P_L(s)\right\} \right\| + \left\| \int L^{-1}\left\{\left\|\sum_{j=1}^{m_2} \frac{\Delta f(s)}{\Delta v_j(s)}\right\| \sum_{j=1}^{m_2} d\Delta v_i(s)\right\} \right\| + \cdots \\
& \cdots \left\| \int L^{-1}\left\{\left\|\left[\frac{\Delta f(s)}{\Delta P_L(s) - \sum_{j=1}^{m_2} G_{wv,j}\Delta v_j(s)}\right]^2\right\| \left[\Delta P_L(s) - \sum_{j=1}^{m_2} G_{wv,j}\Delta v_j(s)\right] \cdot 2sdH_i\right\} \right\| \\
\leq & \left\| \int L^{-1}\left[\kappa\, d\Delta P_L(s)\right] \right\| + \left\| \int L^{-1}\left[\eta \sum_{j=1}^{m_2} d\Delta v_j(s)\right] \right\| + \cdots \\
& \cdots \left\| \int L^{-1}\left\{\rho^2 \left[\Delta P_L(s) - \sum_{j=1}^{m_2} G_{wv,j}\Delta v_j(s)\right] \cdot 2sdH_i\right\} \right\| \\
\leq & \left\| \int \kappa \cdot 1(t - t_0) d\Delta P_L \right\| + \left\| \int \eta \cdot 1(t - t_0) \sum_{j=1}^{m_2} d\Delta v_j \right\| + \cdots \\
& \cdots \left\| \int \rho^2 \cdot \left[1(t - t_0)\Delta P_L - \sum_{j=1}^{m_2} \left(\frac{g_j}{q_j}\right) \cdot e^{-\left(\frac{1}{q_j}\right)t} \Delta v_j\right] 2sdH_i \right\| \\
\leq & \kappa \left\|\Delta P_L\right\| + \eta \left\|\sum_{j=1}^{m_2} \Delta v_j\right\| + \rho^2 \left\|H_i\right\| \left\|\left[\Delta P_L - \sum_{j=1}^{m_2} \left(\frac{g_j}{q_j}\right) \cdot e^{-\left(\frac{1}{q_j}\right)t} \Delta v_j\right]\right\|
\end{aligned}
\tag{37}
$$

where $\kappa = \left\|\frac{\Delta f(s)}{\Delta P_L(s)}\right\|$, $\eta = \left\|\sum_{j=1}^{N} \frac{\Delta f(s)}{\Delta v_j(s)}\right\|$ and $\rho = \left\|\left[\frac{\Delta f(s)}{\Delta P_L(s) - \sum_{j=1}^{N} G_{wv,j}\Delta v_j(s)}\right]\right\|$ represent the respective bounded magnitudes of the transfer functions established in Eq. (36). Also, $L^{-1}[1/s] = 1(t - t_0)$, with $1(t - t_0) = 1$ when $t \geq t_0$. In this way, $||1(t - t_0)|| = 1$. Therefore, the system remains stable when the effects of inertia variations are considered.

In the same way, the output frequency deviation Δf of the traditional power system model without considering inertia effects is given as [23]:

$$
\|\Delta f_{base}(t)\| \leq \kappa_{base} \left\|\Delta P_L\right\| \tag{38}
$$

The expressions in Eqs. (37) and (38) show that the frequency deviations in both cases remain bounded when a stable power system is considered. However, it is also evident that the boundaries are different. The limits of the conventional model depend on the boundaries of load disturbance ΔP_L. However, the consideration of inertia effects and wind turbine integration to frequency regulation affects the boundaries of frequency deviations. In this case, the limits depend on factors such as load disturbance, variations in wind speed, system configuration, and the specific value of H_i.

5 Simulation Results and Discussion

A modified version of the WSCC 9-bus power system [1] was employed for simulating wind turbine contribution in the LFC for a multi-area power system. For the system of Fig. 3, 20% of conventional generation in Area III was replaced by a wind farm. The wind farm was formed by 32 DFIG WT of 2 MW each, whose model parameters were shown in Table 2. The system was modeled as indicated in Fig. 1, and several case studies were analyzed.

5.1 Case 1: Frequency Response for a Load Step Change

A step change of 10% is applied to the simulated system at $t = 0$ s. The system with a reduced inertia was simulated without wind, and with increasing constant wind speed (5 and 10 m/s) as disturbance. Contribution of wind turbines in frequency

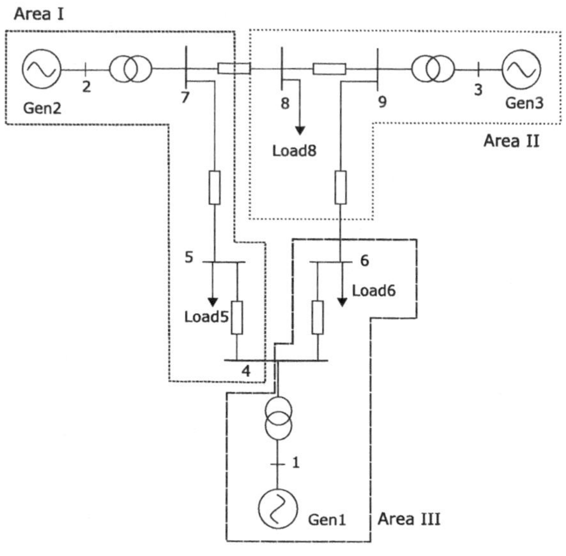

Fig. 3 WSCC 9-bus system multi-area partitioning. The modified system parameters are summarized in Table 1, considering an hydraulic machine and two gas units. The Power Base was set at 100 MVA

Table 1 WSCC 9 bus system parameters [1]

Parameter	Value	Parameter	Value	Parameter	Value
H_1	23.64 s	T_{12}	2.064 p.u.	R_1	2 p.u.
H_2	6.4 s	T_{13}	6.1191 p.u.	R_2	10 p.u.
H_3	1.505 s	T_{23}	14.4353 p.u.	R_3	7.5019 p.u.
MVA_{nom1}	247.5	$D1, D2, D3$	0.8	B_1	2.8 s
MVA_{nom2}	192	$Tg1, Tg2, Tg3$	0.2	B_2	10.8 s
MVA_{nom3}	128	$T_{\tau 1}, T\tau 2, T\tau 3$	0.3	B_3	8.3 s

Table 2 Wind-turbine model simulation parameters [25]

Parameter	Value	Parameter	Value
P_{nom}	2 MW	R_s	0.00491 p.u.
V_{nom}	966 V	X_{ls}	0.09273 p.u.
K_1	5000 Nm	X_m	3.96545 p.u.
K_2	2000 Nm	R_r	0.00552 p.u.
T_w	1	X_{lr}	0.1 p.u.
K_a	500	H	4.5 s
T_a	20	J	506.61 Kg m^2

regulation is not being considered. Fig. 4 presents the grid frequency deviations for this case. Frequency nadir is lower for the system with reduced inertia and no wind. Despite the wind being considered exclusively as disturbance, the effects of the wind power injections help to improve the frequency characteristic of the system. Wind power production increases with higher speeds. However, higher speeds may lead to a more oscillatory response, as seen from the curve for a speed of 10 m/s.

5.2 Case 2: Frequency Response for a Load Step Change and WT Contribution to LFC

For the previous example, contribution of wind turbines in frequency regulation is now being considered as constant in every case. Figure 5 presents the grid frequency deviations for this case. Again, we can see how the lower frequency nadir is given for the case without active power injections from RES. Response is similar as the one shown in Fig. 4; however, the contribution of wind turbines to frequency regulation provided a smoother response in grid frequency deviation. Again, a higher speed and oscillatory response even with inertia emulation from wind turbines are concerning.

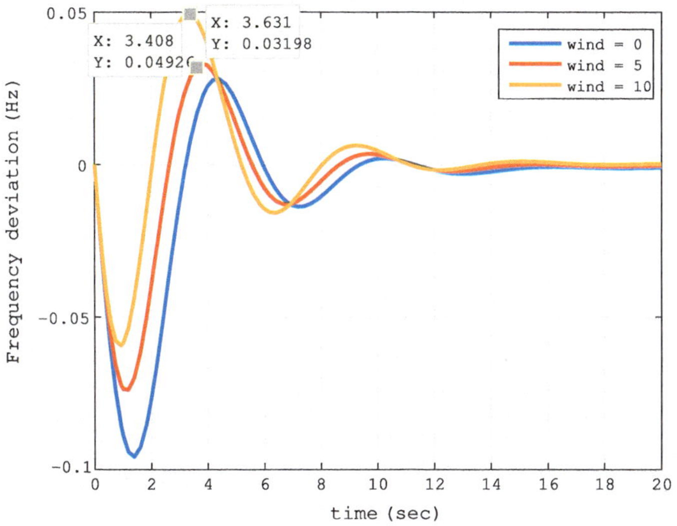

Fig. 4 Case 1: Frequency response without contribution of WT and increasing wind speed

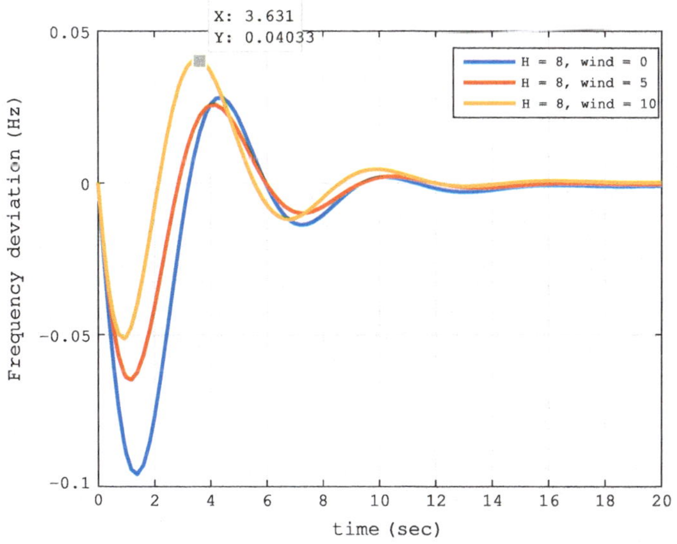

Fig. 5 Case 2: Frequency response with constant contribution of WT and increasing wind speed

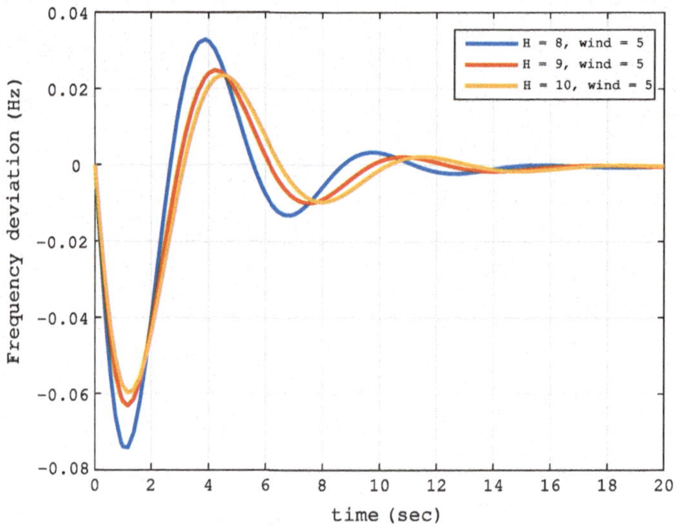

Fig. 6 Case 3: Frequency response with constant wind speed and increasing contribution of WT

5.3 *Case 3: Frequency Response for a Load Step Change and Increasing Wind Contribution to LFC*

A step change of 10% is applied to the simulated system at $t = 0$ s. Contribution of wind turbines in frequency regulation is being increased (0–10% and 20%) in every case by modifying loop parameters. The system with a reduced inertia was simulated with a constant wind speed of 5 m/s. Figure 6 presents the grid frequency deviations for this case. Here, the effects of the increased inertia with the contribution of wind turbines are evident. However, these effects are being shown with the assumption of constant wind speed.

5.4 *Case 4: Frequency Response with Constant Wind Speed and Increasing Contribution of WT After Unstable Conditions*

For this case, one of the system poles was changed to a value in the right half-plane of the s-domain. This caused an unstable response in frequency deviations. The same conditions of Sect. 5.3 are applied in this case and the resulting responses are plotted in Fig. 7. The system took longer time in reaching unstable conditions when wind turbines contributed to grid frequency regulation. Wind speed was kept as constant value of 5 m/s. Variability in wind speed may lead to a faster unstable mode.

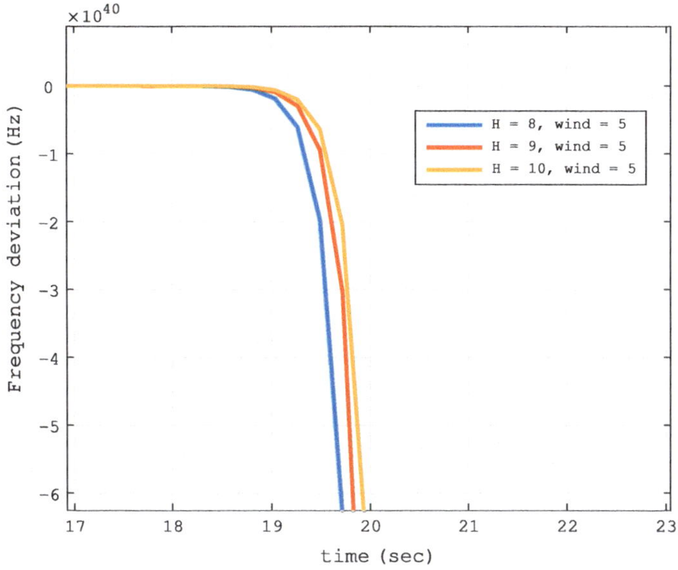

Fig. 7 Case 4: Frequency response with constant wind speed and increasing contribution of WT, starting from unstable case

5.5 *Case 5:Frequency Response with a Simulated Wind Profile and Increasing Contribution of WT*

The power generation in wind units depends on the wind profile. Wind speed is highly variable and unpredictable, causing fluctuations on wind power generation and frequency deviations. A wind speed profile was simulated with data obtained from the database of Virgin Islands [24]. Load disturbance was not considered, just wind speed. The resultant responses are plotted in Fig. 8, while contribution of wind turbines in frequency regulation is increased (0–10% and 20%). According to Fig. 8, the curves with higher inertia contributions from wind turbines presented smaller peaks than the purely conventional system. In consequence, frequency deviation had a better performance with contribution of RES despite the inherent variability of the wind speed.

5.6 *Case 6: Frequency Response with a Simulated Wind Profile and Load Disturbance*

The same conditions of the immediately previous case were replicated, but now a step load disturbance of 10% was applied at $t = 50$ s. The contributions of wind turbines make the system more robust to disturbance action, even under the effects

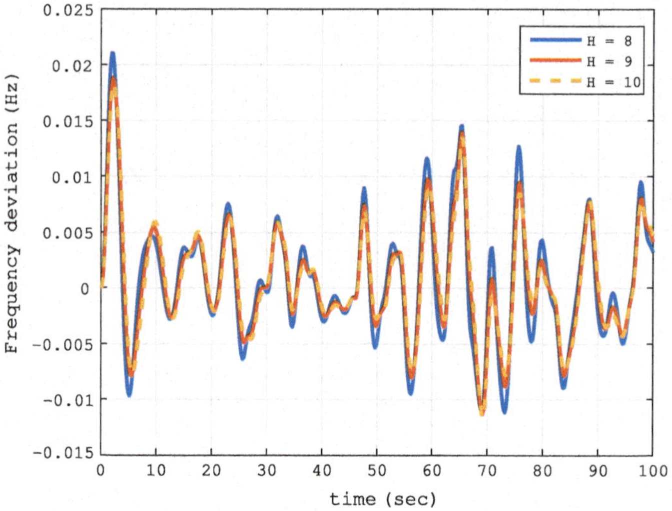

Fig. 8 Case 5: Frequency response with a simulated wind profile and increasing contribution of WT

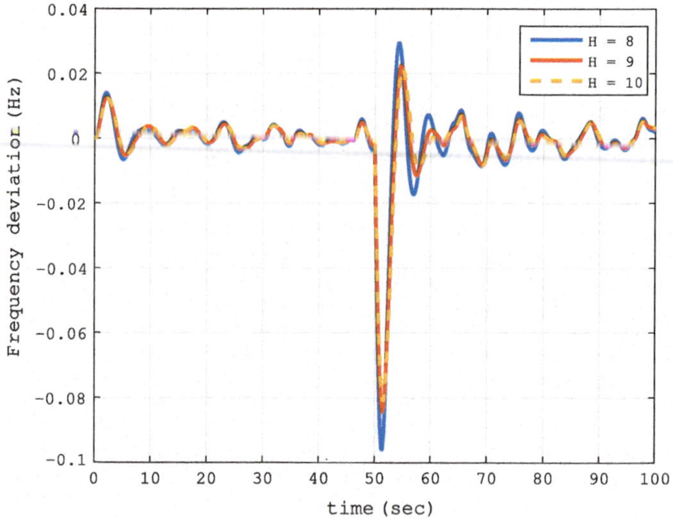

Fig. 9 Case 6: Frequency response with a simulated wind profile and increasing contribution of WT, and load disturbance $\Delta P_L = 0.1$ at $t = 50$ s

of variable wind speed (see Fig. 9). Starting from an stable case, the consideration of inertia variations maintains the stability of the system. This is expected from Eq. (35), and the analysis of Sect. 4.2. The dynamic characteristics of grid frequency deviations can actually improve with a higher value of inertia coefficient.

6 Conclusions

This chapter addressed the effects of inertia variations for power systems with integrated wind units. The system transfer functions were obtained from a linearized wind turbine model. The mathematical relationships were formulated to analyze the sensitivity and stability regarding inertia coefficient H. These expressions were verified through simulation of several cases under different stability conditions and disturbances in wind speed and load.

Simulations have shown a better and smoother response of frequency deviations for the cases with contributions of wind turbines in comparison with the basic case of the purely conventional system. These improvements on frequency response provide a better power quality. Actually, results indicate a better performance of the frequency response for higher contributions of wind turbines to system inertia. Moreover, results suggest more robustness to parameter variations of systems with wind turbine participation. However, the frequency deviation rate increases when the uncertainty of system parameters grows. This behavior would lead to instability scenarios under frequency disturbances for the power system.

With the growing development of intermittent renewable energy sources and its integration of the electrical grid, renewable energy will take more responsibility for frequency regulation tasks in the foreseeable future. Therefore, the impact of changes in generator inertia coefficient H on the power system frequency regulation must be accounted. We showed, both with theoretic and simulation analyses, that when including the wind generation system into the control loop, an inaccurate generator inertia coefficient H has a relatively small impact when the power system is inherently stable; while the system frequency deviation may be accelerated when the power system is indeed unstable after disturbance.

In future works, the analysis of the impacts of the combination of the frequency sensitivities with respect to the main system parameters should be explored. This would include parameter such as generator inertia coefficient H, the governor speed coefficient R and load-damping coefficient D. Their effects on the regulation of power system frequency response and stability studies in LFC should be explored together rather than considering them individually.

Acknowledgements Colciencias supported contributions of J. Patiño through the program "Convocatoria 528—Convocatoria Nacional para Estudios de Doctorados en Colombia 2011".

References

1. Anderson PM, Fouad AA (2002) Power system control and stability. IEEE Press Power Engineering Series. Wiley-IEEE Press
2. DeMarco CL, Baone CA (2014) Chapter 29—Control of power systems with high penetration variable generation. In: Jones LE (ed) Renewable energy integration. Academic Press, Boston, pp 369–379
3. REN21 (ed) (2017) Renewables 2017 global status report. REN21 Secretariat, Paris

4. Vidyanandan KV, Senroy N (2013) Primary frequency regulation by deloaded wind turbines using variable droop. IEEE Trans Power Syst 28(2):837–846
5. Díaz-González Francisco, Hau Melanie, Sumper Andreas, Gomis-Bellmunt Oriol (2014) Participation of wind power plants in system frequency control: Review of grid code requirements and control methods. Renew Sustain Energy Rev 34:551–564
6. Dreidy M, Mokhlis H, Mekhilef S (2017) Inertia response and frequency control techniques for renewable energy sources: a review. Renew Sustain Energy Rev 69:144–155
7. Pradhan C, Bhende CN (2017) Frequency sensitivity analysis of load damping coefficient in wind farm-integrated power system. IEEE Trans Power Syst 32(2):1016–1029
8. Ruiz S, Patino J, Espinosa J (2018) PI and LQR controllers for frequency regulation including wind generation. Int J Electr Comput Eng (IJECE) 8(6)
9. Nanda J, Kaul BL (1978) Automatic generation control of an interconnected power system. Proc Inst Electr Eng 125(5):385–390(5)
10. Pereira L, Kosterev D, Mackin P, Davies D, Undrill J, Zhu W (2002) An interim dynamic induction motor model for stability studies in the WSCC. IEEE Trans Power Syst 17(4):1108–1115
11. Borsche TS, Liu T, Hill DJ (2015) Effects of rotational Inertia on power system damping and frequency transients. In: 2015 54th IEEE conference on decision and control (CDC), Dec 2015, pp 5940–5946
12. Gautam D, Vittal V, Harbour T (2009) Impact of increased penetration of DFIG-based wind turbine generators on transient and small signal stability of power systems. IEEE Trans Power Syst 24(3):1426–1434
13. Huang Hao, Li Fangxing (2013) Sensitivity analysis of load-damping characteristic in power system frequency regulation. IEEE Trans Power Syst 28(2):1324–1335
14. Huang H, Li F (2014) Sensitivity analysis of load-damping, generator inertia and governor speed characteristics in hydraulic power system frequency regulation. In: 2014 Australasian Universities power engineering conference (AUPEC), Sept 2014, pp 1–6
15. Patino J, Valencia F, Espinosa J (2017) Sensitivity analysis for frequency regulation in a two-area power system. Int J Renew Energy Res 7(2):700–706
16. Patiño J, López JD, Espinosa J (2018) Analysis of control sensitivity functions for power system frequency regulation. In: Figueroa-García JC, López-Santana ER, Rodriguez-Molano JI (eds) Applied computer sciences in engineering, vol 915. Springer International Publishing, Cham, pp 606–617
17. Thomsen SC (2006) Nonlinear control of a wind turbine, ME thesis. PhD thesis, Technical University of Denmark
18. Delille G, Francois B, Malarange G (2012) Dynamic frequency control support by energy storage to reduce the impact of wind and solar generation on isolated power system's inertia. IEEE Trans Sustain Energy 3(4):931–939
19. Kundur P (1994) Power system stability and control. McGraw-Hill Professional
20. Bevrani H (2014) Robust power system frequency control. Power electronics and power systems, 2 edn. Springer
21. Ulbig Andreas, Borsche Theodor S, Andersson Gran (2014) Impact of low rotational inertia on power system stability and operation. IFAC Proc Vol 47(3):7290–7297
22. Tielens Pieter, Van Hertem Dirk (2016) The relevance of inertia in power systems. Renew Sustain Energy Rev 55:999–1009
23. Huang Hao (2014) Studies of economics and stability with variable generation and controllable load. PhD thesis, University of Tennessee

24. Roberts O, Andreas A (1997) United States Virgin Islands: St. Thomas (Bovoni); St. Croix (Longford)
25. Moore IF (2012) Inertial response from wind turbines. PhD thesis, Cardiff University, Cardiff

Wind Turbines Integration into Power Systems: Advanced Control Strategy for Harmonics Mitigation

Alex Reis, José Carlos de Oliveira and Leandro Pains Moura

Abstract When considering the topic of grid integration, it is recognized that wind farms can bring about negative power quality impacts at the connection point. A lot of attention has been given to this type of generation system, aiming at achieving the required level of compatibility between the wind turbine and the AC grid. Amongst the various existing power quality phenomena, harmonic distortion emerges as a relevant challenge once power electronics converters that currently exist on wind energy conversion systems have a direct effect upon the Voltage Harmonic Total Distortion (THD). The search for solutions that reduce the harmonic current injection emerge as an issue that needs ample consideration. In this context, this chapter is focused on two main points directed to full converter wind turbines (type IV). The first exploits an analysis and an experimental validation of the physical factors responsible for harmonic current emissions. The second part contemplates an advanced control strategy aimed at reducing the level of harmonic currents

1 Introduction

Electric power systems are currently going through great paradigm breakups and modernization of processes, because of scientific and technological advances on generation, transmission, distribution and use of energy. This is mainly due to growing concerns about climate change, sustainability and efficiency, high-energy prices, energy reliability, among other issues. In this scenario, over the last decades, wind power stations have reached significant participation on electrical generation in several countries and the connection of wind turbines to the electrical grid has motivated

A. Reis (✉)
University of Brasília (UnB), Brasília, Brazil
e-mail: reialex@gmail.com

J. C. de Oliveira · L. Pains Moura
Federal University of Uberlândia (UFU), Uberlândia, Brazil
e-mail: jcoliveira@ufu.br

L. Pains Moura
e-mail: leandropm91@gmail.com

R.-E. Precup et al. (eds.), *Advanced Control and Optimization Paradigms for Wind Energy Systems*, Power Systems, https://doi.org/10.1007/978-981-13-5995-8_4

the need for international standards or electrical grid codes [1]. These include requirements related to steady state, transients, dynamics and power quality issues of the system.

An IEEE Task Force on distributed generation (DG) planning and optimization [2] shows that there are numerous challenges when integrating wind power systems. Among such aspects, harmonic current emissions by wind turbines emerges as a relevant topic [3]. This fact is aggravated when considering the large number of units that make up a wind farm [4] and the ever-growing number of installations. This phenomenon can cause serious effects on the connection grid, which may result in hosting capacity limitation to preserve the overall performance of the network.

Considering the aforementioned scenario, the development of techniques to mitigate the impacts of harmonics on the point of common coupling (PCC) emerges as a relevant engineering matter in need of resolution. To this end, passive [5] and active [6] filters appear as traditional solutions. These devices can be installed to allow the connection of wind farms to existing grids without causing standard distortion limit violations. Although these techniques are effective, the rated power of the filter is very sensitive to background harmonic distortions on the coupling point. Consequently, extra financial investments can be required so that the filter can handle the harmonics produced by the wind generation plant and the grid. Noteworthy here is that there are still no acceptable procedures for the sharing of the harmonic responsibility neither the costs of the solution between the utility and wind farm [7].

In addition, the control of the switching pattern of the converter characterizes an alternative approach for harmonic current control. This methodology allows the harmonic current to be fixed to preset values without hardware modifications. The idea is based on the fact that, by setting the switching instants to predefined values, it is possible to eliminate individual harmonic components and the total distortion produced by the inverters [8–11].

Based on the above-mentioned strategy, this chapter deals with the physical factors responsible for harmonic current emissions by full-converter wind turbines, as well as presenting a proposal of an advanced control function to mitigate the harmonics in a wind power generator. The methodology is directed toward guaranteeing the integration of a large-scale wind farm, through minor changes to the background harmonic distortion at the busbar common coupling, onto the existing electrical grid.

2 Wind Turbine Harmonic Generation

The output voltage of the inverter unit is a determining factor in establishing the harmonic flow between the wind turbine and the electrical system. The harmonics generated by the wind farm are influenced by the converter control, as well as electrical characteristics at the coupling point. In this context, details are provided below to elucidate the main factors that influence the generation of harmonics currents associated with full converter wind turbines.

2.1 Harmonic Distortions Inherent to the PWM Switching Pattern

Harmonic Distortions produced by the PWM switching process are essentially related to two factors: the frequency modulation index (m_f) and the amplitude modulation index (m_a). Such magnitudes are defined by (1) and (2), respectively.

$$m_f = \frac{f_p}{f_m} \quad (1)$$

$$m_a = \frac{V_m}{V_p} \quad (2)$$

here: f_p is a carrier frequency, f_m is a modulator frequency, V_m is the amplitude of the modulator wave and V_p is the amplitude of the carrier wave.

The index m_a is directly related to the amplitude of the converter output voltage in fundamental frequency [12]. The correlation between such magnitudes is indicated in Fig. 1, which details three regions of operation, namely: linear, overmodulation and saturation range.

- Linear region: This range is related to an amplitude modulation index less than the unit. As a result, the inverter output voltage linearly varies with the amplitude modulation index. The harmonic content of the voltage is exemplified in Fig. 2, which presents the expected waveform of the operation in the linear region for a 2-level voltage source converter;
- Overmodulation region: This occurs when the amplitude modulation index is greater than the unit and less than 3.24, which results in a non-linear relationship between the output voltage of the converter and the m_a index. Under this condition

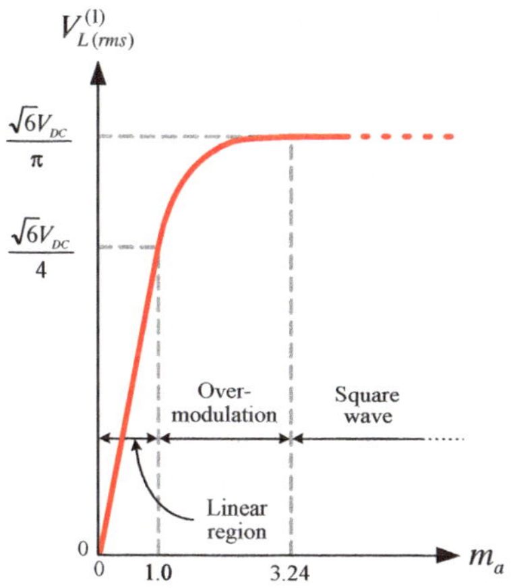

Fig. 1 Inverter operating regions

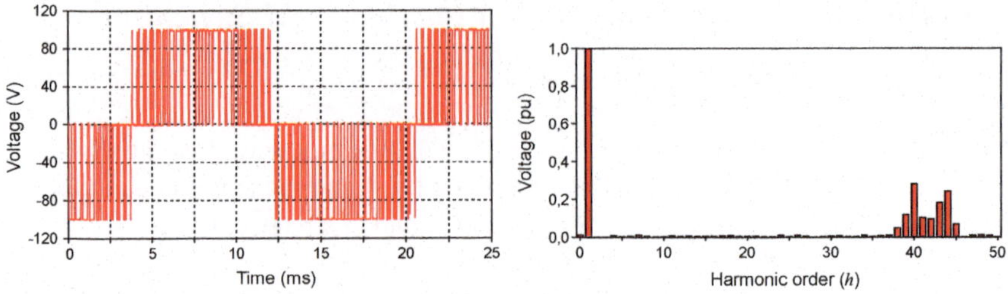

Fig. 2 Inverter output voltage($m_f = 41.67$): linear region ($m_a = 1$)

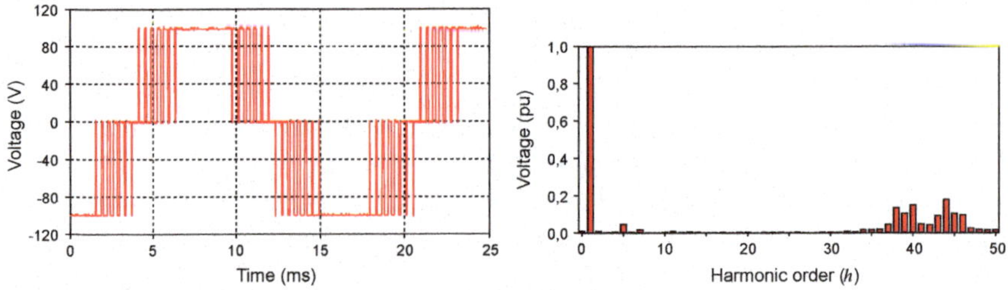

Fig. 3 Inverter output voltage ($m_f = 41.67$): overmodulation region ($m_a = 2$)

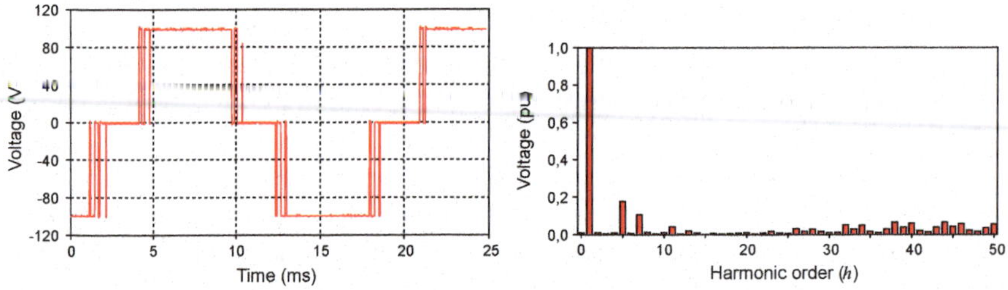

Fig. 4 Inverter output voltage ($m_f = 41.67$): saturation region ($m_a = 5$)

of operation, the converter produces low frequency harmonic components on the output frequency, as presented in Fig. 3;

- Saturation region (operation with square wave): This comprises the functionality of the inverter at offering the highest levels for output voltages. Under such conditions, Fig. 4 shows that the voltage waveform produced by the inverter contains a predominant range of low order components.

In line with the above, it is important to highlight that the constructive details (switching frequency, coupling filters, etc.) and the rated voltages of both AC and DC sides of the inverter directly affect the harmonic distortions produced by the equipment.

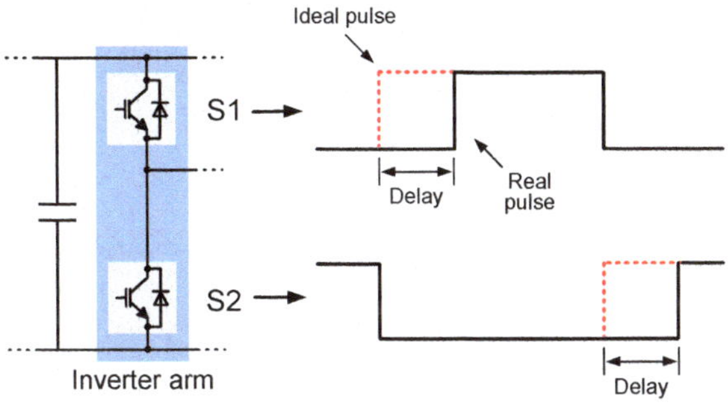

Fig. 5 Drive pulses considering the dead time of the switches

2.2 *Harmonic Distortions Inherent to Switches Dead Time*

A second aspect of upmost importance to the characterization of the converter harmonic current emissivity is the switches dead time. Such a mechanism can be established as a time delays that are programmed between switches drive pulses, as shown in Fig. 5.

Although the programming of time delays guarantees a secure converter operation, the output voltage waveform is changed due to the modification in the switching process. During the dead time, both power electronics switches on an inverter arm remain temporarily inactive and this causes an increase/reduction in the pulse width of the converter output voltage which changes the voltage output harmonic content.

Reference [13] presents a quantitative evaluation of the effects through the inclusion of dead time into the converter switching process. In a simplified manner, the voltage harmonic distortions caused by this procedure can be calculated by superimposing the fundamental voltage wave with a square wave. This wave is 180 out of phase with the current circulating in the system and results in a ΔV amplitude defined by (3) [13].

$$\Delta V = \frac{M t_m}{T} V_{cc} \tag{3}$$

Here: M is the switching number in a cycle, t_m is the dead time, T is the switching period, V_{cc} is the DC voltage.

Figure 6 shows the aforementioned effects, which demonstrates that the output current of the inverter is changed.

2.3 *Harmonic Distortions Inherent to the Network Coupling Point*

The harmonic flow established between the wind turbine and the AC network are affected by background distortion existing on the point of common coupling (PCC). Such relationship can be justified by the interaction of the PCC voltages and currents

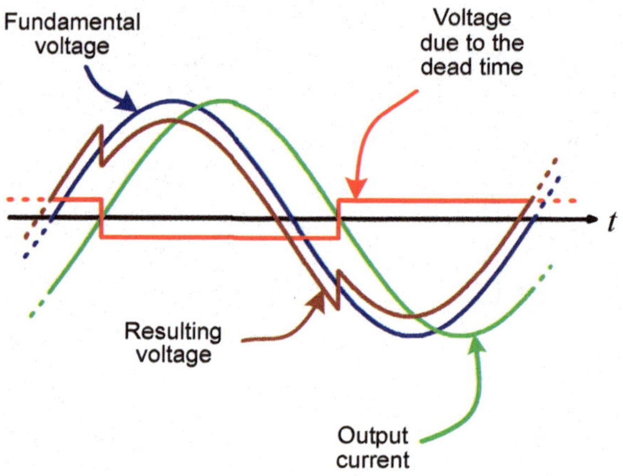

Fig. 6 Dead time effect on the output voltage/current

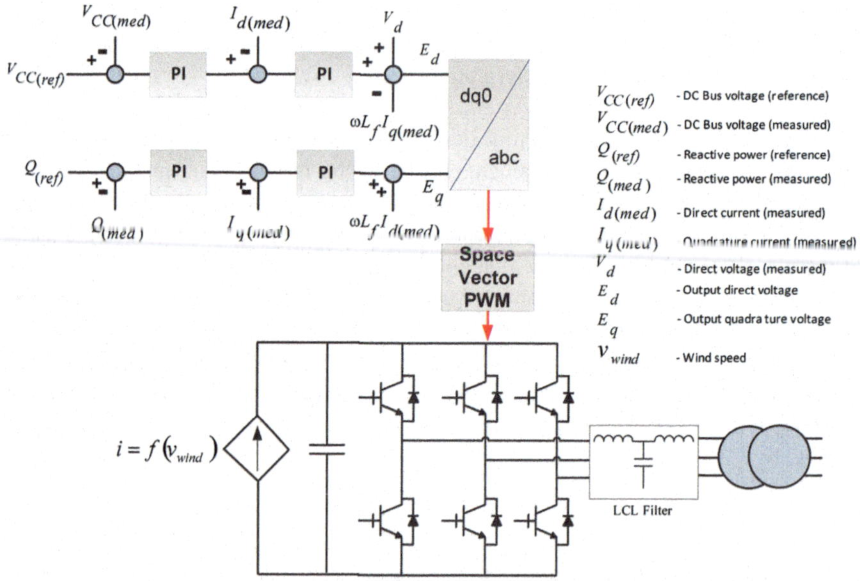

Fig. 7 Wind turbine inverter control loop

on the feedback of the control system of the connection converters. Figure 7 presents a classical control loop of a wind turbine inverter and clearly highlights such physical correlation.

Primarily, it is worth noting that the control loop of the inverter requires the measurement of AC voltages, which are necessary to establish the references for synchronization. In this way, eventual voltage background distortions that exist on the connection bus will affect the operation of the PLL (Phase Locked Loop) or

FLL (Frequency Locked Loop) algorithms. However, it must be stressed that the structures currently employed by commercial inverters, such as DDSRF-PLL or DSOGI-FLL, are not significantly affected by the mentioned association. If this is the case, aforementioned interaction has shown to be of minor relevance upon the synchronization requirements [14].

On the other hand, the currents that will be established between the wind turbine and the connection point are shown as more sensitive. For a better understanding of the mechanism, the balanced three-phase currents between the power converter and the electric network can be represented through Eqs. (4), (5) and (6). These currents are used as a feedback to the control system. In order to simplify the mathematical process, only a single harmonic component related to a generic order is added to the respective fundamental component.

$$i_a = I_1 cos(\omega t - \phi) + I_h cos[h(\omega t - \phi)] \tag{4}$$

$$i_b = I_1 cos(\omega t - 2\pi/3 - \phi) + I_h cos[h(\omega t - 2\pi/3 - \phi)] \tag{5}$$

$$i_c = I_1 cos(\omega t + 2\pi/3 - \phi) + I_h cos[h(\omega t + 2\pi/3 - \phi)] \tag{6}$$

where: i_a, i_b and i_c are instantaneous currents on phases a, b and c, respectively; I_1 is the amplitude of the fundamental current; h is the harmonic order; I_h is the amplitude of the harmonic current; ω is the angular frequency; ϕ is the phase angle difference between voltage and current.

By applying Park transformation, which uses the fundamental voltage angle as a reference, the above currents can be referred to the *dq0* system, for which the result is given by (7) and (8). These expressions show the existence of an oscillatory term that is proportional to the amplitude of the harmonic components, and which is added to the continuous signal arising from the fundamental frequency components. These variables will possess oscillating terms due to the presence of the harmonic component of order *h*, thus influencing the voltage to be synthesized on the inverter terminals.

$$i_d = I_1 cos(\phi) + I_h cos[(h - 1)(\omega t - \phi)] \tag{7}$$

$$i_q = I_1 sin(\phi) + I_h sin[(h - 1)(\omega t - \phi)] \tag{8}$$

where i_d and i_q are the direct axis currents and the quadrature axis, respectively.

2.4 Experimental Analysis of the Influence Factors

This section is aimed at validating the aforementioned dependence between the harmonic generation by the inverters and the operational condition imposed by the control and the connecting busbar. The experimental arrangement set up to highlight the described factors capable of modifying the harmonic spectrum is shown in Fig. 8.

The laboratory structure is constituted by a 2-level voltage source inverter, a LCL (Inductive-Capacitive-Inductive) coupling filter, a transformer and a programmable voltage source CSW5550 from California Instruments. This latter component was programmed to provide a voltage that represents the wind turbine, the synchronous generator and the rectifier. These components are necessary to represent the transference of the active energy produced by the synchronous generator to the electrical network. Additionally, there are still the voltage/currents conditioning boards, a DSP F28335 from Texas Instruments to provide the integration of the control algorithm and real time calculations. The inverter main parameters are given in Table 1.

In order to illustrate the harmonic generation dependence with the interference factors herein discussed, the following case studies were investigated:

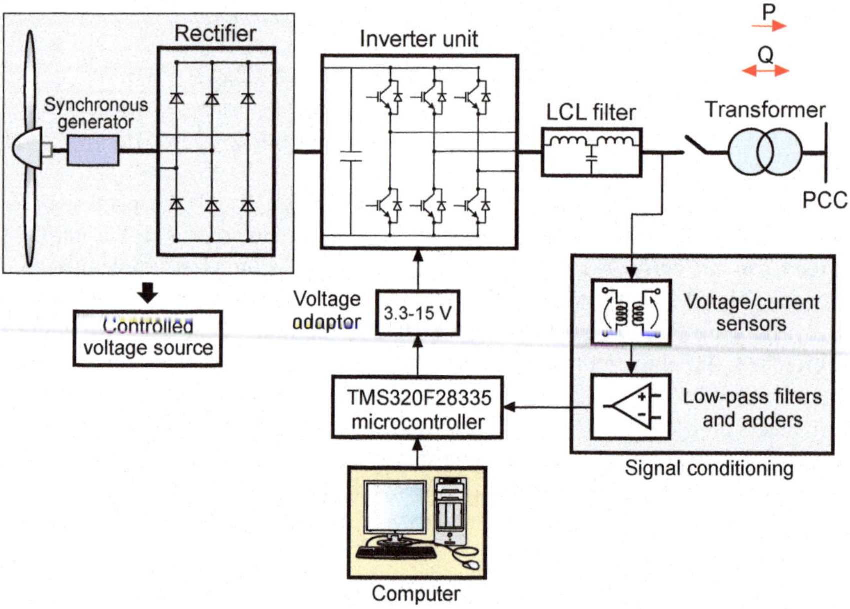

Fig. 8 Laboratory physical structure

Table 1 Inverter parameters

Parameters	Values
Input voltage (V)	450.0
Output voltage (V)	220.0
Output current (A)	8.0
Rated power (kVA)	3.0
Switching frequency (kHz)	5.0
Dead time (μs)	1.5

Table 2 Inverter input voltages and modulation indexes Case 1

Subcase	V_{dc} (V)	Modulation index	Active power (kW)
Sub-case 1.1	450.0	0.93	2.763
Sub-case 1.2	200.0	1.05	2.754
Sub-case 1.3	370.0	1.26	2.748

- Case 1: This case was carried out to emphasize the relationship between the harmonic current generation with the modulation index of the inverter unit;
- Case 2: The second study was focused on the effect of the inverter switching dead time upon the harmonic content;
- Case 3: This study was aimed at demonstrating the dependence between the harmonic currents with the AC busbar background distortions.

2.4.1 Results and Analysis—Case 1

The relationship between DC inverter voltage, modulation index and active power injected into the network are presented in Table 2. The results show that, to transfer a constant active power, with distinct levels of DC voltage, the inverter switching control employs different modulation indexes. Therefore, with such strategy, the inverter is required to work within three distinct conditions: the first in the linear region, the second in the overmodulation region, and the third, at the border of these two regions.

Figure 9 and Fig. 10 presents, respectively, the currents injected at the PCC over the three aforementioned regions. It can be noted that the low order harmonic distortion of the output currents is considerably higher when the inverter unit is operating in the overmodulation region. This fact is in line with theoretical expectations.

2.4.2 Results and Analysis—Case 2

Table 3 summarizes the case studies herein carried out with distinct time delays with the same modulation indexes (linear region) and constant active power delivered to the PCC.

The frequency spectrum of the currents associated to the two dead time conditions is presented in Fig. 11. It can be noted that with the dead time of the inverter being increased, the harmonics will be affected. The corresponding current waveforms are given in Fig. 12.

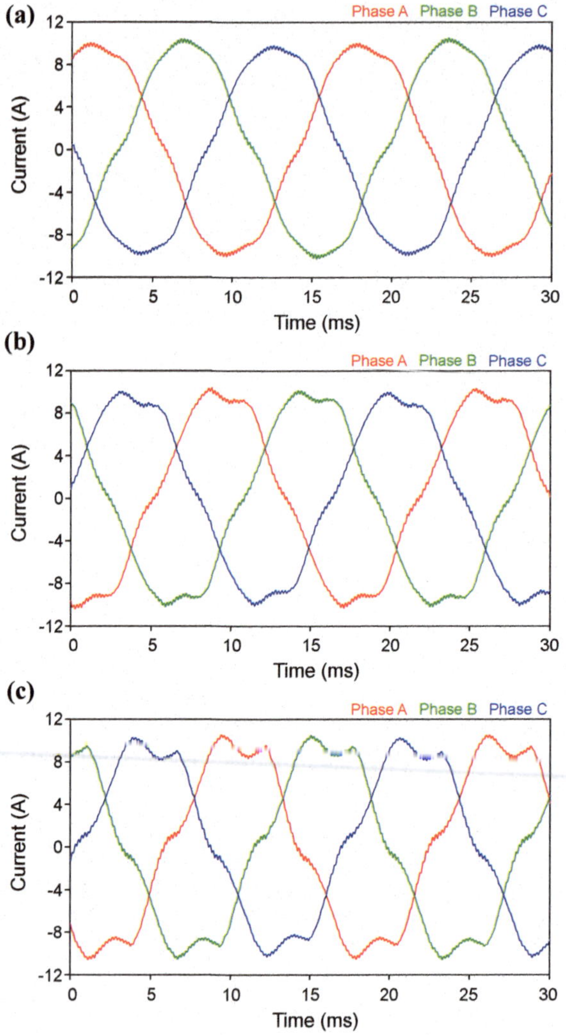

Fig. 9 Current waveforms at the inverter output Case 1; **a** Sub-case 1.1—$V_{cc} = 450$ V (linear region); **b** Sub-case 1.2—$V_{cc} = 400$ V (overmodulation region); **c** Sub-case 1.3—$V_{cc} = 370$ V (higher overmodulation region)

2.4.3 Results and Analysis—Case 3

In order to investigate the relationship between the PCC background distortion effect upon the inverter harmonic generation, the experimental arrangement was considered as having two situations associated to the busbar voltage. The first was admitted as having a background distortion close to zero and the second with an existing THD of 6.36%. Table 4 gives the imposed busbar voltage conditions and the corresponding total harmonic current distortion. Both situations are carried out with a modulation index in the linear region and constant switching dead time of 3.0 μs.

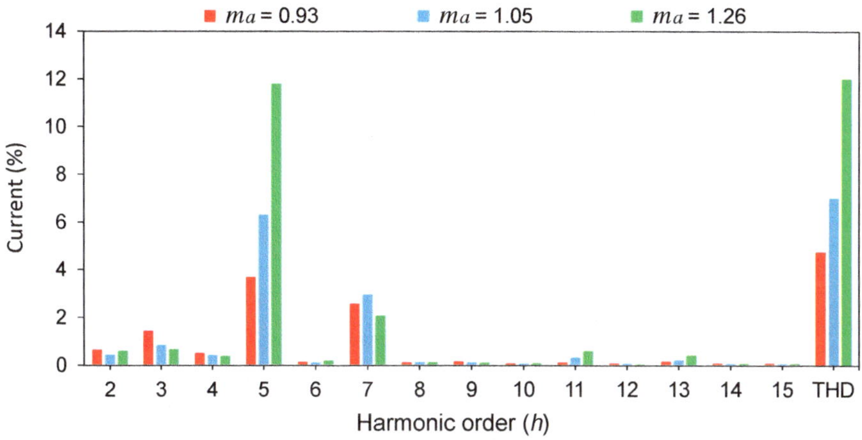

Fig. 10 Current harmonic spectrum with the three modulation indexes Case 1

Table 3 Dead time values—case 2

Subcase	t_m (μs) (V)	Modulation index	Active power (kW)
Sub-case 2.1	3.0	0.93	2.754
Sub-case 2.2	4.5	0.95	2.753

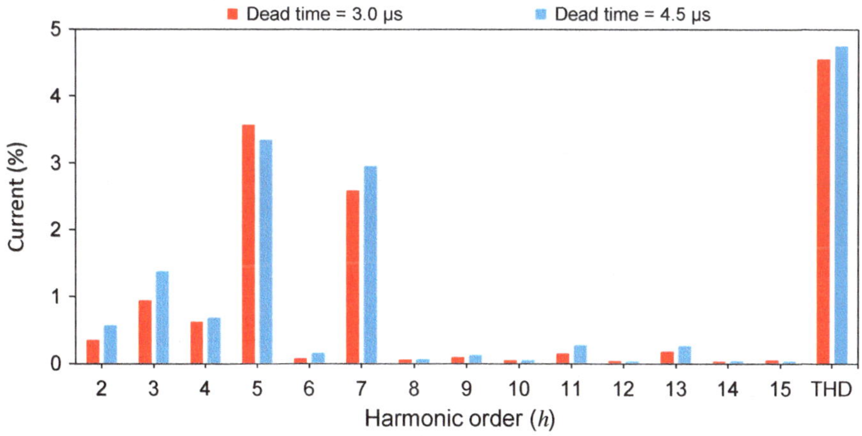

Fig. 11 Currents harmonic spectrum with different dead time Case 2

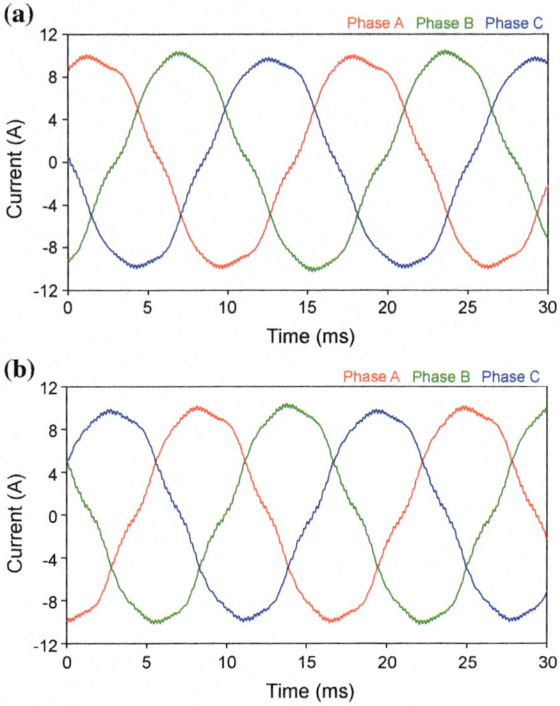

Fig. 12 Current waveforms at the inverter output Case 2; **a** Sub-case 2.1 dead time of 3.0 μs; **b** Sub-case 2.2 dead time of 4.5 μs

Table 4 Relationship between the PCC voltage background distortion and the inverter harmonic current emission Case 3

Subcase	Voltage THD (%)	Current THD (%)	Active power (kW)
Sub-case 3.1	≈0.00	3.36	2.293
Sub-case 3.2	6.36	17.81	2.302

As for the sub-case 3.2, Fig. 13 indicates the PCC background voltage components.

Figure 14 indicates the current waveforms with the busbar background distortion close to zero. In addition, Fig. 15 shows the current waveforms supplied by the inverter when the PCC background voltage distortion exists. In order to compare the effect of the given background PCC voltage distortion, Fig. 16 shows the harmonic current spectrum related to both situations. It can be seen that the harmonic content is highly increased with the previous PCC voltage distortion. Again, this is in accordance with the theoretical expectations.

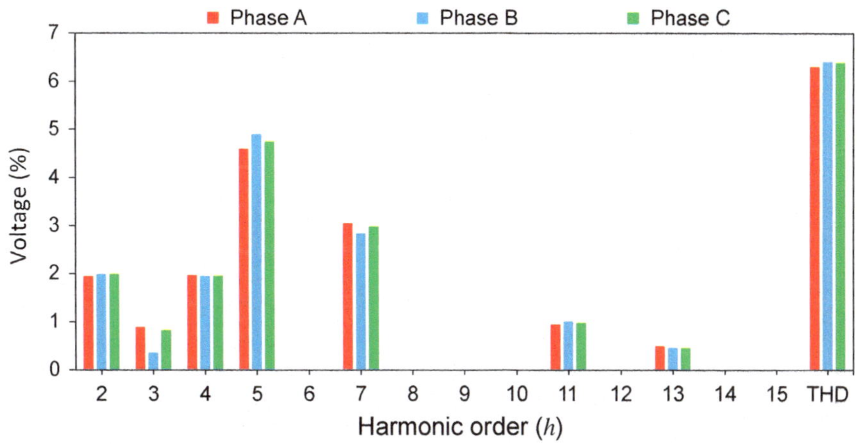

Fig. 13 PCC background voltage harmonic frequency components THD = 6.36% Sub-case 3.2

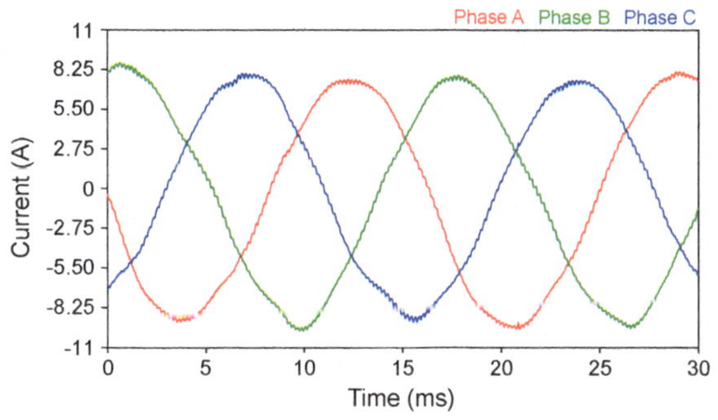

Fig. 14 Inverter output currents (without PCC background distortion) Sub-case 3.1

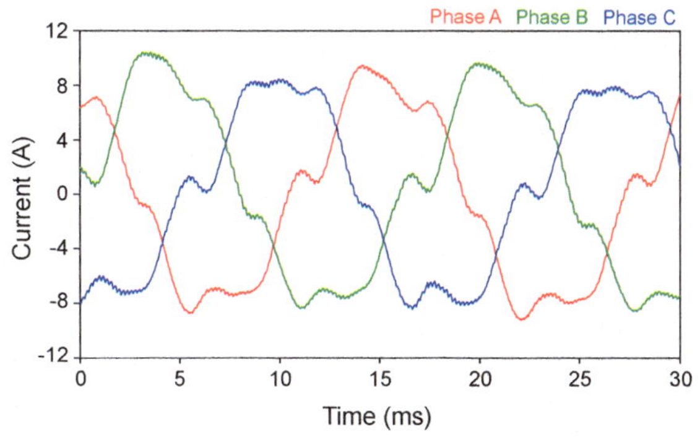

Fig. 15 Inverter output current waveforms (with background distortion) Sub-case 3.2

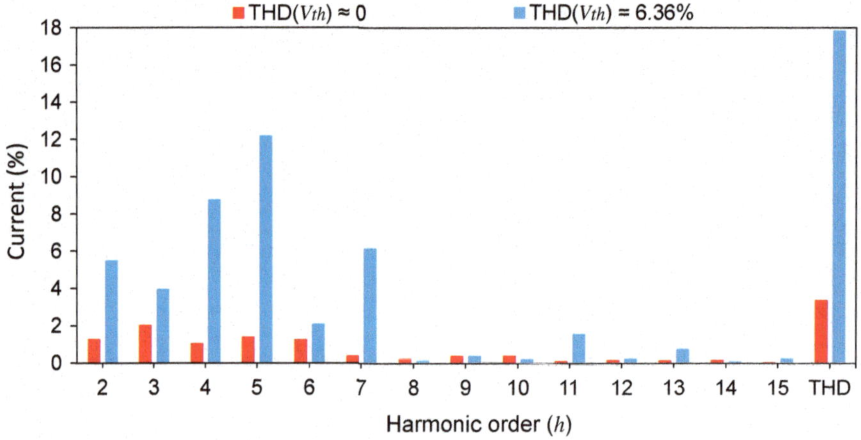

Fig. 16 Current harmonic spectrum at the PCC Sub-cases 3.1 and 3.2

3 Wind Turbine Grid Integration: Advanced Control Strategy

Once the wind turbine harmonic generation concepts have been understood, this section is focused on the establishment of an advanced control strategy applied to full converter wind systems, aiming at reducing the level of harmonic currents throughout a switching control technique inspired on the classical active filters. The physical and mathematical principles of such methodology as well as computational modelling and performance evaluation are carried out in next sections.

3.1 Harmonic Current Mitigation Technique

Figure 17 shows the full-converter wind turbine circuit and its connection to a generic AC grid at the PCC busbar. As previously stated, the inverter unit is the main non-linear component that is responsible for the harmonic content injected onto the existing AC network. Through highlighting the importance of the effects presented in the previous section, Eq. (9) is obtained through a Fourier analysis of a two-level inverter output voltage. The acquired relationship clearly shows that deviations on the switching angle α_k will promptly affect the inverter output voltage waveform and its harmonic content.

$$e_n = \frac{4}{n\pi}\left\{1 + 2\sum_{k=1}^{M}(-1)^k \cos(n\alpha_k)\right\}\frac{V_{cc}}{2} \tag{9}$$

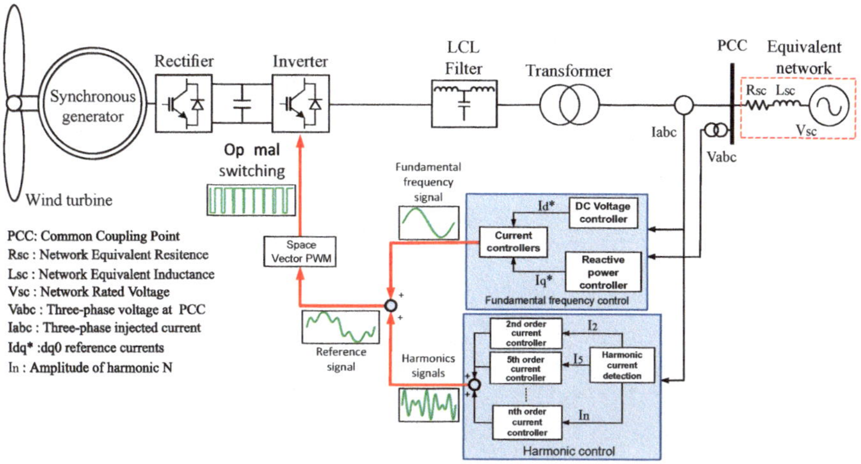

Fig. 17 Full converter wind turbine control block diagram

where e_n is the amplitude of the output voltage of the nth order ($n = 1, 2, \ldots, n$), V_{cc} is the inverter DC voltage and M is the number of switching per half cycle.

Concerning the harmonic compensation approach focused herein, this technique is feasible through controllers that operate in parallel with the fundamental frequency mechanism by superposing harmonic signals onto the reference that was originally applied to the classical inverter PWM device [15]. This condition modifies the PWM pattern to allow both active power supply and one or more harmonic orders mitigation.

In order to exemplify the fundamentals of such technique, Fig. 18 shows three different switching conditions applied to the inverter unit. They are:

- Case 1—operation with switching pulses produced by just the fundamental frequency control signal;
- Case 2—operation with switching pulses produced by a control signal containing the previous fundamental waveform added to a 2nd harmonic component;
- Case 3—operation with switching pulses produced by a control signal containing a waveform given by: fundamental, 2nd harmonic and 5th order component.

Figure 18a is associated to the PWM reference signal containing only the fundamental waveform. The switching pattern resulting from this procedure produces a set of pulses that results on the inverter harmonic currents generation in accordance with Fig. 18b. The 2nd harmonic current shows the highest level. Other even components are of minor relevance.

By adding a 2nd harmonic component to the fundamental control signal, as given in Fig. 18c, such condition modifies the inverter pattern pulses in such a way to produce the inverter harmonic generation seen in Fig. 18d. This shows that the corresponding 2nd harmonic current previously injected by the inverter has been drastically reduced.

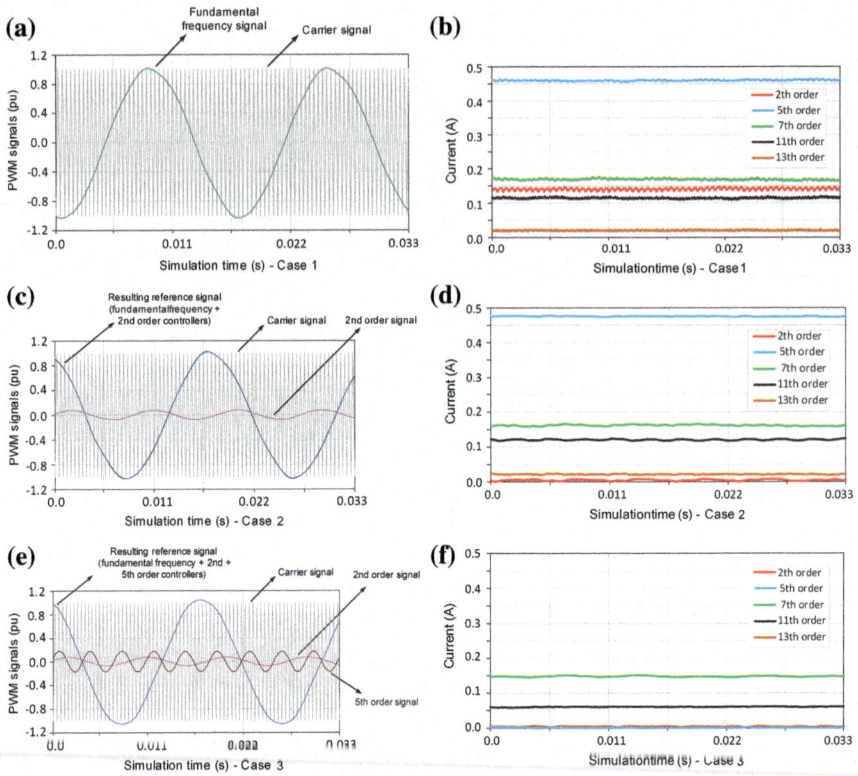

Fig. 18 Inverter harmonic current generation as a function of the switching pulses pattern

Finally, Fig. 18e indicates that the control signal given by the addition of a fundamental, a 2nd and a 5th component will cause the inverter harmonic current generation set in Fig. 18f.

The results are clear enough to show that the methodology herein considered dynamically changes the inverter PWM switching patterns.

Based on the above, Fig. 19 shows the control strategy block diagram required to implement the harmonic compensation technique herein considered [16].

3.2 Computational Studies

In order to demonstrate the applicability of the harmonic compensation technique discussed, this section is dedicated to present computational studies carried out to highlight the methodology. This was realized using a typical distribution system with a wind farm similar to an actual application. The studies are carried out with the aim of eliminating some pre-defined harmonic components. Figure 20 presents a

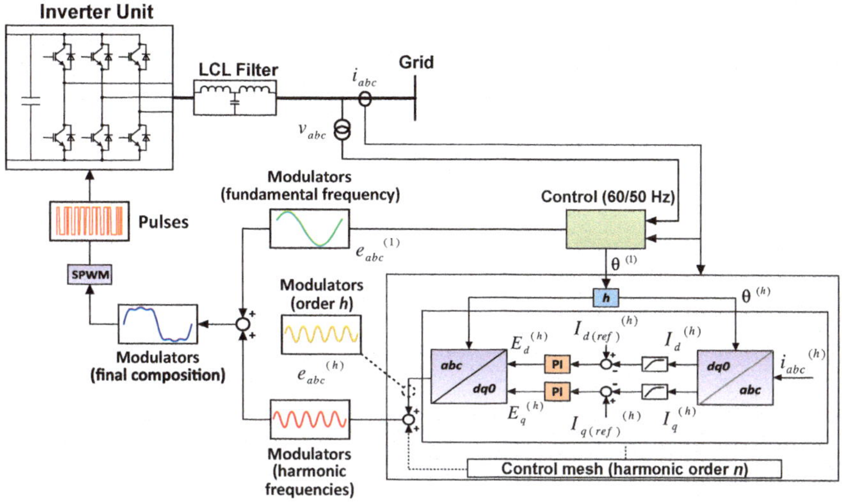

Fig. 19 Inverter control strategy for harmonic mitigation

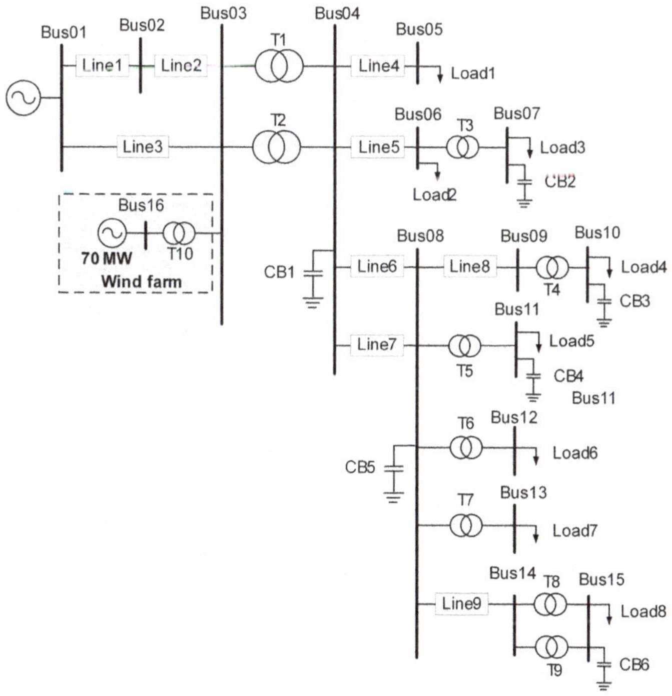

Fig. 20 The simulated AC grid with wind farm generation

Table 5 Busbar 01—equivalent generator

Generator	Nominal voltage (kV)	Short circuit power (MVA)
G1	138.00	680.0

Table 6 Capacitor bank parameters

Busbar	Rated power (MVAr)
4	13.80
7	1.02
10	1.02
11	4.00
8	5.08
15	5.05

Table 7 Line parameters

Line	Resistance (Ω)	Inductance (mH)	Capacitance (μF)
1	17.05	152.55	1.14
2	2.51	22.43	0.17
3	16.72	172.51	1.33
4	9.88	60.34	0.48
5	2.51	22.43	0.17
6	2.37	11.18	0.11
7	2.09	18.19	0.14
8	13.05	54.52	0.37
9	0.70	6.06	0.05

single-line diagram of the electrical arrangement [17] that has been modelled in the ATP/EMTP software.

Tables 5, 6, 7, 8 and 9 emphasize the electrical parameters for the individual components that perform the arrangement. For greater similarity with that encountered in the field, a group of nonlinear loads are adopted to produce the AC system background distortion. The corresponding harmonic currents related to the loads are given on Table 9.

Figure 21 presents the single-line diagram of the wind farm. It comprises of 35 wind turbine units that are represented in accordance with [18]. Table 10 presents the main parameters of the inverter and Table 11 the data of the medium voltage cables.

The simulation considers a constant wind speed corresponding to an active power production of a 100% of the installed capacity of the wind farm. Similar results were verified for other situations and the conclusions were similar to those presented herein.

Table 8 Transformers parameters

Transformer	Rated power (MVA)	Rated voltage (kV)	Impedance (%)
1	60.0	138.0/69.0	15.60
2	60.0	138.0/69.0	15.60
3	12.0	69.0/13.8	4.64
4	5.0	69.0/13.8	6.01
5	12.0	69.0/13.8	8.35
6	12.0	69.0/13.8	8.35
7	12.0	69.0/13.8	8.35
8	12.0	69.0/13.8	6.15
9	12.0	69.0/13.8	6.15
10	70.0	138.0/34.5	12.0

Table 9 Load parameters

	Rated power—fundamental frequency		Harmonic currents (A)			
Load	Active (MW)	Reactive (MVAr)	5th	7th	11th	13th
1	20.8	8.9	7.0	5.0	11.6	2.3
2	1.8	0.8	1.2	0.9	0.3	0.2
3	5.4	2.3	9.0	7.0	4.7	3.0
4	2.1	0.9	7.1	5.1	1.8	1.2
5	7.1	3.0	12.0	9.0	6.2	3.9
6	0.9	0.4	3.0	2.2	0.8	0.5
7	3.0	1.3	10.1	7.3	2.6	1.7
8	18.9	8.1	10.1	7.3	2.6	1.7

Figure 22 presents the amplitude of the harmonic currents established between wind farm and AC grid. During the first 2 seconds, the wind generator inverters operate only with fundamental frequency control loops, which are aimed at injecting the required active and reactive power. The harmonic content produced within this period is in total agreement with those theoretically expected, both in order and value. Following this, at $t = 2\,s$, the harmonic compensation strategy is activated and the 2nd, 5th, 7th, 11th and 13th harmonic components were eliminated showing, computationally, the success of the proposed strategy.

In a complementary way, Fig. 23 shows the harmonic voltages at the PCC. Both the individual harmonic distortions and the THD are shown under the following conditions:

- AC grid with no wind farm operation;
- AC grid with wind farm at full power and no harmonic mitigation technique activation;

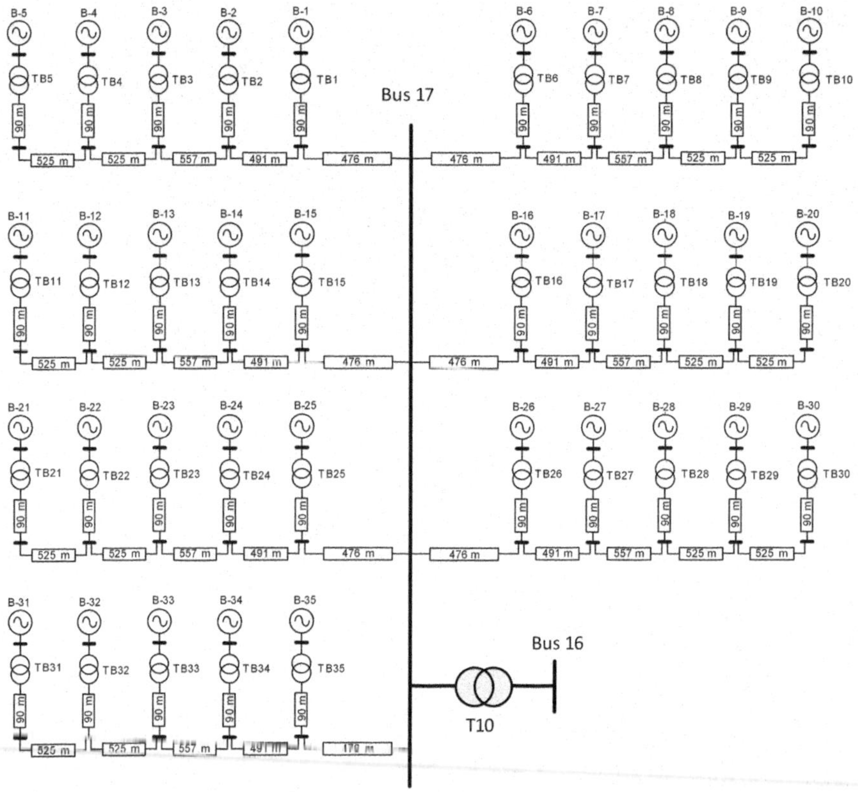

Fig. 21 Wind farm—Single-line diagram

Table 10 Wind system inverter—main parameters

Parameter	Value
Rated power	2.00 MVA
DC inverter input voltage	1200 V
Switching frequency	2500.00 Hz
LCL Filter—Converter Side Inductance	0.14 mH
LCL Filter—Capacitance	780.00 μF
LCL Filter—Damping Resistance	0.10 Ω
LCL Filter—Grid Side Inductance	0.14 mH
Switches dead time	8.00 μs

Table 11 Wind farm—Cables parameters

Parameter	Value
Resistance	0.1932 Ω/km
Inductive reactance	0.1444 Ω/km
Capacitance	0.21 nF/km

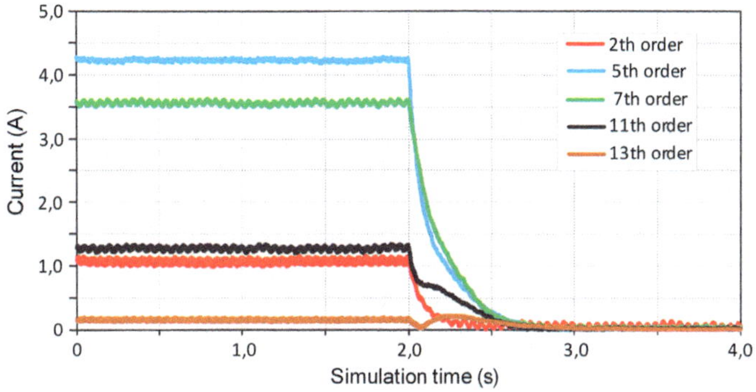

Fig. 22 Harmonic currents circulating between wind farm and power system—without and with harmonic compensator

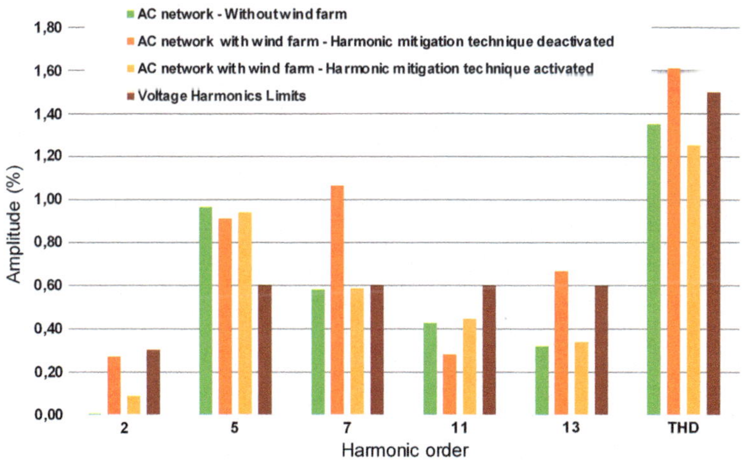

Fig. 23 Harmonic voltages at the PCC without and with harmonic compensator—AC grid under normal conditions

- AC grid with wind farm at full power and harmonic mitigation technique activated;
- Typical harmonic level limits established by regulatory agencies [19].

The results emphasize the mitigation approach effectiveness at compensating the individual and total harmonic distortion to the required levels. The PCC total harmonic distortion of 1.22% is approximately the same already existing and produced by the AC loads, with or without the wind farm operation. With no harmonic mitigation control activation, the corresponding THD achieved a value of 1.81%. This is above the given standard value.

4 Conclusions

This chapter emphasized issues associated with the problems inherent to electrical power quality challenges, with emphasis placed upon the harmonic distortions produced when integrating wind farms to the connecting electricity networks. Within this context, theoretical and experimental considerations related to the factors that influence and determine the orders and magnitudes of the harmonic currents produced by the full converter type generation units were established. Following on, the principles that guide a strategy for an intrinsic harmonic compensation technique aimed at mitigating current components were presented.

Based on the aforementioned generation structures that is based on synchronous generators and a full power connection converter, three influencing factors were pointed out that play a decisive role in the generation of the harmonic currents injected by such installations. One such factor is the switching operating principle of the inverters and their respective PWM technologies for the control of the pulses for the electronic keys (modulation index); another one is related to the so-called "dead time" in order to avoid any eventual undesirable short-circuit between switches assembled on the same branch of the inverter and finally, the existence of pre-distortions in the AC busbar to which the wind park will be connected (PCC background distortion).

Concerning the modulation index, this is presented as a factor that exercises major influence over the spectrum content. It became evident that when operating the inverter in its linear region, the harmonic frequencies present higher orders. In the overmodulation and saturation region, there already exists the emergence of low order harmonics in addition to those of higher frequencies. Regarding the switch dead-time, which guarantees the safe operation of the inverter semiconductors, this effect was shown through adequate verification to be of minor significance. On the other hand, the existence of background distortions are shown to be a major factor that impacts on the operation of the complex and in its generation of harmonics.

In succession, the chapter was directed to the proposition and exploitation of a control strategy that integrates functionalities into wind turbines to mitigate specific harmonic currents. The methodology is an intrinsic compensation mechanism, which dynamically changes the inverter PWM switching patterns, and as such aims at mitigating pre-defined harmonic orders.

The main idea put forward relies on the use of controllers that operate parallel to the fundamental frequency control. A PI controller was used, which superimposes harmonic frequencies onto the reference signal used by the Space Vector PWM. This strategy changes the amplitude and phase of the inverter output harmonic voltage, allowing for harmonic emission control.

In order to verify the efficacy of this proposal a case study associated to a real site grid was used and two situations were considered. Both of which have demonstrated that by activating the compensation mechanism, the existing voltage distortion present at the connection busbar were kept constant, i.e. no further changes on the distortions other than those already existing were noted.

Acknowledgements The authors are grateful to CAPES, FAPEMIG (Project: TEC APQ 2381-13) and FAPDF (Project: 193.001.505/2017) for the financial support for this research.

References

1. International Electrotechnical Commission: IEC 61400-21—wind turbines Part 21: (2008) Measurement and assessment of power quality characteristics of grid connected wind turbines
2. Keane A et al (2013) State-of-the-art techniques and challenges ahead for distributed generation planning and optimization. IEEE Trans Power Syst. https://doi.org/10.1109/TPWRS.2012.2214406
3. Rosentino AJP, Gondim IN, Reis A, Oliveira JC (2016) A critical analysis of the harmonic distortion procedure applied to wind farm connection. In: 17th international conference on harmonics and quality of power (ICHQP). https://doi.org/10.1109/ICHQP.2016.7783349
4. Yang K, Bollen MHJ, Larsson EOA (2015) Aggregation and amplification of wind-turbine harmonic emission in a wind park. IEEE Trans Power Deliv. https://doi.org/10.1109/PESGM.2015.7285778
5. Tentzerakis ST, Papathanassiou SA (2007) An investigation of the harmonic emissions of wind turbines. IEEE Trans Energy Convers. https://doi.org/10.1109/TEC.2006.889607
6. Sharaf AM, Wang W, Altas IH (2010) A novel hybrid active filter compensator for stabilization of wind-utility grid interface scheme. Eur Trans Electr Power. https://doi.org/10.1002/etep.313
7. Wilsun X, Yilu L (2000) A method for determining customer and utility harmonic contributions at the point of common coupling. IEEE Trans Power Deliv. https://doi.org/10.1109/61.853023
8. Napoles J, Leon JI, Portillo R, Franquelo LG, Aguirre MA (2010) Selective harmonic mitigation technique for high-power converters. IEEE Trans Ind Electron. https://doi.org/10.1109/TIE.2009.2026759
9. Liserre M, Teodorescu R, Blaabjerg F (2006) Multiple harmonics control for three-phase grid converter systems with the use of PI-RES current controller in a rotating frame. IEEE Trans Power Electron. https://doi.org/10.1109/TPEL.2006.875566
10. Cantarellas AM, Sintamarean C, Rakhshani E, Rodriguez P, Teodorescu R (2013) Control of an active filter based three-level grid connected converter for wind turbine application. Renew Energy Power Qual J 1:1–6
11. Sha D, Wu D, Liao X (2011) Analysis of a hybrid controlled three-phase grid-connected inverter with harmonics compensation in synchronous reference frame. IET Power Electron. https://doi.org/10.1049/iet-pel.2010.0231
12. Holmes DG, Lipo TA (2003) Pulse width modulation for power converters—principles and practice. Wiley
13. Jeong S, Park MH (1991) The analysis and compensation of dead-time effects in PWM inverters. IEEE Trans Ind Electron. https://doi.org/10.1109/41.88903
14. Rodriguez P, Luna A, Munoz-Aguilar RS, Etxeberria-Otadui I, Teodorescu R, Blaabjerg F (2012) A stationary reference frame grid synchronization system for three-phase grid-connected power converters under adverse grid conditions. IEEE Trans Power Electron. https://doi.org/10.1109/TPEL.2011.2159242
15. Teodorescu R, Liserre M, Rodriguez P (2011) Grid converters for photovoltaic and wind power systems. Wiley

16. Blaabjerg F, Teodorescu R, Liserre M, Timbus AV (2006) Overview of control and grid synchronization for distributed power generation systems. IEEE Trans Ind Electron. https://doi.org/10.1109/TIE.2006.881997
17. Santos IN, Cuk V, Almeida PM, Bollen MHJ, Ribeiro PF (2015) Considerations on hosting capacity for harmonic distortions on transmission and distribution systems. Electric Power Syst Res. https://doi.org/10.1016/j.epsr.2014.09.020
18. Reis A, Moura LP, Oliveira JC (2018) Computational studies of voltage regulation provided by wind farms through reactive power control. Int J Emerg Electric Power Syst. https://doi.org/10.1515/ijeeps-2017-0234
19. Brazilian National Grid Operator (Operador Nacional do Sistema ONS): (2010) Brazilian transmission system grid code (Procedimentos de Rede—Submódulo 2.8)

Power Conversion and Predictive Control of Wind Energy Conversion Systems

Venkata Yaramasu, Samir Kouro, Apparao Dekka, Salvador Alepuz, Jose Rodriguez and Mario Duran

Abstract Wind energy conversion systems have become mature and competitive with conventional and other renewable energy sources. Electric generators, power converters, and control systems ensure safe, efficient, reliable, and high-performance operation for wind energy systems, while meeting the strict grid codes. This book chapter presents power conversion systems and predictive control strategies for permanent magnet synchronous generator, squirrel cage induction generator, and doubly-fed induction generator-based wind energy conversion systems. Various forms of predictive control techniques including predictive current control, predictive torque control, and predictive power control are discussed with case studies. The discrete-time models of overall system are presented in terms of power converter switching states. The predictive control strategies fulfill the control requirements such as maximum power point tracking, regulation of DC-link voltage, grid synchronization, generation of reactive power to a three-phase grid, and fault-ride through operation. The intuitiveness of material presented in this book chapter allows readers extending the predictive control strategies for other power conversion applications.

V. Yaramasu (✉)
Northern Arizona University, Flagstaff, AZ, USA
e-mail: Venkata.Yaramasu@nau.edu

S. Kouro
Universidad Tecnica Federico Santa Maria, Valaraiso, Chile
e-mail: samir.kouro@ieee.org

A. Dekka
Ryerson University, Toronto, Ontario, Canada
e-mail: dapparao@ieee.org

S. Alepuz
Universitat Pompeu Fabra, Mataro, Spain
e-mail: alepuz@tecnocampus.cat

J. Rodriguez
Universidad Andres Bello, Santiago, Chile
e-mail: jose.rodriguez@unab.cl

M. Duran
University of Malaga, Andalusia, Spain
e-mail: mjduran@uma.es

R.-E. Precup et al. (eds.), *Advanced Control and Optimization Paradigms for Wind Energy Systems*, Power Systems, https://doi.org/10.1007/978-981-13-5995-8_5

1 Introduction

Wind energy installed capacity has skyrocketed from 6,100 MW in 1996 to 539,123 MW by 2017 [1]. The wind power capacity worldwide accounts for approximately 4% of the global electricity production. Besides the installed capacity, the power rating of wind turbines (WTs) increased exponentially to 10 MW [2]. This trend continues to increase, and future WT projects in the 10–15 MW class are anticipated from global WT manufacturers such as *Clipper, Mitsubishi, Sway Turbine, GoldWind, Sinovel, Mecal, MingYang, GE Energy, United Power,* and *Gamesa* [3].

The past three decades have witnessed fixed-speed, semi-variable speed, and full-variable speed operation of wind energy conversion systems (WECS) through different combinations of electric generators and power electronic converters [4]. The present WT manufacturers primarily use doubly-fed induction generator (DFIG) for semi-variable speed operation, and squirrel-cage induction generator (SCIG), permanent magnet synchronous generator (PMSG), wound rotor synchronous generator (WRSG), and high-temperature-superconducting synchronous generator (HTS-SG) for full-variable speed operation. Among the classes of wind generators for full-variable speed operation, PMSG is the most popular because it offers high power density and reliability, does not require rotor field excitation and gearbox, and operates with high efficiency [5].

The control system design ensures safe, reliable, and efficient operation of WECS. The present industry practices high performance vector control techniques such as field-oriented control (FOC) and direct torque control (DTC) to control wind generators [6]. Grid-side converters (GSCs) in variable-speed WECS are controlled by decoupled voltage-oriented control (VOC) and direct power control (DPC) [7]. In recent years, the finite control-set model predictive control (FCS-MPC) strategy has emerged as a simple and promising digital control tool for electric power conversion systems [8]. During the low-switching frequency operation of megawatt-level WECS, the predictive control mitigates several technical and operational disadvantages associated with the classical control techniques [3].

Considering the present and emerging trends in wind energy industry, the authors present a comprehensive survey on variable-speed WECS with respect to power conversion and control schemes. The principles of FOC/VOC, DTC, and DPC schemes are used to design predictive current control (PCC), predictive torque control (PTC), and predictive power control (PPC), respectively. The PCC, PTC, and PPC schemes are discussed in detail for PMSG, SCIG, and DFIG WECS, respectively. The design procedure for the calculation and extrapolation of reference control variables, prediction of control variables' behavior, and cost function is provided. The PCC, PTC, and PPC schemes accomplish control objectives similar to the FOC/VOC, DTC, and DPC schemes, respectively, but without any PI controllers, modulation stage, hysteresis controllers, and look-up tables. The predictive control schemes operate with variable switching frequency as they do not employ any modulation stage. The state-of-the-art and emerging technologies presented in this chapter on variable-speed WECS will help the eventual readers to update their knowledge on this specific topic.

This chapter is organized as follows:

- The WT power converters are discussed in Sect. 2 with respect to low-voltage (LV) and medium-voltage (MV) configurations.
- An overview of digital control schemes for WTs is given in Sect. 3.
- PCC of PMSG WECS, PTC of SCIG WECS, and PPC of DFIG WECS are discussed in Sects. 4, 5, and 6, respectively.
- Section 7 provides concluding remarks for this book chapter.

2 Power Converter Topologies for WECS

The two-stage (AC/DC + DC/AC) and three-stage (AC/DC + DC/DC + DC/AC) converters are employed in WECS, and the former class is very popular in wind industry. The power converters are also classified into LV (<1000 V) and MV (3000–4000 V) categories.

2.1 LV Converters

A full-variable speed WECS with SCIG/SG and full-power (100%) back-to-back (BTB) connected two-level voltage source converter (2L-VSC) is shown in Fig. 1 [9]. The 2L voltage source rectifier (2L-VSR) and 2L voltage source inverter (2L-VSI) are identical in terms of topology, construction, modulation, and control. Due to the nature of connection, the 2L-VSR and 2L-VSI are also referred to as machine-side converter (MSC) and grid-side converter (GSC), respectively. The MVA capacity of GSC is usually higher than the MSC such that the WT meets the grid codes. The DC-link provides decoupling between the wind generator and grid. The DC chopper consisting of an IGBT and a resistor is used to dissipate surplus energy during grid

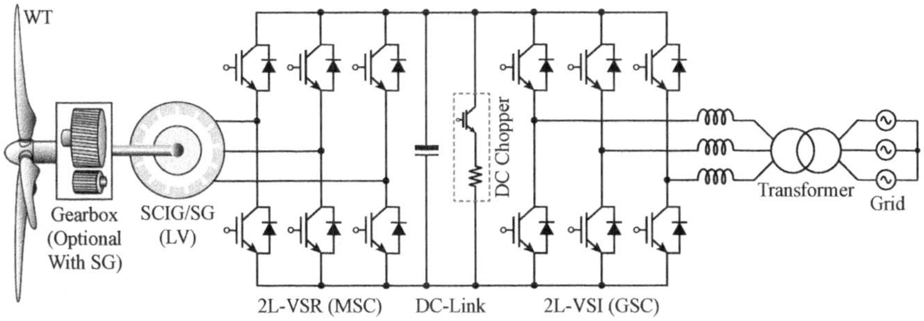

Fig. 1 Full-variable speed WECS with SCIG/SG and full-power BTB connected 2L-VSC

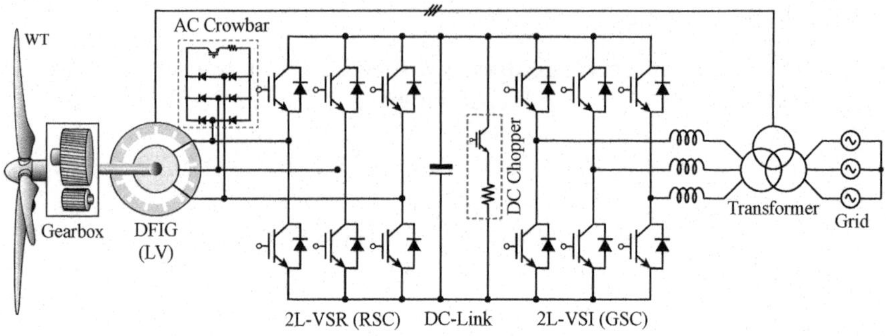

Fig. 2 Semi-variable speed WECS with DFIG and partial-power BTB connected 2L-VSC

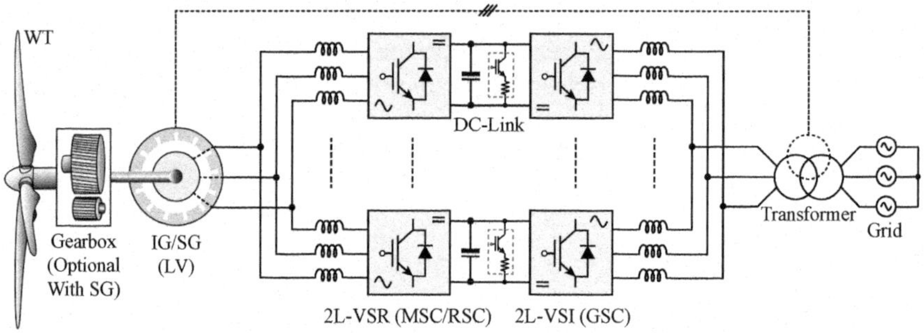

Fig. 3 Variable-speed WECS with IG/SG and parallel connected BTB 2L-VSCs

voltage dips. This configuration is very popular for SCIG/SG WTs rated below 0.75 MW power level.

Figure 2 shows the semi-variable speed WECS with DFIG and partial-power (30%) BTB 2L-VSC in rotor circuit. The DFIG rotor is connected to the grid through rotor-side converter (RSC), DC-link, and GSC [10], and the stator is directly connected to the grid. The partial-power converter in DFIG WECS controls the generator active/reactive power, DC-link voltage and grid power factor. Due to 30% rated capacity of converter, the DFIG speed is controlled in ±30% range. The use of partial-power converter decreases the size and weight of DFIG WTs in comparison to the full-variable speed WTs. During grid faults, the rotor over-currents and DC over-voltages are suppressed by AC crowbar and DC chopper, respectively.

For power ratings greater than 0.75 MW in full-variable speed WTs and 2.5 MW in semi-variable speed WTs, the 2L-VSCs are connected in parallel to increase the power handling capability (Fig. 3) [11]. For example, 10 converters are connected in parallel in the *Enercon* E-126 WT to increase power capacity to 7.5 MW. The parallel connection of converters lead to energy efficiency and redundancy. The system efficiency can be improved by turning off one or more converters when the wind speed is low. Higher converter cost, converter derating, circulating currents, and

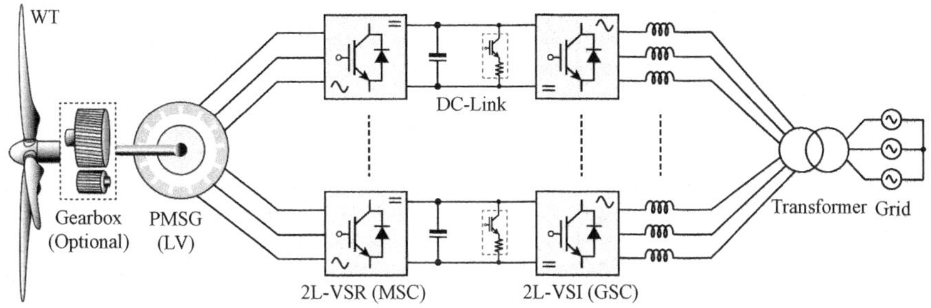

Fig. 4 Full-variable speed WECS with multiphase PMSG and distributed 2L-VSCs

complex configuration and control are the major drawbacks of this power conversion configuration.

Issues with the circulating currents in converters can be solved by employing multiphase PMSG, distributed converters, and open-winding transformer as shown in Fig. 4. For example, the *Gamesa* G10x WT uses 4 sets of three-phase windings and 4 BTB 2L-VSC modules in parallel to handle 5 MW power. Due to the redundancy of phases, fault tolerance against power converter and generator faults is improved [12]. To reach system voltage levels of 10–35 kV and perform transformerless grid connection, the power converters can be connected in cascade [13].

The SG WECS does not require excitation current from the grid. Therefore, diode-bridge rectifiers can be used for AC/DC conversion with unidirectional power flow. The diode bridge rectifiers are reliable and less expensive than the VSR. However, due to the use of diode-bridge rectifiers, significant amount of 5th (14%) and 7th (7%) harmonics appear in the generator currents; and as a consequence the generator electromagnetic torque contains 6th harmonic distortion (10%) [14]. A six-phase SG with 30° phase shift solves the issues with high torque ripples by cancelling the 5th and 7th harmonic components in generator currents. Figure 5 shows the six-phase configuration that has been commercialized by WT manufacturers such as *Vensys* and *GoldWind* for 1.5 MW output power [15]. The boost converter performs the maximum power point tracking (MPPT) and DC voltage step-up operations. The interleaving operation of boost converter and 2L-VSI decreases the inductor and grid current ripples, respectively.

2.2 *MV Converters*

For power ratings greater than 3.0 MW, the MV operation of WTs is more efficient, reliable, and economical. The BTB connected neutral-point-clamped (NPC) converters have been used in the commercial WTs for 6.0 MW power rating and MV operation in 3–4 kV range [16]. The WT manufacturers such as *ADVEN, Shandong, XEMC-Darwind,* and *Zephyros* have used NPC converters in conjunction with

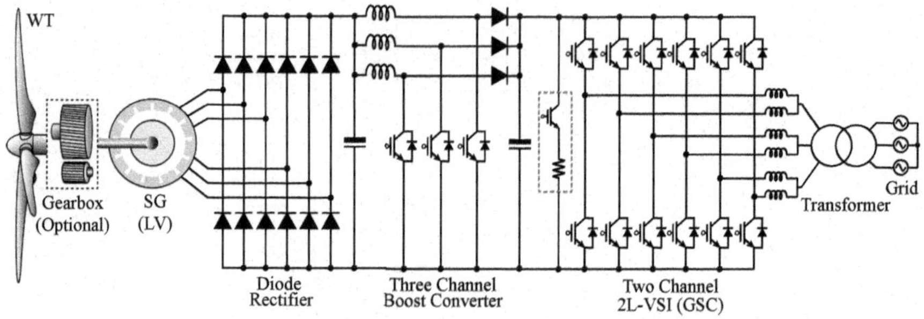

Fig. 5 SG WECS with six-phase diode rectifier, parallel boost converters, and parallel 2L-VSIs

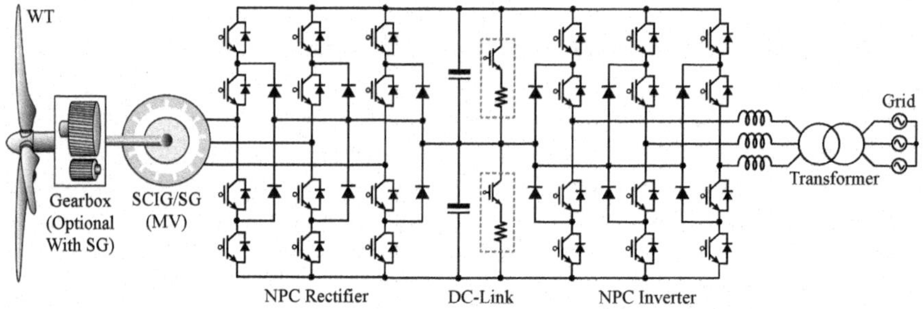

Fig. 6 Full-variable speed WECS with MV-SCIG/SG and BTB connected NPC converter

MV-PMSG. In comparison to the 2L-VSCs, the NPC converters offer several advantages such as reduced *dv/dt* in the output voltage, reduced switching losses, lower output current ripple, smaller harmonic filter requirement, and reduced electromagnetic interference [17]. The neutral-point voltage control is a major issue with the NPC converters; and this issue can be solved through the digital control scheme design (Fig. 6).

3 Control Schemes for WECS

There are two type of control systems used simultaneously to control a WT: aerodynamic control (by adjusting the pitch angle of blades) and electromechanical control (by adjusting the WT speed). The focus of this chapter is on the later. The electromechanical control is performed by the power converter stages, and has the following control goals: MPPT by controlling the generator torque/speed through the MSC/RSC; grid synchronization, DC-link voltage control and active/reactive power control by the GSC. The classic or mainstream control methods used for these goals are briefly described in the following subsections.

3.1 MPPT Control

The MPPT schemes are designed to extract the peak power during varying wind speed conditions. Some of the popular MPPT schemes are optimal tip speed ratio (OTSR) control, optimal torque (OT) control, and WT power curves (WTPC)-based control [15]. In OTSR method, the ω_m^* is obtained from the measured wind speed v_w and WT parameters, that is,

$$\omega_m^* = K_1^{\text{op}}\, v_w. \tag{1}$$

In the case of OT method, the reference electromagnetic torque T_e^* is calculated from measured wind generator mechanical speed ω_m, that is,

$$T_e^* = K_2^{\text{op}}\, \omega_m^2 \tag{2}$$

where K_1^{op} and K_2^{op} are the proportional constants and obtained from the offline calculations at the rated conditions. In WTPC-based control, the generator output power P_s versus v_w curves are derived from the initial experimental tests. Unlike OTSR and WTPC-based control, the OT control does not require anemometer and easy to implement.

3.2 Control of PMSG

In PMSG wind energy conversion systems, the vector control strategies are adopted to achieve the decoupled torque and flux control [6]. Some of the popular vector control strategies are zero d-axis current (ZDC) control and maximum torque per ampere (MTPA) control [15]. The ZDC control is applied to surface-mount PMSG, whereas the MTPA control is applied to interior PMSG to develop a maximum torque with a minimum value of stator current , thereby reducing the power losses and improving the system efficiency. In ZDC scheme, the reference d-axis stator current i_{ds}^* is set to zero, whereas the reference q-axis stator current i_{qs}^* is calculated from T_e^* as shown below:

$$i_{ds}^* = 0, \qquad i_{qs}^* = \frac{T_e^*}{1.5\, P_p\, \psi_r} \tag{3}$$

where i_{ds} and i_{qs} are d- and q-axis stator current components, ψ_r represents the peak value of rotor flux linkage, and P_p represents the number of pole pairs.

In MTPA scheme, the reference d- and q-axis current components are calculated from the measured i_{ds} and are given as,

$$i_{qs}^* = \frac{T_e^*}{1.5\, P_p \left[\psi_r + \left(L_{ds} - L_{qs}\right) i_{ds}\right]} \tag{4}$$

$$i_{ds}^* = -\frac{\psi_r}{2\,(L_{ds} - L_{qs})} - \sqrt{\frac{\psi_r^2}{4\,(L_{ds} - L_{qs})^2} + [i_{qs}^*]^2} \tag{5}$$

where L_{ds} and L_{qs} are stator dq-axis inductances.

3.3 Control of SCIG

Similar to SG, the SCIG can be controlled with FOC and DTC schemes [6]. In FOC scheme, the rotor flux vector is aligned with the d-axis of the synchronous reference frame. As a result, ψ_{dr} becomes equal to ψ_r and ψ_{qr} becomes zero. Furthermore, the stator d-axis and q-axis current components (i_{ds} and i_{qs}) are aligned with the d-axis and q-axis of the synchronous reference frame. As a result, the decoupled control of rotor flux and torque are possible in SCIG, and the resultant torque expression is given by,

$$T_e = K_T\, \psi_r\, i_{qs} \quad (\text{for } \psi_{dr} = \psi_r \text{ and } \psi_{qr} = 0), \quad K_T = \frac{3\, P_p\, L_m}{2\, L_r} \tag{6}$$

where L_m and L_r are magnetizing and rotor self inductances, respectively.

The indirect FOC scheme estimates rotor flux value and its position from the stator currents and rotor electrical speed ω_r. The ω_r is added to the estimated slip frequency ω_{sl}, which gives the stator angular frequency ω_s. The ω_s is integrated to obtain the rotor flux angle θ_e. By using this angle, the stator currents in abc-frame are converted into the dq-frame. The DTC scheme eliminates the inner current control loop and directly controls the stator flux and torque. The electromagnetic torque of SCIG is given by,

$$T_e = \frac{3\, P_p\, L_m}{2\, \sigma\, L_s\, L_r}\, \psi_s\, \psi_r \sin \theta_T \tag{7}$$

where σ is total leakage coefficient, L_s is stator inductance, and θ_T is torque angle.

In DTC approach, the stator flux magnitude is maintained constant such that the torque T_e varies linearly with θ_T. During this process, there is a change in rotor flux magnitude ψ_r, but its variation is small and negligible. The main features of DTC scheme are simplicity in the design and fast dynamic response in comparison to the FOC method, but the DTC operates at variable switching frequency and produces a higher current ripple.

3.4 Control of DFIG

The DFIG can be controlled by FOC, DTC, or DPC schemes, which are implemented in synchronous reference frame with stator voltage orientation [10]. Hence, the q-axis stator voltage component becomes zero, and the three-phase rotor currents are decoupled into active and reactive current components. By neglecting stator winding resistance, the d- and q-axis rotor currents in terms of stator active and reactive powers are given as follows [15]:

$$i_{dr} = -\left(\frac{2\,L_s}{3\,v_{ds}\,L_m}\right) P_s \tag{8}$$

$$i_{qr} = \left(\frac{2\,L_s}{3\,v_{ds}\,L_m}\right) Q_s - \left(\frac{1}{\omega_s\,L_m}\right) v_{ds}. \tag{9}$$

3.5 Grid Synchronization

The grid synchronization of GSC is achieved by using the grid voltage angle θ_g. Therefore, the active and reactive power flow between the grid and GSC is effectively regulated. Furthermore, the θ_g simplifies the digital controller design and allows the decoupled control of electrical variables by transforming them from abc to dq, and vice-versa. In reality, the AC-grids are weak and requires a sophisticated method to obtain the grid voltage angle θ_g. Some of the popular methods are PI-based phase-locked loop (PLL) and dual second order generalized integrator (DSOGI) PLL [7].

3.6 Control of Grid-Side Converter

In WECS, the GSC regulates the DC-link voltage v_{dc} by controlling the grid active power, whereas the reactive power Q_g is controlled to obtain the required grid power factor. To achieve these objectives, the VOC and DPC methods are widely employed [3]. To obtain the decoupled control of active and reactive powers, the grid voltage vector is aligned with the d-axis of SRF; as a result, v_{qg} becomes zero. Hence, the P_g varies linearly with i_{dg} whereas the Q_g is proportional to i_{qg} under a constant v_{dg} value [18], that is,

$$P_g = +1.5\,v_{dg}\,i_{dg} \tag{10}$$

$$Q_g = -1.5\,v_{dg}\,i_{qg}. \tag{11}$$

From the above equations, the VOC scheme regulates the active power (P_g) and reactive power (Q_g) through the control of dq-axis grid currents. On the other hand, the P_g and Q_g are controlled by hysteresis comparators and a lookup table in DPC method.

3.7 Fault-Ride Through Control

Often, the wind energy conversion systems will be disconnected from the grid due to the external disturbances including the faults and voltage sags and swells, which leads to instability of the power system network. Under such scenarios, the GSC supplies reactive power at its full capacity and zero active power to the grid. However, due to large moment of inertia and mechanical time constant, the wind generators still supply active power to the GSC. This injected active power increases the DC-link voltage during the fault and damages the complete power converter system. Hence, to dissipate the surplus energy, a DC chopper is connected across DC-link of WECS. In the case of DFIG WECS, the rotor over currents due to the surplus energy is suppressed by using AC crowbar. In the direct drive PMSG WECS, the surplus energy is stored in the inertia of rotating masses such that the FRT requirements will be met [19].

3.8 Reactive Power Generation

The WECS should be able to generate reactive power during normal and abnormal conditions. The RPG is not only improves the grid power factor, but also compensate the voltage drop in transmission lines. Also, the RPG helps to stabilize the grid and compile with the grid codes and FRT requirements. Hence, the RPG is one of the important grid codes, which ensures efficient and reliable grids. This requirement can be easily fulfilled by designing the WT GSC with high MVA capacity or high-performance control scheme for MSC/RSC and GSC.

4 Predictive Current Control of PMSG WECS

The control structure of PCC scheme for the MSC and GSC in PMSG WECS is shown in Fig. 7. The MSC perform MPPT operation, whereas the GSC regulates the DC-link voltage and reactive power. These control requirements are translated into reference generator and grid currents in dq-frame. The PCC schemes for the MSC and GSC ensure that the actual currents follow their reference currents during all operating conditions. The PCC scheme for MSC and GSC mainly consists of feedback measurements, calculation and extrapolation of reference currents, prediction of future behavior of currents, and optimization process of cost function to select the switching signals.

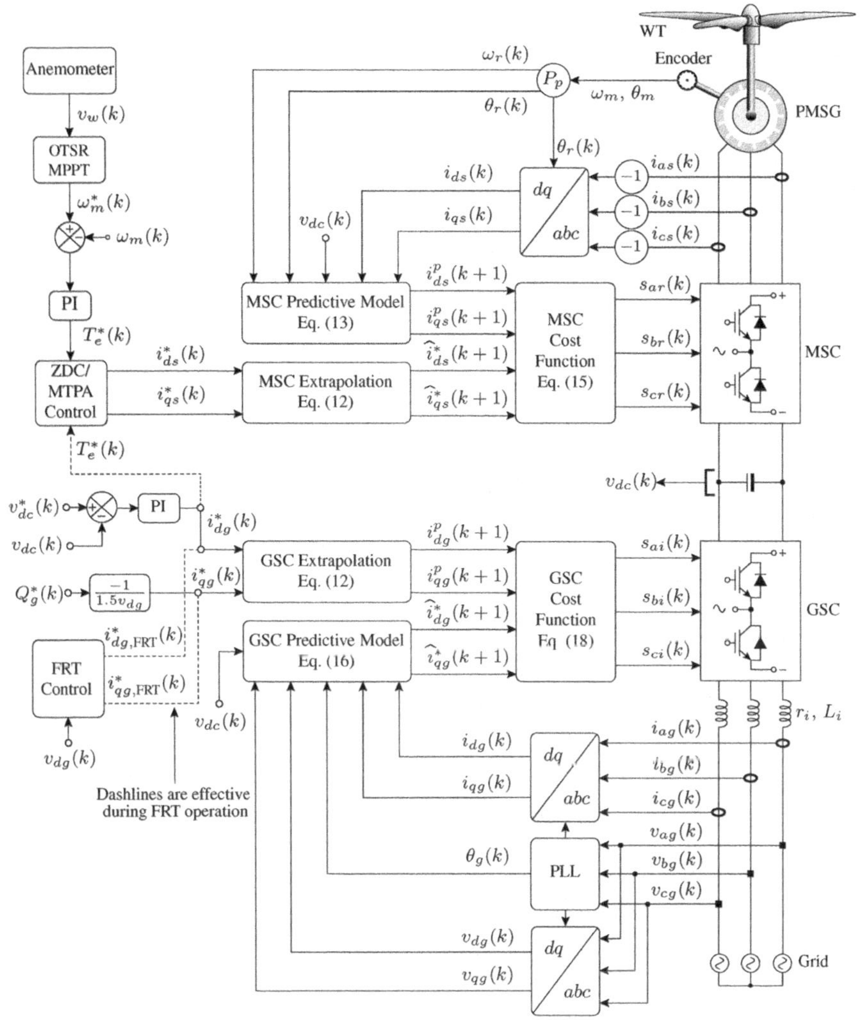

Fig. 7 Block diagram of PCC scheme for MSC and GSC in PMSG WECS

4.1 Control of MSC

The classical control scheme of MSC has an outer speed control loop and an inner current control loop. However, the PCC scheme eliminates the PI controllers in inner current control loops and improves the dynamic response. To implement the PCC in SRF, the measured generator currents will be transformed into SRF by using the electrical speed and position (ω_r and θ_r) values at kth sampling instant. The ω_r and θ_r values are obtained by multiplying the measured mechanical speed and position (ω_m and θ_m) values with the pole pairs P_p. Furthermore, to adopt the motor models for generator operation, the measured stator currents of PMSG are multiplied with -1. The outer speed control loop requires a mechanical speed ω_r, which is obtained from the speed encoder. On the other hand, the reference mechanical speed ω_m^* is calculated by using OTSR MPPT scheme given in (1) and wind speed v_w. A PI controller minimizes the mechanical speed error and generates the reference torque T_e^*. By using the T_e^* information, the ZDC or MTPA control schemes generate d- and q-axis reference current components based on the models given in (3)–(5). Finally, the reference currents are extrapolated to $(k+1)$ sampling instant by using first order Lagrange extrapolation technique as shown below:

$$\left.\begin{aligned} \widehat{i_{ds}^*}(k+1) &= 2\,i_{ds}^*(k) - i_{ds}^*(k-1) \\ \widehat{i_{qs}^*}(k+1) &= 2\,i_{qs}^*(k) - i_{qs}^*(k-1) \end{aligned}\right\} \tag{12}$$

The PCC scheme uses the mathematical model of PMSG in discrete-time domain, which is obtained from the forward Euler approximation technique. The discrete-time model of PMSG stator currents is given as,

$$\left.\begin{aligned} \begin{bmatrix} i_{ds}^p(k+1) \\ i_{qs}^p(k+1) \end{bmatrix} &= \begin{bmatrix} 1-\frac{R_s\,T_s}{L_{ds}} & \frac{\omega_r(k)\,L_{qs}\,T_s}{L_{ds}} \\ -\frac{\omega_r(k)\,L_{ds}\,T_s}{L_{qs}} & 1-\frac{R_s\,T_s}{L_{qs}} \end{bmatrix} \begin{bmatrix} i_{ds}(k) \\ i_{qs}(k) \end{bmatrix} \\ &+ \begin{bmatrix} \frac{T_s}{L_{ds}} & 0 \\ 0 & \frac{T_s}{L_{qs}} \end{bmatrix} \begin{bmatrix} v_{ds}^p(k) \\ v_{qs}^p(k) \end{bmatrix} + \begin{bmatrix} 0 \\ -\frac{\omega_r(k)\,\psi_r\,T_s}{L_{qs}} \end{bmatrix} \end{aligned}\right\} \tag{13}$$

where p denotes the predicted variable. v_{ds}^p and v_{qs}^p are predicted d- and q-axis stator voltage components, R_s is the stator winding resistance, and ω_r is the electrical speed of the rotor.

The MSC output voltages are predicted by using $\alpha\beta$-frame switching states ($s_{\alpha r}$ and $s_{\beta r}$), DC-link voltage v_{dc}, and rotor electrical angle θ_r, that is,

$$\begin{bmatrix} v_{ds}^p(k) \\ v_{qs}^p(k) \end{bmatrix} = \begin{bmatrix} \cos\theta_r(k) & \sin\theta_r(k) \\ -\sin\theta_r(k) & \cos\theta_r(k) \end{bmatrix} v_{dc}(k) \begin{bmatrix} s_{\alpha r}^p(k) \\ s_{\beta r}^p(k) \end{bmatrix}. \tag{14}$$

The MSC cost function designed to minimize the error between extrapolated reference and predicted d and q-axis stator current components and switching frequency [20], that is,

$$\begin{aligned} g_r(k) &= \left[\widehat{i_{ds}^*}(k+1) - i_{ds}^p(k+1)\right]^2 + \left[\widehat{i_{qs}^*}(k+1) - i_{qs}^p(k+1)\right]^2 \\ &+ \lambda_{sw,r} \sum_{x=a,\,b,\,c} \left[s_{xr}^p(k) - s_{xr}^{\mathrm{op}}(k)\right]^2 \end{aligned} \tag{15}$$

where $\lambda_{sw,r}$ is the weight factor for switching frequency minimization in high-power WECS. The $\lambda_{sw,r}$ value is heuristically selected to achieve the desired switching frequency operation.

The MSC has a total of eight switching states, which gives eight possible predictions for v_{ds}^{p} and v_{qs}^{p}. These voltage components are used to predict the PMSG currents, i_{ds}^{p} and i_{qs}^{p}. The cost function $g_r(k)$ is employed to evaluate the difference between predicted and extrapolated reference currents. The switching state combination corresponding to the minimum cost function value is chosen as an optimal actuation. The corresponding switching signals s_{ar}, s_{br}, and s_{cr} are applied to the MSC at the $(k+1)$ sampling instant.

4.2 Control of GSC

The classical control scheme of GSC also has an inner current control loop along with an outer DC-link voltage control loop. The PCC scheme eliminates the PI controllers in inner current control loops as shown in Fig. 7. The grid voltage angle θ_g is used to transform the grid voltages and currents from abc-frame to synchronous dq-frame. The reference DC-link voltage v_{dc}^{*} is set to 3.062 times grid phase voltage [15]. A PI controller minimizes the error between measured and reference DC-link voltage and generates reference d-axis grid current component (i_{dg}^{*}). The grid operator defines the reference reactive power command Q_g^{*}, and it is set to zero to achieve the unity power factor operation. The reference q-axis grid current component i_{qg}^{*} is obtained from Q_g^{*} as shown in Fig. 7. Similar to the models in (12), the reference d- and q-axis grid current components are extrapolated to $(k+1)$ sampling instant by using first-order Lagrange extrapolation.

The predicted d- and q-axis grid current components in discrete-time domain are given as,

$$\left.\begin{aligned} \begin{bmatrix} i_{dg}^{p}(k+1) \\ i_{qg}^{p}(k+1) \end{bmatrix} &= \begin{bmatrix} 1-\frac{r_i T_s}{L_i} & \omega_g T_s \\ -\omega_g T_s & 1-\frac{r_i T_s}{L_i} \end{bmatrix} \begin{bmatrix} i_{dg}(k) \\ i_{qg}(k) \end{bmatrix} \\ &+ \begin{bmatrix} \frac{T_s}{L_i} & 0 \\ 0 & \frac{T_s}{L_i} \end{bmatrix} \begin{bmatrix} v_{di}^{p}(k) \\ v_{qi}^{p}(k) \end{bmatrix} + \begin{bmatrix} -\frac{T_s}{L_i} & 0 \\ 0 & -\frac{T_s}{L_i} \end{bmatrix} \begin{bmatrix} v_{dg}(k) \\ v_{qg}(k) \end{bmatrix} \end{aligned}\right\} \tag{16}$$

where v_{dg} and v_{qg} are the d- and q-axis grid voltage components. i_{dg} and i_{qg} are the d- and q-axis grid current components. r_i and L_i are output filter internal resistance and inductance, respectively.

The d- and q-axis voltage components of GSC are predicted by using $\alpha\beta$-frame GSC switching states, DC-link voltage, and grid voltage angle as shown below:

$$\begin{bmatrix} v_{di}^{p}(k) \\ v_{qi}^{p}(k) \end{bmatrix} = \begin{bmatrix} \cos\theta_g(k) & \sin\theta_g(k) \\ -\sin\theta_g(k) & \cos\theta_g(k) \end{bmatrix} v_{dc}(k) \begin{bmatrix} s_{\alpha i}^{p}(k) \\ s_{\beta i}^{p}(k) \end{bmatrix}. \tag{17}$$

The GSC cost function designed to minimize the error between extrapolated reference and predicted d and q-axis grid current components and switching frequency, that is,

$$\begin{aligned} g_i(k) = &\left[\widehat{i^*_{dg}}(k+1) - i^p_{dg}(k+1)\right]^2 + \left[\widehat{i^*_{qg}}(k+1) - i^p_{qg}(k+1)\right]^2 \\ &+ \lambda_{sw,i} \sum_{x=a,b,c} \left[s^p_{xi}(k) - s^{\mathrm{op}}_{xi}(k)\right]^2 \end{aligned} \tag{18}$$

The GSC also has a total of eight switching states, which gives eight possible predictions for v^p_{di} and v^p_{qi}. These voltage components are used to predict the d- and q-axis grid current components i^p_{dg} and i^p_{qg}. The cost function $g_i(k)$ is employed to evaluate the difference between predicted and extrapolated reference currents ($\widehat{i^*_{dg}}$ and $\widehat{i^*_{qg}}$). The switching state combination corresponding to the minimum cost function value is chosen as an optimal actuation. The corresponding switching signals s_{ai}, s_{bi}, and s_{ci} are applied to the GSC at the $(k+1)$ sampling instant.

The mechanical components of WECS have large moment of inertia and are used to store the surplus energy during grid disturbances, therefore, no DC chopper is used [21]. This method is highly efficient and cost-effective compared with the DC chopper. During grid disturbance, FRT control subsystem starts operating when the grid voltage falls below 90% (0.9 pu) of the rated value and provides rated (1.0 pu) reactive power (i_{qg} component), whereas the FRT control subsystem provides 2% reactive power for each 1% voltage dip, when the grid voltage is in the range of 50% (0.5 pu) to 90% (0.9 pu). The dashed lines for $T_e^*(k)$, $i^*_{dg,\mathrm{FRT}}(k)$, and $i^*_{qg,\mathrm{FRT}}(k)$ correspond to the control variables during grid faults.

4.3 Results and Analysis

The startup transient performance of surface-mount PMSG WECS is analyzed through MATLAB simulation results (Fig. 8). The rated specifications of a 750 kW, 690 V, 9.75 Hz direct drive PMSG are as follows: $R_s = 6.521$ mΩ, $L_{ds} = 3.848$ mH, $L_{qs} = 3.848$ mH, $\psi_r = 8.532$ Wb (*peak*), $P_p = 26$, and $\omega_r = 61.26$ rad/s. The GSC operates with unity power factor (UPF), that is $Q_g^* = 0$. $T_s = 100$ μs.

From $t = 0$ to 0.5 s, the wind speed v_w is maintained at a constant value of 6 m/s. From $t = 0.5$ s, the v_w is increased linearly so that the v_w reaches to the rated value of 12 m/s at $t = 1.5$ s. Furthermore, the reference mechanical speed ω^*_m varies proportional to the v_w (Fig. 8a). The speed controller regulates the actual mechanical speed at its reference value (Fig. 8a). The speed controller gives a reference torque command T_e^*, and the PCC adjust the gating pattern of MSC such that the generator developed torque T_e follows the reference torque command T_e^* (Fig. 8b). As the PMSG accelerates, the stator currents magnitude and frequency increases with the speed (Fig. 8c). The PCC scheme maintains the d-axis current component at its

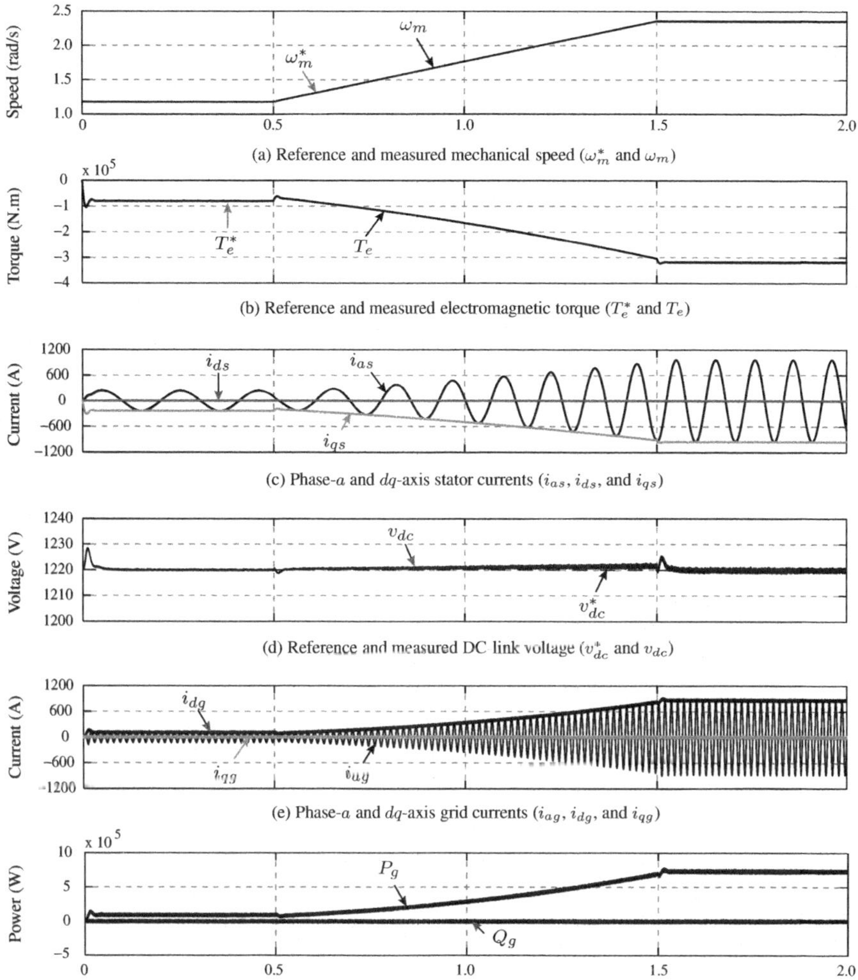

(a) Reference and measured mechanical speed (ω_m^* and ω_m)

(b) Reference and measured electromagnetic torque (T_e^* and T_e)

(c) Phase-a and dq-axis stator currents (i_{as}, i_{ds}, and i_{qs})

(d) Reference and measured DC link voltage (v_{dc}^* and v_{dc})

(e) Phase-a and dq-axis grid currents (i_{ag}, i_{dg}, and i_{qg})

(f) Grid active and reactive power (P_g and Q_g)

Fig. 8 Simulated waveforms of the PCC scheme for PMSG WECS during startup

reference value of zero. The q-axis current i_{qs} maintains linear relationship with the PMSG torque T_e. The DC voltage control loop of GSC regulates the average DC-link voltage v_{dc} at its reference value of 1220 V (Fig. 8d). As expected, the ripple in v_{dc} increases with the increase in the current (generator and grid) magnitudes. The grid d-axis current component follows the peak value of the grid current waveform in natural reference frame. On the other hand, the grid q-axis current component is regulated at zero value to achieve the unity power factor operation on the grid side (Fig. 8e). The total harmonic distortion (THD) of stator currents, THD of grid currents, average switching frequency of MSC, and average switching frequency of GSC during rated operating condition are 0.85%, 7.5%, 1800 Hz, and 1700 Hz, respectively. These results prove the superiority of the PCC scheme in achieving a fast dynamic performance and perfect current tracking in a PMSG based WECS.

5 Predictive Torque Control of SCIG WECS

Figure 9 shows the structure of PTC scheme for MSC and PCC scheme for GSC in SCIG WECS. The PTC scheme regulates the SCIG torque T_e and stator flux by controlling the rotor speed and stator currents. The PTC scheme offers simple implementation and fast dynamic response in the complete operational range [22]. The generator-side reference torque and flux are estimated to achieve the MPPT.

5.1 *Control of MSC*

The optimum torque (OT) MPPT scheme in (2) provides the reference torque command ($T_e^*(k)$), whereas the reference stator flux $\psi_s^*(k)$ is maintained constant at its rated value. These reference commands are extrapolated to $(k+1)$ state similar to the models in (12). The $\alpha\beta$-axis rotor flux linkages are estimated from three-phase stator currents and measured electrical rotor speed $\omega_r(k)$ (refer to Sect. 3.3).

The discrete-time model of SCIG is given by [23],

$$\begin{bmatrix} i_{\alpha s}^p(k+1) \\ i_{\beta s}^p(k+1) \\ \psi_{\alpha r}^p(k+1) \\ \psi_{\beta r}^p(k+1) \end{bmatrix} = \begin{bmatrix} 1-\frac{T_s}{\tau_\sigma} & 0 & \frac{k_r\,T_s}{\sigma\,L_s\,\tau_r} & \frac{k_r\,\omega_r(k)\,T_s}{\sigma\,L_s} \\ 0 & 1-\frac{T_s}{\tau_\sigma} & -\frac{k_r\,\omega_r(k)\,T_s}{\sigma\,L_s} & \frac{k_r\,T_s}{\sigma\,L_s\,\tau_r} \\ \frac{L_m\,T_s}{\tau_r} & 0 & 1-\frac{T_s}{\tau_r} & -\omega_r(k)\,T_s \\ 0 & \frac{L_m\,T_s}{\tau_r} & \omega_r(k)\,T_s & 1-\frac{T_s}{\tau_r} \end{bmatrix} \begin{bmatrix} i_{\alpha s}(k) \\ i_{\beta s}(k) \\ \psi_{\alpha r}(k) \\ \psi_{\beta r}(k) \end{bmatrix} + \begin{bmatrix} \frac{T_s}{\sigma\,L_s} & 0 \\ 0 & \frac{T_s}{\sigma\,L_s} \\ 0 & 0 \\ 0 & 0 \end{bmatrix} \begin{bmatrix} v_{\alpha s}^p(k) \\ v_{\beta s}^p(k) \end{bmatrix} \tag{19}$$

where $i_{\alpha s}$ and $i_{\beta s}$ are the α- and β-axis stator current components. $v_{\alpha s}$ and $v_{\beta s}$ are the α- and β-axis stator voltage components. $\psi_{\alpha r}$ and $\psi_{\beta r}$ are the α- and β-axis rotor flux linkages. T_s is the sampling time.

The SCIG variables in the above model are defined as follows [24]:

- Coupling coefficients of stator and rotor: $k_s = \frac{L_m}{L_s}$ and $k_r = \frac{L_m}{L_r}$.
- Total leakage coefficient, $\sigma = 1 - k_s\,k_r = 1 - \frac{L_m^2}{L_s\,L_r}$.
- Equivalent resistance, $r_\sigma = R_s + k_r^2\,R_r$ (Ω).
- Stator transient time constant, $\tau_\sigma = \frac{\sigma\,L_s}{r_\sigma}$ (s).

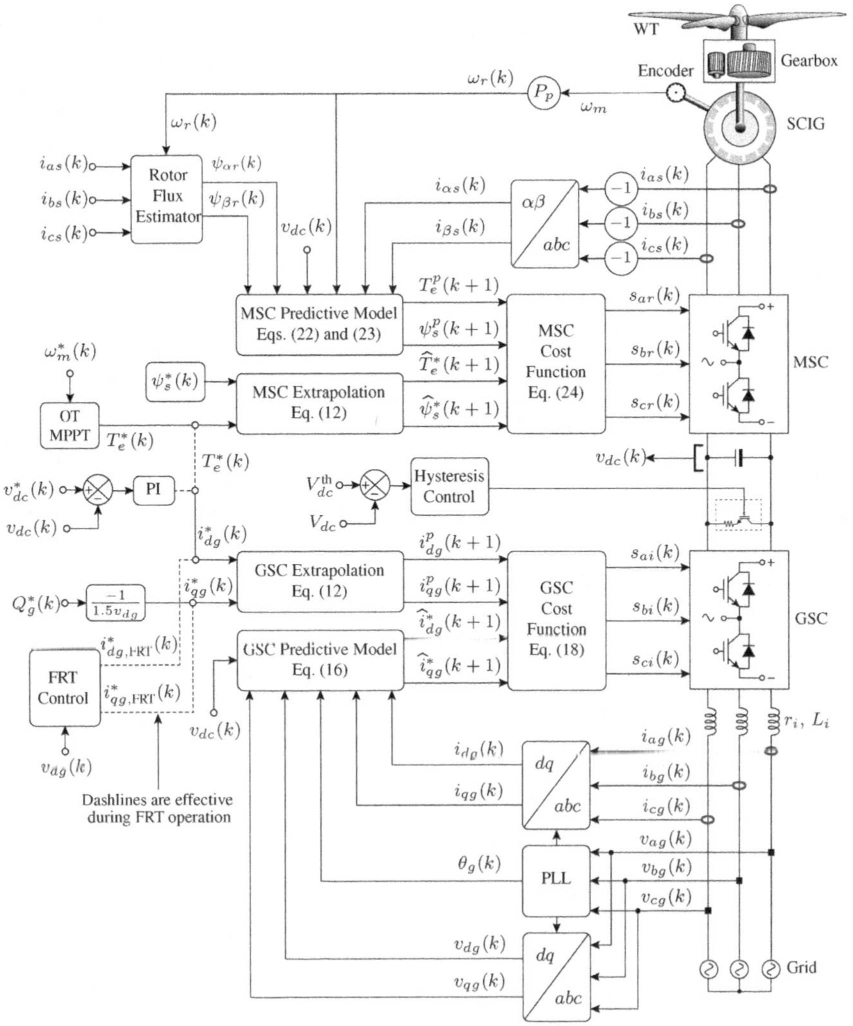

Fig. 9 Block diagram of PTC scheme for MSC and PCC scheme GSC in SCIG WECS

- Stator and rotor time constants: $\tau_s = \frac{L_s}{R_s}$ (s) and $\tau_r = \frac{L_r}{R_r}$ (s)
- Total leakage inductance, $\sigma L_s = \frac{L_s L_r - L_m^2}{L_r}$.

The stator (or MSC) voltages are expressed in terms of DC-link voltage and switching signals in $\alpha\beta$-frame as follows:

$$\begin{bmatrix} v^p_{\alpha s}(k) \\ v^p_{\beta s}(k) \end{bmatrix} = v_{dc}(k) \begin{bmatrix} s^p_{\alpha r}(k) \\ s^p_{\beta r}(k) \end{bmatrix}. \tag{20}$$

The predicted α- and β-axis stator flux linkages at $(k+1)$ sampling instant are obtained from the predicted α- and β-axis stator current components and α- and β-axis rotor flux linkage components in (19), that is,

$$\begin{bmatrix} \psi^p_{\alpha s}(k+1) \\ \psi^p_{\beta s}(k+1) \end{bmatrix} = \begin{bmatrix} \sigma L_s & 0 & k_r & 0 \\ 0 & \sigma L_s & 0 & k_r \end{bmatrix} \begin{bmatrix} i^p_{\alpha s}(k+1) \\ i^p_{\beta s}(k+1) \\ \psi^p_{\alpha r}(k+1) \\ \psi^p_{\beta r}(k+1) \end{bmatrix}. \tag{21}$$

From the above equation, the peak value of stator flux linkage is given as

$$\psi^p_s(k+1) = \sqrt{[\psi^p_{\alpha s}(k+1)]^2 + [\psi^p_{\beta s}(k+1)]^2}. \tag{22}$$

Similarly, the predicted electromagnetic torque is calculated by using the predicted α- and β-axis stator current components and stator flux linkage components in (19) and (21), respectively. The model of $T^p_e(k+1)$ is given as,

$$T^p_e(k+1) = 1.5\, P_p \left[\psi^p_{\alpha s}(k+1)\, i^p_{\beta s}(k+1) - \psi^p_{\beta s}(k+1)\, i^p_{\alpha s}(k+1)\right]. \tag{23}$$

A single cost function is employed to control the generator electromagnetic torque, generator stator flux, and the MSC switching frequency simultaneously as,

$$\begin{aligned} g_r(k) = {} & \lambda_T \left[T^*_e(k+1) - T^p_e(k+1)\right]^2 + \lambda_\psi \left[\widehat{\psi}^*_s(k+1) - \psi^p_s(k+1)\right]^2 \\ & + \lambda_{sw,r} \sum_{x=a,b,c} \left[s^p_{xr}(k) - s^{\mathrm{op}}_{xr}(k)\right]^2 \end{aligned} \tag{24}$$

where λ_T, λ_ψ, and $\lambda_{sw,r}$ are the weight factors for torque, flux, and switching frequency control, respectively. These weight factors are defined as,

$$\lambda_T = \frac{T_{e,R}}{T_{e,R}} = 1, \qquad \lambda_\psi = \frac{T_{e,R}}{\psi_{s,R}} \tag{25}$$

where $T_{e,R}$ and $\psi_{s,R}$ represent the rated torque and rated stator flux, respectively.

Finally, the cost function is evaluated for all possible eight switching states and the optimum switching states which gives the least cost value is selected and applied to the MSC.

5.2 *Control of GSC*

The control structure of GSC in SCIG WECS and PMSG WECS is quite identical (Fig. 7). A comprehensive analysis and description on GSC control has already been presented in Sect. 4.2. The SCIGs require three-stage gearbox to operate at higher-speeds. Therefore, the storage of surplus energy in WT-SCIG rotor inertia during grid faults is not feasible. The DC chopper dissipates surplus energy in DC-link during grid faults. Hysteresis control is used to ensure that average DC voltage V_{dc} never exceeds the maximum threshold voltage V_{dc}^{th}.

5.3 *Results and Analysis*

The transient performance of SCIG WECS with the PTC scheme is verified during a step change in wind speed (Fig. 10). A 3.0 MW, 3000 V, and 60 Hz SCIG WECS with $R_s = 18.885$ mΩ, $R_r = 21.404$ mΩ, $L_{ls} = L_{lr} = 0.815$ mH, $L_m = 27.168$ mH, $\psi_r = 6.198$ Wb (*peak*), $P_p = 2$, and $n_m = 1814$ rpm is considered. The GSC and control parameters are $r_i = 30$ mΩ (0.01 pu), $L_i = 1.6$ mH (0.2 pu), $v_{dc}^* = 5304$ V (3.062 pu), $Q_g^* = 0$ MVAR, and $T_s = 100$ μs.

The transient performance of SCIG WECS with PTC scheme is validated by changing the wind speed from 12 m/s (1.0 pu) to 7.2 m/s (0.6 pu). This change reflects in the generate reference speed ω_m^* and it varies from 1.0 pu to 0.6 pu. Due to the large moment of inertia, the generator actual speed will not change instantaneously due to the slow response of the speed control loop (Fig. 10a). The speed controller provides a reference torque command T_e^* corresponding to the square of the reference speed. The PTC scheme ensures that T_e follows T_e^* with a small steady state error (Fig. 10b), whereas the stator flux is maintained constant at its rated value (Fig. 10c). The torque and flux values vary proportional to the q- and d-axis stator current components (Fig. 10d). The change in wind speed also effects the DC-link voltage control loop and exhibits oscillation in the DC-link voltage v_{dc} of GSC (Fig. 10e). The magnitude of GSC current decreases with the reduction in generator active power output (Fig. 10f). During rated operating condition, the THD of stator currents, THD of grid currents, average switching frequency of MSC, and average switching frequency of GSC are noted as 6.9%, 5.8%, 1560 Hz, and 1780 Hz, respectively. The presented results validate the fast dynamic response and decoupled control of the PTC scheme.

6 Predictive Power Control of DFIG WECS

Figure 11 shows the structure of PPC scheme for RSC and PCC scheme for GSC in DFIG WECS. For simplicity, the DFIG stator terminals are assumed to be directly connected to the wind farm collection point (WFCP) and the rotor terminals are connected to the WFCP by BTB VSC and a step-up transformer with a step-up ratio of $1{:}V_s/V_r$. V_s and V_r are the rated *rms* stator and rotor voltages, respectively. The stator reactive power reference Q_s^* is defined such that the RSC controls the overall WECS reactive power Q_o. For unity power factor operation of WFCP, the Q_s^* value

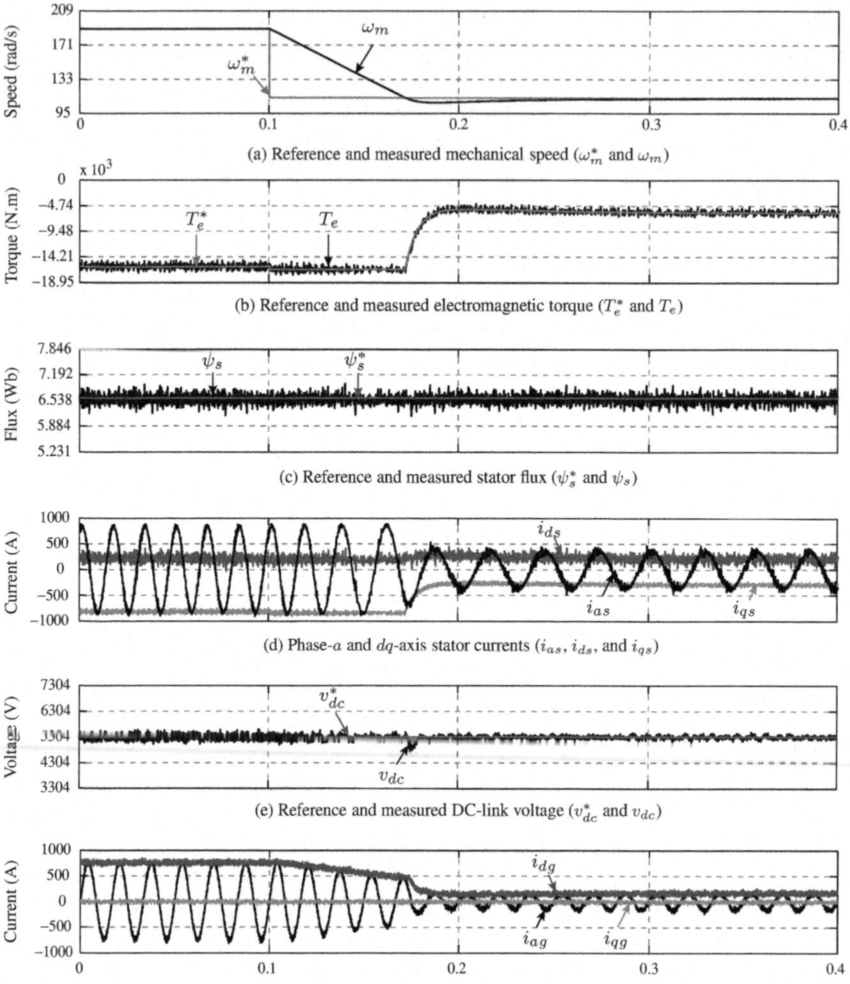

(a) Reference and measured mechanical speed (ω_m^* and ω_m)

(b) Reference and measured electromagnetic torque (T_e^* and T_e)

(c) Reference and measured stator flux (ψ_s^* and ψ_s)

(d) Phase-a and dq-axis stator currents (i_{as}, i_{ds}, and i_{qs})

(e) Reference and measured DC-link voltage (v_{dc}^* and v_{dc})

(f) Phase-a and dq-axis grid currents (i_{ag}, i_{dg}, and i_{qg})

Fig. 10 Simulated waveforms for PTC scheme of SCIG WECS during step change in wind speed

is set to zero. On the other hand, the active power is regulated through the MPPT algorithm.

6.1 Control of RSC

The RSC control is implemented in dq-reference frame and uses stator voltage angle θ_s to transform the stator voltage and current variables from abc-frame to dq-frame.

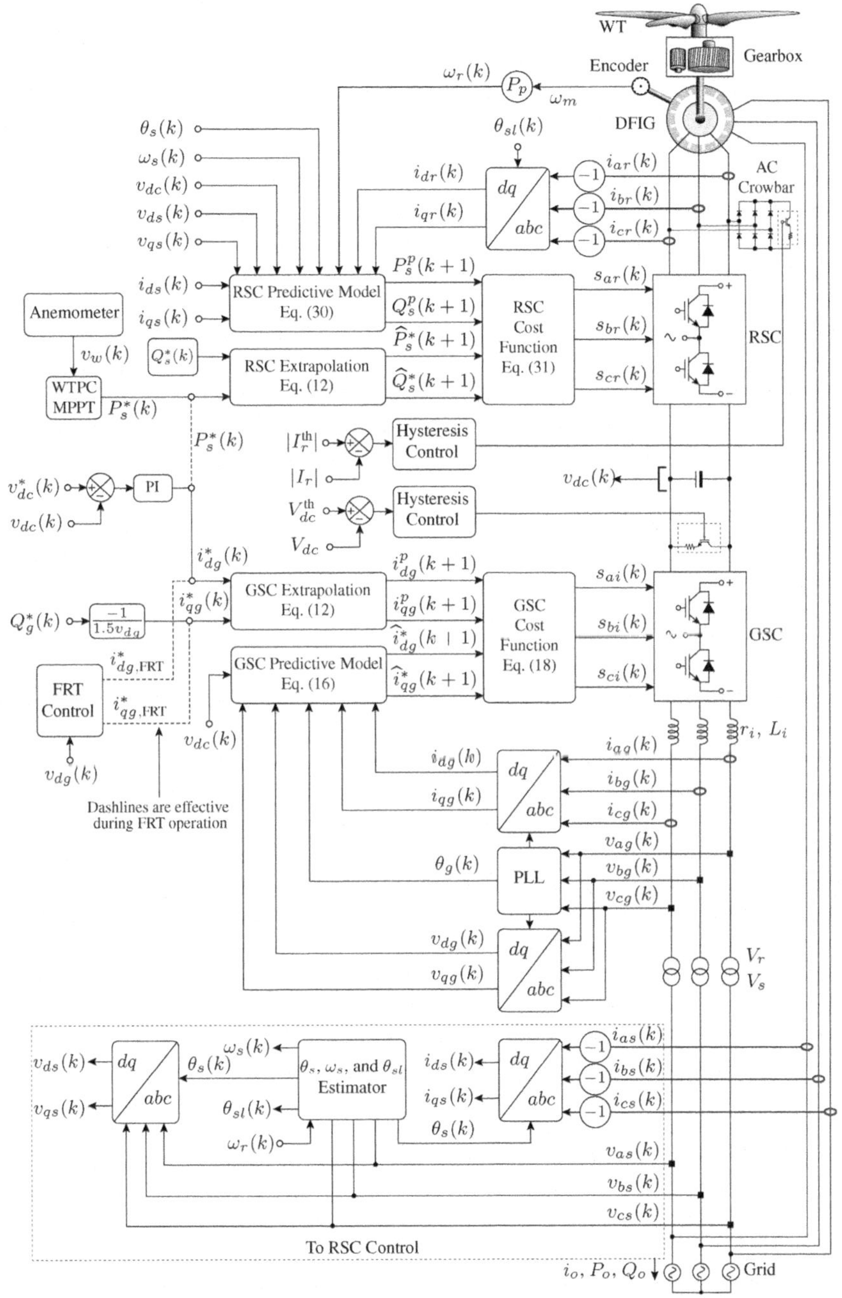

Fig. 11 Block diagram of PPC scheme for RSC and PCC scheme GSC in DFIG WECS

On the other hand, the rotor circuit variables are converted into dq-frame by using the slip angle θ_{sl}. The reference active power P_s^* is calculated by WTPC MPPT control. The reference active and reactive powers are extrapolated to $(k+1)$ state using the models in (12). The rotor current dynamic model is given as [3]:

$$\begin{bmatrix} i_{dr}^p(k+1) \\ i_{qr}^p(k+1) \end{bmatrix} = \Phi(k) \begin{bmatrix} i_{ds}(k) \\ i_{qs}(k) \\ i_{dr}(k) \\ i_{qr}(k) \end{bmatrix} + \Gamma_{\mathbf{b}} \begin{bmatrix} v_{ds}(k) \\ v_{qs}(k) \\ v_{dr}^p(k) \\ v_{qr}^p(k) \end{bmatrix} \tag{26}$$

where

$$\begin{aligned} \Phi(k) &\approx k_\sigma \begin{bmatrix} R_s L_m T_s & -\omega_r(k) L_s L_m T_s & 1 - R_r L_s T_s & -\omega_r(k) L_r L_s T_s + k_\phi \\ \omega_r(k) L_s L_m T_s & R_s L_m T_s & \omega_r(k) L_r L_s T_s - k_\phi & 1 - R_r L_s T_s \end{bmatrix} \\ \Gamma_{\mathbf{b}} &\approx k_\sigma \begin{bmatrix} -L_m T_s & 0 & L_s T_s & 0 \\ 0 & -L_m T_s & 0 & L_s T_s \end{bmatrix} \end{aligned} \tag{27}$$

with

$$k_\sigma = \left(\frac{1}{\sigma L_s L_r} \right), \quad k_\omega = \omega_s \sigma L_s L_r, \quad k_\phi \approx k_\omega T_s \approx \omega_s \sigma L_s L_r T_s. \tag{28}$$

The rotor voltages in the aforementioned expression are predicted by using the switching states in dq-frame and actual DC-link voltage. The model is defined by the following:

$$\begin{bmatrix} v_{dr}^p(k) \\ v_{qr}^p(k) \end{bmatrix} = \begin{bmatrix} \cos\theta_{sl}(k) & \sin\theta_{sl}(k) \\ -\sin\theta_{sl}(k) & \cos\theta_{sl}(k) \end{bmatrix} v_{dc}(k) \begin{bmatrix} s_{\alpha r}^p(k) \\ s_{\beta r}^p(k) \end{bmatrix} \tag{29}$$

The future value of stator powers are predicted in terms of rotor currents and stator voltages, that is,

$$\begin{aligned} P_s^p(k+1) &= -\left(\frac{3\, v_{ds}(k)\, L_m}{2\, L_s} \right) i_{dr}^p(k+1) \\ Q_s^p(k+1) &= \left(i_{qr}^p(k+1) + \frac{v_{ds}(k)}{\omega_s L_m} \right) \frac{2\, L_s}{3\, v_{ds}(k)\, L_m}. \end{aligned} \tag{30}$$

In the PPC scheme for RSC, the stator active and reactive powers and switching frequency reduction are the main control objectives. Accordingly, the cost function is formulated as follows:

$$\begin{aligned} g_r(k) &= \left[\widehat{P}_s^*(k+1) - P_s^p(k+1) \right]^2 + \left[\widehat{Q}_s^*(k+1) - Q_s^p(k+1) \right]^2 \\ &+ \lambda_{sw,r} \sum_{x=a,\,b,\,c} \left[s_{xr}^p(k) - s_{xr}^{\mathrm{op}}(k) \right]^2. \end{aligned} \tag{31}$$

The cost function is evaluated for all possible eight switching states, which leads to eight predictions of i_{dr}, i_{qr} correspondingly P_s and Q_s. During each sampling interval, the best suited switching states for RSC are selected on the basis of the minimum cost function value, which ensures the accurate tracking of stator powers to their respective reference powers. The optimal switching states is applied to the RSC during the $(k + 1)$ state.

6.2 Control of GSC

The control of GSC in DFIG WECS is similar to the control of GSC in PMSG and SCIG WECS (Figs. 7 and 9). The average DC-link voltage is maintained at its reference through the control of GSC-side currents, whereas the Q_g^* is set to zero. The voltage in the rotor circuit is only a fraction of the WFCP (or DFIG stator) voltage. The DC chopper is employed to dissipate surplus energy in DC-link during grid faults, and it is controlled using a hysteresis control. The AC crowbar uses another hysteresis controller to ensure that rotor currents do not exceed the threshold value during grid faults.

6.3 Results and Analysis

The operation of DFIG WECS during a transition from 0.7 pu rotor speed (sub-synchronous mode) to 1.0 pu rotor speed (super-synchronous mode) is analyzed. 3.0 MW, 690 V, 60 Hz DFIG WECS parameters are given as follows: V_s = 398.4 V, V_r = 91.63 V, R_s = 1.443 mΩ, R_r = 1.125 mΩ, L_{ls} = 0.094 mH, L_{lr} = 0.085 mH, L_m = 0.802 mH, ψ_s = 1.5057 Wb (*peak*), ψ_r = 1.7576 Wb (*peak*), P_p = 2, and n_m = 2160 rpm. The GSC parameters are as follows: r_i = 0.34 mΩ, L_i = 0.02 mH, and v_{dc}^* = 281 V. The reference stator reactive power is zero ($Q_s^* = 0$). The sampling time used in the implementation of PPC scheme is T_s = 100 μs.

In this study, the wind speed v_w profile with a gradual change from 8.4 m/s (0.7 pu) to 12 m/s (1.0 pu) is considered. Accordingly, the rotor speed linearly increased from 0.7 pu to 1.0 pu with in the time duration of 1 s. The WTPC MPPT produces a cubic reference power P_s^* (cubic of v_w) for the RSC PPC scheme. The outer speed control loops maintains the rotor speed ω_m at its reference value ω_m^* (Fig. 12a). The change in speed effects the torque T_e developed by the DFIG, and it varies from –0.49 pu to –1.0 pu and follows the T_e^* trajectory (Fig. 12b). During transient interval, the mechanical input torque T_m is greater than T_e. Hence, the DFIG accelerates and its the rotor speed ω_m increases until T_e is equal to T_m.

The DFIG slip dynamically changes from 0.16 to –0.2 during the transient interval and reach zero at t = 0.63 s. The changes in d-axis rotor current are in proportion to the T_e profile, whereas the q-axis rotor current varies proportionally with the Q_s^* value. The PPC method ensures that measured dq-axis rotor currents to follow the

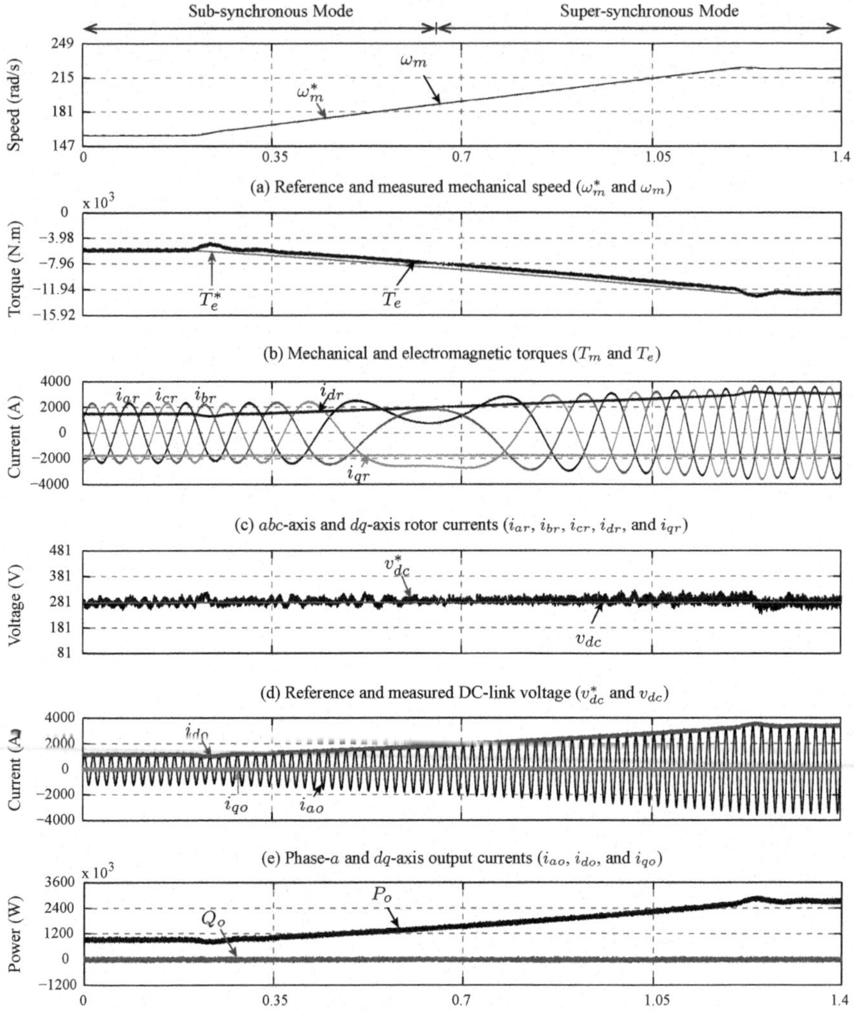

(a) Reference and measured mechanical speed (ω_m^* and ω_m)

(b) Mechanical and electromagnetic torques (T_m and T_e)

(c) *abc*-axis and *dq*-axis rotor currents (i_{ar}, i_{br}, i_{cr}, i_{dr}, and i_{qr})

(d) Reference and measured DC-link voltage (v_{dc}^* and v_{dc})

(e) Phase-*a* and *dq*-axis output currents (i_{ao}, i_{do}, and i_{qo})

(f) Output active and reactive power (P_o and Q_o)

Fig. 12 Transient waveforms for DFIG WECS during transition from sub-synchronous to super-synchronous mode of operation

dynamically changing reference rotor currents such that the rotor speed and torque are controlled. The PPC scheme exhibits an excellent dynamic performance with low tracking error. During sub-synchronous mode, the slip value becomes positive which can be seen in the phase relation between three-phase rotor currents (i_{ar} leads i_{br}, and i_{br} leads i_{cr}), whereas the slip value becomes negative in the super-synchronous mode. Hence, the phase relation becomes reverse (Fig. 12c).

During the entire transient interval, the DC-link voltage v_{dc} is maintained constant at the v_{dc}^* value by the DC voltage PI controller (Fig. 12d). In PMSG and SCIG WECS, the GSC controls the WECS reactive power but the RSC influences the reactive power in DFIG WECS. The magnitude of generator output current increases in proportion to the active power extracted from wind. The output d-axis current i_{do} follows the envelop of three-phase output currents, whereas the q-axis output current i_{qo} is regulated at zero value to obtain the UPF operation (Fig. 12e). The WT output active power P_o follows the ω_m spectrum and the WT output reactive power Q_o is maintained zero by the PPC scheme employed for RSC (Fig. 12f). During sub-synchronous mode, the THD of rotor currents, THD of output currents, average switching frequency of RSC, and average switching frequency of GSC are noted as 1.4%, 6.4%, 1960 Hz, and 1770 Hz, respectively. The THD of rotor currents, THD of output currents, average switching frequency of RSC, and average switching frequency of GSC during suer-synchronous mode are 0.9%, 2.4%, 1730 Hz, and 1690 Hz, respectively. The presented results prove the effectiveness of the PPC scheme for DFIG WECS.

7 Summary

This book chapter provides a comprehensive analysis on the power converters and predictive control schemes for PMSG, SCIG, and DFIG based high-power WECS. The PCC scheme for PMSG WECS, PTC scheme for SCIG WECS, and PPC scheme for DFIG WECS are analyzed. The discrete-time models of complete WECS are developed by combining the models of wind generators, MSC/RSC, GSC, harmonic filter, and a three-phase grid. The predictive control schemes eliminate the PI controllers, modulation stage, hysteresis controllers and lookup tables in the inner control loop of MSC/RSC and GSC, thereby providing a fast dynamic response for control variables such as generator and grid currents, generator electromagnetic torque and stator flux, and generator active and reactive powers. The discrete and nonlinear nature of power converters in high-power WECS is naturally handled by the MPC, thus leading to an excellent steady-state and fast dynamic response. The MPC scheme has been demonstrated to be simple and intuitive yet powerful tool to control high-power WECS. The technical details given in this chapter are also applicable for other power conversion applications, such as active power filters, variable-speed electric drives, and photovoltaic energy, which employ a digital control structure with an inner current control loop.

Acknowledgements This work was supported in part by the Northern Arizona University, in part by the AC3E Conicyt/FB0008 Project, and in part by Spanish Ministry of Science and Innovation under Project ENE2014-52536-C2-1-R.

References

1. Global Wind Energy Council (GWEC) (2017) Global wind report: annual market update. http://www.gwec.net. Accessed June 2018
2. Yaramasu V, Wu B, Sen PC, Kouro S, Narimani M (2015) High-power wind energy conversion systems: state-of-the-art and emerging technologies. Proc. IEEE 103(5):740–788
3. Yaramasu V, Wu B (2016) Model predictive control of wind energy conversion systems. Wiley-IEEE Press, Hoboken, NJ, first edition
4. Yaramasu V, Wu B (2017) Encyclopedia of sustainable technologies. In: Power electronics for high-power wind energy conversion systems, pp 37–49, 1st edn. Elsevier, Amsterdam, Netherlands
5. Yaramasu V, Dekka A, Durn MJ, Kouro S, Wu B (2017) PMSG-based wind energy conversion systems: survey on power converters and controls. IEE Proc Electr Power Appl 11(6):956–968
6. Casadei D, Profumo F, Serra G, Tani A (2002) FOC and DTC: two viable schemes for induction motors torque control. IEEE Trans Power Electron 17(5):779–787
7. Teodorescu R, Liserre M, Rodriguez P (2011) Grid converters for photovoltaic and wind power systems. Wiley-IEEE Press, Chichester, UK
8. Vazquez S, Rodriguez J, Rivera M, Franquelo LG, Norambuena M (2017) Model predictive control for power converters and drives: advances and trends. IEEE Trans Ind Electron 64(2):935–947
9. Chinchilla M, Arnaltes S, Burgos JC (2006) Control of permanent-magnet generators applied to variable-speed wind-energy systems connected to the grid. IEEE Trans Energy Convers 21(1):130–135
10. Cardenas R, Pena R, Alepuz S, Asher G (2013) Overview of control systems for the operation of DFIGs in wind energy applications. IEEE Trans Ind Electron 60(7):2776–2798
11. Xu Z, Li R, Zhu H, Xu D, Zhang CCH (2012) Control of parallel multiple converters for direct-drive permanent-magnet wind power generation systems. IEEE Trans Power Electron 27(3):1259–1270
12. Duran MJ, Barrero F (2016) Recent advances in the design, modeling, and control of multiphase machines—part II. IEEE Trans Ind Electron 63(1):459–468
13. Iov F, Blaabjerg F, Clare J, Wheeler P, Rufer A, Hyde A (2009) UNIFLEX-PM—a key-enabling technology for future European electricity networks. Europ Power Electron Drives Assoc J 19(4):6–16
14. Xia YY, Fletcher JE, Finney SJ, Ahmed KH, Williams BW (2011) Torque ripple analysis and reduction for wind energy conversion systems using uncontrolled rectifier and boost converter. IET Renew Power Gener 5(5):377–386
15. Wu B, Lang Y, Zargari N, Kouro S (2011) Power conversion and control of wind energy systems. IEEE Press series on power engineering, 1st edn. Wiley-IEEE Press, Hoboken, NJ
16. Faulstich A, Stinke JK, Wittwer F (2005) Medium voltage converter for permanent magnet wind power generators up to 5 MW. In: European Conference on Power Electronics and Applications (EPE), pp. 9–P.9. Dresden, Germany
17. Kouro S, Malinowski M, Gopakumar K, Pou J, Franquelo LG, Wu B, Rodríguez J, Perez MA, Leon JI (2010) Recent advances and industrial applications of multilevel converters. IEEE Trans Ind Electron 57(8):2553–2580
18. Yaramasu V, Wu B (2014) Model predictive decoupled active and reactive power control for high-power grid-connected four-level diode-clamped inverters. IEEE Trans Ind Electron 61(7):3407–3416
19. Yaramasu V, Wu B, Alepuz S, Kouro S (2014) Predictive control for low-voltage ride-through enhancement of three-level-boost and NPC-converter-based PMSG wind turbine. IEEE Trans Ind Electron 61(12):6832–6843
20. Yaramasu V (2017) Advances in energy research. In: Acosta MJ (ed) MATLAB/simulink implementation of predictive current control for PMSG wind energy conversion systems, vol 27, pp 187–230. Nova Science Publishers Inc., Hauppauge, NY

21. Calle-Prado A, Alepuz S, Bordonau J, Nicolas-Apruzzese J, Cortes P, Rodriguez J (2015) Model predictive current control of grid-connected neutral-point-clamped converters to meet low-voltage ride-through requirements. IEEE Trans Ind Electron 62(3):1503–1514
22. Rodríguez J, Kennel RM, Espinoza JR, Trincado M, Silva CA, Rojas CA (2012) High-performance control strategies for electrical drives: an experimental assessment. IEEE Trans Ind Electron 59(2):812–820
23. Miranda H., Cortes P, Yuz JI, Rodríguez J (2009) Predictive torque control of induction machines based on state-spacemodels. IEEE Trans Ind Electron 56(6):1916–1924
24. Holtz J (1995) The representation of AC machine dynamics by complex signal flow graphs. IEEE Trans Ind Electron 42(3):263–271

Adaptive Guaranteed Performance Control of Wind Energy Systems

Wenchao Meng and Qinmin Yang

Abstract In this chapter, we present an adaptive guaranteed performance controller for wind energy conversion system (WECS) equipped with doubly fed induction generator (DFIG). The proposed controller consists of outer loop control concerning the aeroturbine mechanical subsystem, and inner loop control concerning the electrical subsystem. As opposed to most existing studies, we are capable of quantifying and further guaranteeing the system performance on both transient and steady state stages with the help of error transformation techniques. The stability is guaranteed via standard Lyapunov synthesis. Finally, the effectiveness of the proposed scheme is validated on a 1.5 MW DFIG-based wind turbine using the FAST (Fatigue, Aerodynamics, Structures, and Turbulence) simulator developed by the National Renewable Energy Laboratory (NREL).

Nomenclature

ρ	Air density
V	Wind speed
R	Rotor radius
C_p	Power coefficient
$C_q(\lambda, \beta)$	Torque coefficient
β	Pitch angle
λ	Tip-speed ratio
P_r	Rotor power
T_a	Aerodynamic torque
J_r, J_g	Rotor and generator inertias
K_r, K_g	Rotor and generator external damping
T_{hs}, T_{ls}	High-speed and low-speed torque

W. Meng · Q. Yang (✉)
College of Control Science and Engineering, Zhejiang University,
Hangzhou Zhejiang 310027, People's Republic of China
e-mail: qmyang@zju.edu.cn

W. Meng
e-mail: wmengzju@gmail.com

R.-E. Precup et al. (eds.), *Advanced Control and Optimization Paradigms for Wind Energy Systems*, Power Systems, https://doi.org/10.1007/978-981-13-5995-8_6

T_{em}	Electromagnetic torque
n_{ra}	Gearbox ratio
Ω_r	Rotor speed
$\Phi_{d,qs}$	Stator flux
$\Phi_{d,qr}$	Rotor flux
$U_{d,qs}$	Stator winding voltage
$U_{d,qr}$	Rotor winding voltage
$I_{d,qs}, I_{d,rq}$	Stator and rotor winding currents
L_s, L_r	Self-inductance of the stator and rotor
L_m	Mutual inductance between windings
R_s, R_r	Resistance of the stator and rotor
Ω_s	Frequency of the grid
p	Number of pole pairs
$V_{cut\text{-}in}$	Cut-in wind speed
$V_{cut\text{-}off}$	Cut-off wind speed
V_{rated}	Rated wind speed

1 Introduction

In recent years, the wind energy conversion systems have received more and more attentions from both academic and industrial communities due to the depletion of tradition energy source and increasing environment pollution [1, 2]. Because of this, the WECS has experienced the fastest growth and this tendency is expected to endure for a long time [3, 4]. However, it is still provide a very small share in the global energy market due to its high costs, and developing advanced control algorithms is considered to be a promising way to reduce its costs.

Linear control methods have been commonly used for control of wind energy conversion systems [5, 6]. The linear methods only deliver satisfactory performance when the plant works around the corresponding operation point, whereas the operation point of WECS changes frequently because of the random wind. Therefore, the system performance will be impaired if the linear method is enforced.

In order to avoid the drawbacks of linear methods, many nonlinear control methods have been studied [7, 8]. However, in previous studies, only steady state performance of the control system has been taken into account, while the more essential transient performance has been rarely considered.

Therefore, in this chapter, an adaptive guaranteed performance control is proposed for WECS equipped with DFIG. The proposed controller includes two loops [9, 10], i.e., the outer loop control and inner loop control. The outer loop control concerns the aeroturbine mechanical subsystem while the inner loop control concerns the electrical subsystem. Compared with most existing studies, performance indexes including steady-state error, convergence rate and overshoot are guaranteed.

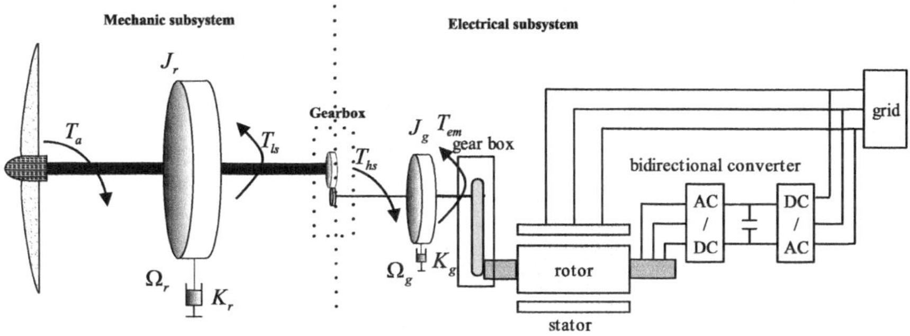

Fig. 1 Wind energy conversion system

2 Wind Energy Conversion System

The wind energy conversion system includes the mechanical subsystem and electrical subsystem as depicted in Fig. 1. For the mechanical subsystem, the rotor power extracted from wind can be formulated as [11]

$$P_r = \frac{1}{2}\rho\pi R^2 C_p(\lambda, \beta) V^3 \tag{1}$$

The tip-speed ratio λ is defined by

$$\lambda = \frac{R\Omega_r}{V} \tag{2}$$

The rotor power P_r can also be formulated as

$$P_r = \Omega_r T_a. \tag{3}$$

with

$$C_q(\lambda) = \frac{C_p(\lambda)}{\lambda} \tag{4}$$

Invoking (3), (4) and (1), we have

$$T_a = \frac{1}{2}\rho\pi R^3 C_q(\lambda) V^2 \tag{5}$$

The rotor dynamics together with the generator dynamics can be written as

$$J_r \dot{\Omega}_r = T_a - K_r \Omega_r - T_{ls} \tag{6}$$

$$J_g \dot{\Omega}_g = T_{hs} - K_g \Omega_g - T_{em} \tag{7}$$

The gearbox ratio n_{ra} is defined as

$$n_{ra} = \frac{\Omega_g}{\Omega_r} = \frac{T_{ls}}{T_{hs}} \tag{8}$$

Substituting (8) into the generator dynamic (7), we have

$$n_{ra}^2 J_g \dot{\Omega}_r = T_{ls} - \left(n_{ra}^2 K_g\right) \Omega_r - n_g T_{em} \tag{9}$$

Thereafter, a simple model of the mechanic subsystem can be obtained as

$$J_o \dot{\Omega}_r = T_a - K_o \Omega_r - T_g \tag{10}$$

where

$$\begin{cases} J_o = J_r + n_{ra}^2 J_g \\ K_o = K_r + n_{ra}^2 K_g \\ T_g = n_{ra} T_{em} \end{cases} \tag{11}$$

The generator power can be formulated as

$$P_g = T_g \Omega_r \tag{12}$$

For the electrical subsystem, we consider the doubly-fed induction generator which connects directly to the grid through the stator, while the rotor winding is interfaced through a bidirectional power electronic converter. In this kind of wound-rotor machine, the power system electrical frequency and the rotor mechanical frequency can be decoupled, which makes a variable speed operation of the wind turbine possible. One of the main advantages is that it can generate and deliver electrical power at the frequency and voltage demanded by the grid. Inspired by [12], the model of DFIG in the Park $d-q$ frame is given by

$$\begin{cases} \dot{\Phi}_{d,s} = U_{d,s} - R_s I_{d,s} + \Omega_s \Phi_{q,s} \\ \dot{\Phi}_{q,s} = U_{q,s} - R_s I_{q,s} - \Omega_s \Phi_{d,s} \\ \dot{\Phi}_{d,r} = U_{d,r} - R_r I_{d,r} + \left(\Omega_s - p\Omega_g\right) \Phi_{q,r} \\ \dot{\Phi}_{q,r} = U_{q,r} - R_r I_{q,r} - \left(\Omega_s - p\Omega_g\right) \Phi_{d,r} \end{cases} \tag{13}$$

with

$$\begin{aligned} \Phi_{d,s} &= L_s I_{d,s} + L_m I_{d,r} \\ \Phi_{q,s} &= L_s I_{q,s} + L_m I_{q,r} \\ \Phi_{d,r} &= L_r I_{d,r} + L_m I_{d,s} \\ \Phi_{q,r} &= L_r I_{q,r} + L_m I_{q,s} \end{aligned} \tag{14}$$

Inspired by [12], a simplified generator model can be given by

$$\dot{I}_{d,r} = \left(\Omega_s - p\Omega_g\right) I_{q,r} - \frac{L_s R_r}{L_e} I_{d,r} + \frac{L_s}{L_e} U_{d,r} + \varpi_d(t) \tag{15}$$

$$\dot{I}_{q,r} = -\frac{L_s R_r}{L_e} I_{q,r} - \left(\Omega_s - p\Omega_g\right)\left(I_{d,r} + \frac{L_m V_s}{L_e \Omega_s}\right) + \frac{L_s}{L_e} U_{q,r} + \varpi_q(t) \tag{16}$$

where V_s is the grid voltage and $L_e = L_s L_r - L_m^2$. The terms $\varpi_d(t), \varpi_q(t)$ are added to represent bounded disturbances [13]. Thereafter, the currents of stator can be algebraically calculated as

$$\begin{aligned} I_{d,s} &= \frac{V_s}{\Omega_s L_s} - \frac{L_m}{L_s} I_{d,r} \\ I_{q,s} &= -\frac{L_m}{L_s} I_{q,r} \end{aligned} \tag{17}$$

The electromagnetic torque along with reactive power is given by

$$T_{em} = -\frac{3}{2} p \frac{V_s L_m}{\Omega_s L_s} I_{q,r} \tag{18}$$

$$Q = \frac{3V_s^2}{2\Omega_s L_s} - \frac{3V_s L_m I_{d,r}}{2L_s} \tag{19}$$

3 Problem Formulation

There are two operation regions for the wind turbine, namely, below the low-speed region and high-speed region as given in Fig. 2 [14, 15].

- Low-speed region: where $V_{cut\text{-}in} \leq V < V_{rated}$ and $P_g < P_{rated}$.
- High speed region: where $V_{rated} \leq V \leq V_{cut\text{-}off}$ and $P_g = P_{rated}$.

In low-speed region, the desired power is given by

$$P_g^* = n_p P_r^{\max} \tag{20}$$

with

$$P_r^{\max} = \frac{1}{2} \rho \pi R^2 C_p^{\max} V^3 \tag{21}$$

Notice that the response of the WT electrical subsystem is much faster than that of the mechanical part of the WT. Hence, the controller design for the electrical subsystem and mechanical subsystem is usually decoupled and a cascaded control structure containing two control loops is usually adopted as shown below

- The outer control loop concerns the aeroturbine mechanical subsystem.
- The inner control loop concerns the electrical subsystem.

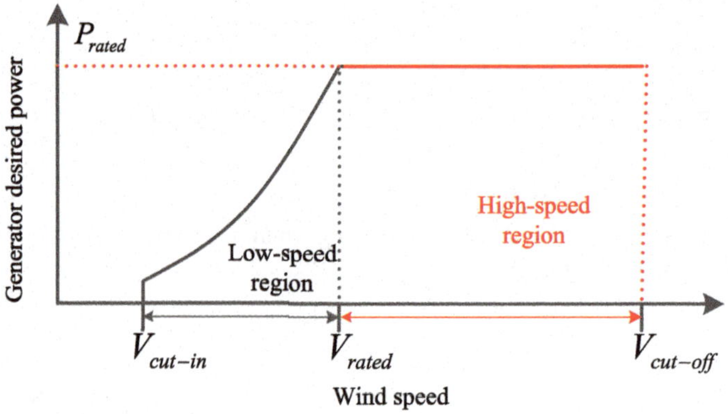

Fig. 2 Generator desired power curve

Because the aeroturbine runs much slower than electrical subsystem, the state of the outer loop can be seen as a slow changing disturbance while calculating the control signal for the inner loop. In the meantime, the outer loop controller is usually designed based on the assumption that the inner electrical control loop is able to track the reference T^* timely. It implies that the stability analysis in the outer control loop and inner control loop can be addressed separately in literature [16].

In this chapter, we consider the low-speed operation region. For the outer control loop, our main goal is to design appropriate generator torque T_g such that P_g can track P_g^*. For the inner control loop, the control objective is to design input voltages $U_{d,r}$, $U_{q,r}$ such that: (1) the electromagnetic torque T_{em} tracks its reference T^*, and (2) the reactive power Q follows its desired value Q^*.

For analysis convenience, we define the following tracking errors

$$\ell_o = P_g^* - P_g \tag{22}$$

$$\ell_{i,T} = T_{em} - T^* \tag{23}$$

$$\ell_{i,Q} = Q - Q^* \tag{24}$$

4 Outer Loop Control

In the outer control loop, both the transient and steady state performance will be considered. Specifically, the imposed performance requirements on $\ell_o(t)$ are

P1:

- The steady tracking error $\ell_o(\infty)$ is required to be within $-\eta_o\hbar_o(\infty) \leq \ell_o(\infty) \leq \hbar_o(\infty)$.
- It converges faster than the signal $\hbar_o(t)$.
- The maximum overshoot is required to be smaller than $\eta_o\hbar_o(0)$.

For evaluating the prescribed performance, the following performance function is firstly introduced.

Definition 1 ([17]) A performance function is a smooth function $\hbar_o(t) : \Re^+ + \{0\} \to \Re^+$ that satisfies $|\ell_o(0)| < \hbar_o(0)$ and $\lim_{t\to\infty} \hbar_o(t) = \hbar_o(\infty) > 0$.

Assume $0 \le \ell_o(0) < \hbar_o(0)$, if the tracking error satisfies

$$-\eta_o \hbar_o(t) < \ell_o(t) < \hbar_o(t) \tag{25}$$

with $0 \le \eta_o \le 1$ being a design parameter, the prescribed performance P1 can be attained.

To proceed the prescribed performance design, an error transformation in [17], which can convert the original error with imposed performance requirements into a new error without imposed performance requirements, will be introduced. Specifically,

$$\ell_o(t) = \hbar_o(t) M_o(\gamma_o) \tag{26}$$

or

$$\gamma_o(t) = M_o^{-1}\left(\frac{\ell_o(t)}{\hbar_o(t)}\right) \tag{27}$$

with γ_o being the new error, and $M_o(\cdot)$ is a function that is smooth and strictly increasing. The function $M_o(\cdot)$ is required to satisfy

$$\begin{cases} \lim_{\gamma_o\to-\infty} M_o(\gamma_o) = -\eta_o \\ \lim_{\gamma_o\to\infty} M_o(\gamma_o) = 1 \end{cases} \tag{28}$$

where $M_o^{-1}(\cdot)$ is the inverse function of $M_o(\cdot)$. There exist many choices for the function $M_o(\gamma_o)$, and a typical choice can be given as

$$M_o(\gamma_o) = \frac{d_o^{\gamma_o} - \eta_o d_o^{-\gamma_o}}{d_o^{\gamma_o} + d_o^{-\gamma_o}} \tag{29}$$

with $d_o > 1$. The function $M_o(\gamma_o)$ is demonstrated in Fig. 3.

The following fact holds as long as $\gamma_o(t)$ exists

$$-\eta_o < M_o(\gamma_o) < 1 \tag{30}$$

The above fact implies (25). It means that the imposed performance requirements in P1 are achieved. Thence, the control task becomes finding a control law to ensure the boundedness of γ_o. For this, by recalling (12), the time derivative of ℓ_o is

$$\dot{\ell}_o = \dot{P}_g^* - \dot{P}_g = \dot{P}_g^* - T_g \dot{\Omega}_r - \dot{T}_g \Omega_r \tag{31}$$

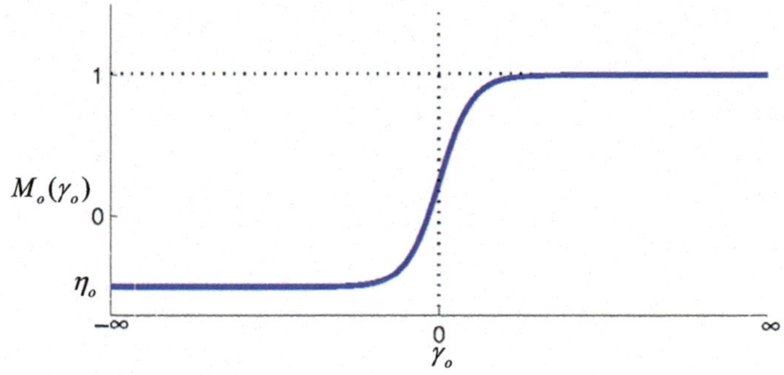

Fig. 3 Graphical illustration of the $M_o(\gamma_o)$ function

Thereafter, by differentiating (27) with respect to time, we have

$$\begin{aligned}
\dot{\gamma}_o &= \frac{\partial M_o{}^{-1}}{\partial\left(\frac{\ell_o(t)}{\hbar_o(t)}\right)} \frac{1}{\hbar_o(t)} \left(\dot{\ell}_o(t) - \frac{\ell_o(t)\dot{\hbar}_o(t)}{\hbar_o(t)}\right) \\
&= \alpha_o \left(\dot{P}_g^* - T_g \dot{\Omega}_r - \dot{T}_g \Omega_r - \beta_o\right) \\
&= \alpha_o \dot{P}_g^* - \alpha_o \dot{T}_g \Omega_r - \alpha_o \frac{T_g}{J_o} \left(T_a - K_o \Omega_r - T_g\right) - \alpha_o \beta_o \\
&= \alpha_o \left[\dot{P}_g^* - \frac{1}{J_o}\left(T_g T_a - T_g^2\right) + \frac{K_o}{J_o} T_g \Omega_r - \dot{T}_g \Omega_r - \beta_o\right]
\end{aligned} \tag{32}$$

where $\alpha_o = \frac{\partial M_o{}^{-1}}{\partial\left(\frac{\ell_o(t)}{\hbar_o(t)}\right)} \frac{1}{\hbar_o(t)}$ and $\beta_o = \frac{\ell_o(t)\dot{\hbar}_o(t)}{\hbar_o(t)}$. Both α_o and β_o are known signals since $\ell_o(t)$, $M_o{}^{-1}(\cdot)$, $\hbar_o(t)$ and $\dot{\hbar}_o(t)$ are all available. An ideal desired control law is firstly presented to assist the controller design. With the known knowledge of the system dynamics, consider the following ideal controller

$$\dot{T}_g = \frac{1}{\Omega_r}\left[\frac{k_o \gamma_o}{\alpha_o} + \frac{K_o}{J_o} T_g \Omega_r + \dot{P}_g^* - \frac{1}{J_o}\left(T_g T_a - T_g^2\right) - \beta_o\right] \tag{33}$$

where $k_o > 0$ is a positive constant. Then, we can easily obtain that

$$\dot{\gamma}_o = -k_o \gamma_o \tag{34}$$

It means that the ideal controller (33) can ensure the exponential convergence of the transformed tracking error γ_o to zero. Notice that the expression of $\dot{P}_d$ is $\dot{P}_d = \frac{1}{2} n_p \rho \pi R^2 C_p^{\max} 3 V^2 \dot{V}$ based on (21) and (20). In order to avoid the knowledge of $\dot{V}$, we use a robust term $\mathrm{sgn}(\alpha_o \gamma_o) B$ to replace $\dot{P}_d$ in (33), and obtain the following desired controller

$$\dot{T}_g = \frac{1}{\Omega_r}\left(\frac{k_o\gamma_o}{\alpha_o} - \beta_o + {\tau_o}^T\xi_o\right) \tag{35}$$

where $\tau_o = [-1/J_o, K_o/J_o, B]^T$, $\xi_o = [T_gT_a - T_g^2, T_g\Omega_r, \ \mathrm{sgn}(\alpha_o\gamma_o)]^T$, and we have used the upper bound of $\dot{P}_d$. The main results of this ideal controller are summarized in the following lemma.

Lemma 1 *For the transformed error dynamics (32), the transformed tracking error γ_o will converge to zero asymptotically if a desired controller is taken as (35).*

Proof A Lyapunov function candidate is built as

$$V_{\gamma_o} = \frac{1}{2}{\gamma_o}^2 \tag{36}$$

By recalling (32) and (54), its time derivative can be given

$$\begin{aligned}
\dot{V}_{\gamma_o} &\le \gamma_o\alpha_o\left[-\dot{T}_g\Omega_r - \frac{T_g}{J_o}\left(T_a - K_o\Omega_r - T_g\right) - \beta_o\right] + |\gamma_o\alpha_o|\,B_o \\
&= \gamma_o\alpha_o\left[-\dot{T}_g\Omega_r + {\tau_o}^T\xi_o - \beta_o\right] \\
&\le -k_o{\gamma_o}^2
\end{aligned} \tag{37}$$

which implies that γ_o converges to zero asymptotically [18].

However, the desired controller has two main defects which should be avoided in practice

- The chattering phenomena may appear because the $\mathrm{sgn}(\cdot)$ function is discontinuous. In WECS, the chattering phenomena is undesirable because it will reduce the lifetime of wind turbines.
- A priori knowledge of τ_o is needed, which may increase the operation costs.

Aimed at mitigating the chattering phenomena, we use the continuous hyperbolic tangent function $\tanh(\alpha_o\gamma_o/\varepsilon_1)$ to replace the discontinuous $\mathrm{sgn}(\alpha_o\gamma_o)$. Notice that the following inequality holds [19]

$$0 \le |\alpha_o\gamma_o| - \alpha_o\gamma_o \tanh\left(\frac{\alpha_o\gamma_o}{\varepsilon_o}\right) \le \kappa\varepsilon_o \quad \text{for} \quad \alpha_o\gamma_o \in \Re \tag{38}$$

where $\kappa = 0.2758$. Furthermore, since τ_o is unknown, let its estimate be $\hat{\tau}_o$, and the following implementable controller is proposed as

$$\dot{T}_g = \frac{1}{\Omega_r}\left(\frac{k_o\gamma_o}{\alpha_o} - \beta_o + \hat{\tau}_o^T\delta_o\right) \tag{39}$$

where $\delta_o = \left[T_gT_a + T_g^2, T_g\Omega_r, \tanh(\frac{\alpha_o\gamma_o}{\varepsilon_o})\right]^T$. The adaptive law for $\hat{\tau}_o$ is given by

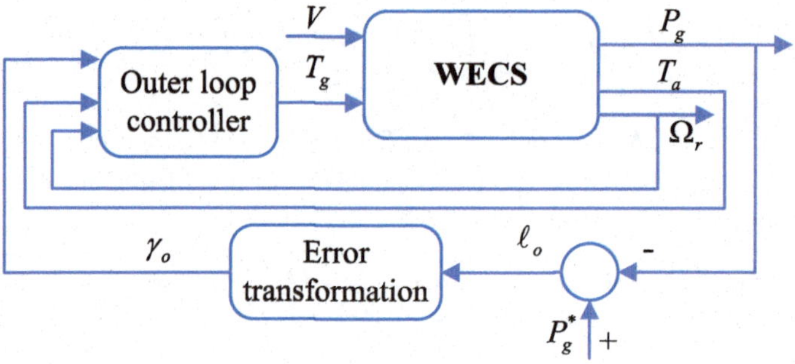

Fig. 4 Controller scheme of the outer loop controller

$$\dot{\hat{\tau}}_o = \Lambda_o \left(\gamma_o \alpha_o \delta_o - \sigma_o \hat{\tau}_o \right) \tag{40}$$

where $\Lambda_o \in \Re^{3\times 3}, \sigma_o > 0$.

Figure 4 depicts the control structure of the outer loop controller.

Theorem 1 *For the mechanical subsystem (10) and (12), if we design the outer controller as (39) while the parameter is updated as (40), the imposed performance given in P1 can be achieved.*

Proof Consider the following Lyapunov function candidate

$$V_o = \frac{1}{2}{\gamma_o}^2 + \frac{1}{2}\tilde{\tau}_o^T {\Lambda_o}^{-1} \tilde{\tau}_o \tag{41}$$

where $\tilde{\tau}_o = \hat{\tau}_o - \tau_o$. By recalling (10), (32) and inequality (38), one has

$$\begin{aligned} \dot{V}_o &= \gamma_o \alpha_o \left(\dot{P}_g^* - T_g \dot{\Omega}_r - \dot{T}_g \Omega_r - \beta_o \right) + \tilde{\tau}_o^T {\Lambda_o}^{-1} \dot{\hat{\tau}}_o \\ &\le \gamma_o \alpha_o \left[-\dot{T}_g \Omega_r - \frac{T_g}{J_o} \left(T_a - K_o \Omega_r - T_g \right) - \beta_o \right] + |\gamma_o \alpha_o| \, B_o + \tilde{\tau}_o^T {\Lambda_o}^{-1} \dot{\hat{\tau}}_o \\ &\le \gamma_o \alpha_o \left[-\dot{T}_g \Omega_r + {\tau_o}^T \delta_o - \beta_o \right] + \kappa \varepsilon_o B_o + \tilde{\tau}_o^T {\Lambda_o}^{-1} \dot{\hat{\tau}}_o \end{aligned} \tag{42}$$

Substituting the outer loop controller (39) and adaptive law (40) into above equation yields

$$\begin{aligned} \dot{V}_o &\le -k_o {\gamma_o}^2 + ({\tau_o}^T - \hat{\tau}_o^T) \gamma_o \alpha_o \delta_o + \tilde{\tau}_o^T \gamma_o \alpha_o \delta_o + \kappa \varepsilon_o B_o - \sigma_o \tilde{\tau}_o^T \hat{\tau}_o \\ &= -k_o {\gamma_o}^2 + \kappa \varepsilon_o B_o - \sigma_o \tilde{\tau}_o^T \hat{\tau}_o \end{aligned} \tag{43}$$

Moreover, by completion of squares, one has

$$\dot{V}_o \le -k_o {\gamma_o}^2 - \frac{\sigma_o \|\tilde{\tau}_o\|^2}{2} + \Delta_o \tag{44}$$

where $\Delta_o = \sigma_o \|\tau_o\|^2/2 + \kappa \varepsilon_o B_o$. Hence, $\dot{V}_o < 0$ when $|\gamma_o| > \sqrt{\frac{\Delta_o}{k_o}}$ or $\|\tilde{\tau}_o\| > \sqrt{\frac{2\Delta_o}{\sigma_o}}$ Therefore, based on the standard Lyapunov extension theorem [20, 21], it can be concluded that γ_o and $\tilde{\tau}_o$ are uniformly ultimately bounded (UUB).

Furthermore, since γ_o is bounded, α_o, β_o are also bounded. $\hat{\tau}_o$ is bounded because of $\hat{\tau}_o = \tilde{\tau}_o + \tau_o$ and the boundedness of τ_o. Hence, from (39), we have that the control input T_g is also bounded. Finally, the boundedness of γ_o implies that the imposed performance requirements as given in P1 are achieved.

5 Inner Loop Control

In the inner loop control, the prescribed transient and steady-state performance are also considered. Specifically, the tracking errors are required to satisfy user-defined conditions as

$$\underline{\ell}_{i,T}(t) < \ell_{i,T} < \bar{\ell}_{i,T}(t) \tag{45}$$

$$\underline{\ell}_{i,Q}(t) < \ell_{i,Q} < \bar{\ell}_{i,Q}(t) \tag{46}$$

where $\bar{\ell}_{i,T}(t)$, $\underline{\ell}_{i,T}(t)$ are lower and upper bounds of the tracking error $\ell_{i,T}$ with $\underline{\ell}_{i,T}(t) < 0 < \bar{\ell}_{i,T}(t)$, and $\bar{\ell}_{i,Q}(t)$, $\underline{\ell}_{i,Q}(t)$ are lower and upper bounds of the tracking error $\ell_{i,Q}$ with $\underline{\ell}_{i,Q}(t) < 0 < \bar{\ell}_{i,Q}(t)$.

Aimed at achieving the goal of guaranteed transient performance, we introduce an improved error transformation technique inspired by [22] that can transform the original constrained errors into new unconstrained errors. Specifically, we define

$$\ell_i = \frac{\bar{\ell}_i(t) - \underline{\ell}_i(t)}{\pi} \arctan(\gamma_i) + \frac{\bar{\ell}_i(t) + \underline{\ell}_i(t)}{2} \tag{47}$$

or

$$\gamma_i(t) = \tan\left(\frac{\pi}{2} \times \frac{2\ell_i - \bar{\ell}_i(t) - \underline{\ell}_i(t)}{\bar{\ell}_i(t) - \underline{\ell}_i(t)}\right), \tag{48}$$

where $\tan(\cdot)$, $\arctan(\cdot)$ are the tangent function and inverse tangent function, respectively, $\gamma_i(t)$ is the transformed error. It can be easily verified that the original tracking error ℓ_i strictly increases with respect to the transformed error γ_i, and thus we have $\frac{\partial \ell_i}{\partial \gamma_i} > 0$. Furthermore, from (47), we have

$$\begin{cases} \lim\limits_{\gamma_i \to -\infty} \ell_i = \underline{\ell}_i(t) \\ \lim\limits_{\gamma_i \to \infty} \ell_i = \bar{\ell}_i(t) \end{cases} \tag{49}$$

From (49), it can be concluded that if γ_i exists, the following fact holds

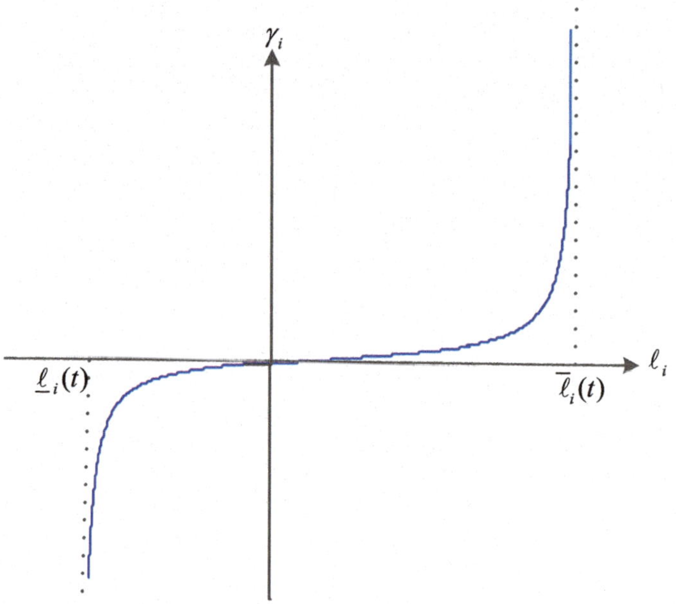

Fig. 5 Graphical illustration of the map from ℓ_i to γ_i

$$\underline{\ell}_i(t) < \ell_i < \bar{\ell}_i(t) \tag{50}$$

which further implies that the guaranteed transient performance in terms of tracking errors is achieved. Therefore, the control objective is converted to finding an inner loop controller that can ensure the boundedness of the transformed error γ_i. The nonlinear mapping between ℓ_i and γ_i is shown in Fig. 5.

Aimed at extracting power from wind as much as possible, the electromagnetic torque should be designed to follow its desired value T^*. The corresponding transformed error of $\ell_{i,T}$ is denoted as $\gamma_{i,T}$. Differentiating $\gamma_{i,T}$ with respect to time and recalling (18), (22) generate

$$\begin{aligned}\dot{\gamma}_{i,T} &= \frac{\partial \gamma_{i,T}}{\partial \ell_{i,T}}\left(-\frac{3}{2}p\frac{U_s L_m}{\Omega_s L_s}\dot{i}_{q,r} - \dot{T}^*\right) + \frac{\partial \gamma_{i,T}}{\partial \bar{\ell}_{i,T}(t)}\dot{\bar{\ell}}_{i,T}(t) + \frac{\partial \gamma_{i,T}}{\partial \underline{\ell}_{i,T}(t)}\dot{\underline{\ell}}_{i,T}(t) \\ &= \alpha_{i,T}\left(-\frac{3}{2}p\frac{U_s L_m}{\Omega_s L_s}\dot{i}_{q,r} - \dot{T}^*\right) + \beta_{i,T}\end{aligned} \tag{51}$$

where $\alpha_{i,T} = \frac{\partial \gamma_{i,T}}{\partial \ell_{i,T}}$ and $\beta_{i,T} = \frac{\partial \gamma_{i,T}}{\partial \bar{\ell}_{i,T}(t)}\dot{\bar{\ell}}_{i,T}(t) + \frac{\partial \gamma_{i,T}}{\partial \underline{\ell}_{i,T}(t)}\dot{\underline{\ell}}_{i,T}(t)$. Because signals $\gamma_{i,T}$, $\ell_{i,T}$, $\bar{\ell}_{i,T}(t)$, $\dot{\bar{\ell}}_{i,T}(t)$, $\underline{\ell}_{i,T}(t)$, $\dot{\underline{\ell}}_{i,T}(t)$ are known, we can easily compute the values of $\alpha_{i,T}$ and $\beta_{i,T}$.

Substituting the dynamics of $I_{q,r}$ given by (16) into (51) and taking the modeling error into account, one has

$$\dot{\gamma}_{i,T} = \alpha_{i,T}\left(f_{i,T}(Z_{i,T}) + g_{i,T}U_{q,r} + d_{i,T}(t)\right) \tag{52}$$

with

$$\begin{aligned} f_{i,T}(Z_{i,T}) &= \frac{3}{2}pL_m\frac{U_sR_r}{\Omega_sL_e}I_{q,r} - \dot{T}^* + \frac{\beta_{i,T}}{\alpha_{i,T}} \\ &\quad + \frac{3}{2}pL_m\frac{U_s}{\Omega_sL_s}\left(\Omega_s - p\Omega_g\right)\left(I_{d,r} + \frac{L_mU_s}{L_e\Omega_s}\right) \\ Z_{i,T} &= [I_{q,r}, I_{d,r}, \Omega_g, \dot{T}^*, \gamma_{i,T}, \alpha_{i,T}]^T \\ g_{i,T} &= -\frac{3}{2}pL_m\frac{U_s}{\Omega_sL_e} \end{aligned} \tag{53}$$

with $d_{i,T}(t) = -\frac{3}{2}p\frac{U_sL_m}{\Omega_sL_s}\varpi_q(t)$ being the system unknown disturbances. Similar to most studies [13], we assume the disturbance term $d_{i,T}$ is bounded.

If the system parameter is available and $d_{i,T}(t) = 0$, a desired control input voltage $U_{q,r}^*$ can be given as

$$U_{q,r}^* = k_{i,T}\frac{\gamma_{i,T}}{\alpha_{i,T}} + {\tau_{i,T}}^T\xi_{i,T} \tag{54}$$

where $\tau_{i,T} = -\left[\frac{1.5pL_mU_sR_r}{\Omega_sL_e}, \frac{1.5pL_mU_s}{L_s}, \frac{1.5pL_m^2U_s^2}{\Omega_sL_eL_s}, -\frac{1.5p^2L_mU_s}{\Omega_sL_s}, \frac{-1.5U_s^2L_m^2p^2}{\Omega_s^2L_sL_e}, -1, 1\right]^T / g_{i,T}$, and $\xi_{i,T} = [I_{q,r}, I_{d,r}, 1, \Omega_gI_{d,r}, \Omega_g, \dot{T}^*, \gamma_{i,T}/\alpha_{i,T}]^T$, $k_{i,T}$ is a positive constant. Thereafter, following lemma shows the system stability with the desired control input $U_{q,r}^*$.

Lemma 2 *Consider the dynamics of $\gamma_{i,T}$ in (52) with $d_{i,T}(t) = 0$. The transformed tracking error $\gamma_{i,T}$ will converge asymptotically to zero if the desired control input $U_{q,r}^*$ is chosen as (54).*

Proof Consider the following Lyapunov candidate

$$V_{i,T}^* = -\frac{1}{2}\frac{\gamma_{i,T}^2}{g_{i,T}} \tag{55}$$

Taking its time derivative and recalling (52) with $d_{i,T} = 0$, we have

$$\dot{V}_{i,T}^* = -\gamma_{i,T}\alpha_{i,T}\left(\frac{f_{i,T}(Z_{i,T})}{g_{i,T}} + U_{q,r}\right) = \gamma_{i,T}\alpha_{i,T}\left({\tau_{i,T}}^T\xi_{i,T} - U_{q,r}\right) \tag{56}$$

Substituting the desired controller (54) into the above equation, we have

$$\dot{V}_{i,T}^* = -k_{i,T}\gamma_{i,T}^2 \tag{57}$$

which implies that $\gamma_{i,T}$ converges to zero asymptotically.

The value of $\tau_{i,T}$ is always distinct for different wind turbine systems, and obtaining its value usually needs substantial human and technological efforts. Because of this, it is not economical to extend the proposed controller to various wind turbines. In order to circumvent such issues, the actual value of $\tau_{i,T}$ is supposed to be unknown in the following adaptive controller design. Further, the bounded disturbance term $d_{i,T}$ is assumed to satisfy

$$\left|\frac{d_{i,T}(t)}{g_{i,T}}\right| \leq B_{i,T} \tag{58}$$

with $B_{i,T}$ being a positive unknown constant and $|\cdot|$ being the absolute value operator. Since both $\tau_{i,T}$ and $B_{i,T}$ are unknown, let their estimates to be $\hat{\tau}_{i,T}$ and $\hat{B}_{i,T}$, and we are ready to present the following adaptive control law

$$U_{q,r} = k_{i,T}\frac{\gamma_{i,T}}{\alpha_{i,T}} + \hat{\tau}_{i,T}^T\xi_{i,T} + \tanh\left(\frac{\alpha_{i,T}\gamma_{i,T}}{\varepsilon_{i,T}}\right)\hat{B}_{i,T} \tag{59}$$

with $\tanh(\cdot)$ being the hyperbolic tangent function. The adaptive law for $\hat{\tau}_{i,T}$ is chosen as

$$\dot{\hat{\tau}}_{i,T} = \Lambda_{i,T}\left(\gamma_{i,T}\alpha_{i,T}\xi_{i,T} - \sigma_{i,T1}\hat{\tau}_{i,T}\right) \tag{60}$$

where the learning rate $\Lambda_{i,T} \in \Re^{7\times 7}$ is a positive definite matrix, and $\sigma_{i,T11}$ is a positive constant. Further, the adaptive law for $\hat{B}_{i,T}$ is chosen as

$$\dot{\hat{B}}_{i,T} = l_{i,T}\left(\gamma_{i,T}\alpha_{i,T}\tanh\left(\frac{\alpha_{i,T}\gamma_{i,T}}{\varepsilon_{i,T}}\right) - \sigma_{i,T2}\hat{B}_{i,T}\right) \tag{61}$$

where $l_{i,T}, \sigma_{i,T2} > 0$.

Theorem 2 *Consider the inner loop dynamics characterized by (13) and (14). If the control input voltage $U_{q,r}$ is selected as (59) with adaptive laws (60) and (61), the electromagnetic torque T_{em} can track its desired value T^* with guaranteed performance in terms of tracking error $e_{i,T}$ satisfying (45).*

Proof Consider the following Lyapunov function candidate

$$V_{i,T}^* = -\frac{1}{2}\frac{\gamma_{i,T}^2}{g_{i,T}} + \frac{1}{2}\tilde{\tau}_{i,T}^T\Lambda_{i,T}^{-1}\tilde{\tau}_{i,T} + \frac{1}{2l_{i,T}}\tilde{B}_{i,T}^2 \tag{62}$$

with $\tilde{\tau}_{i,T} = \hat{\tau}_{i,T} - \tau_{i,T}$, and $\tilde{B}_{i,T} = \hat{B}_{i,T} - B_{i,T}$. Taking the time derivative of $V_{i,T}^*$ generates

$$\dot{V}_{i,T}^* = \gamma_{i,T}\alpha_{i,T}\left(-\frac{f_{i,T}(Z_{i,T})}{g_{i,T}} - U_{q,r} - \frac{d_{i,T}(t)}{g_{i,T}}\right) + \tilde{\tau}_{i,T}^T\Lambda_{i,T}^{-1}\dot{\hat{\tau}}_{i,T} + \frac{1}{l_{i,T}}\tilde{B}_{i,T}\dot{\hat{B}}_{i,T} \tag{63}$$

Using the inequality (38), we have

$$-\gamma_{i,T}\alpha_{i,T}\frac{d_{i,T}(t)}{g_{i,T}} \leq |\gamma_{i,T}\alpha_{i,T}|B_{i,T} \leq \gamma_{i,T}\alpha_{i,T}\tanh\left(\frac{\gamma_{i,T}\alpha_{i,T}}{\varepsilon_{i,T}}\right)B_{i,T} + \kappa\varepsilon_{i,T}B_{i,T} \tag{64}$$

Substituting (64) and (59) into (63) yields

$$\begin{aligned}\dot{V}_{i,T}^{*} &\leq \gamma_{i,T}\alpha_{i,T}\left(\tau_{i,T}{}^{T}\xi_{i,T} - U_{q,r} + \tanh\left(\frac{\gamma_{i,T}\alpha_{i,T}}{\varepsilon_{i,T}}\right)B_{i,T}\right) \\ &\quad + \tilde{\tau}_{i,T}^{T}\Lambda_{i,T}^{-1}\dot{\hat{\tau}}_{i,T} + \frac{1}{l_{i,T}}\tilde{B}_{i,T}\dot{\hat{B}}_{i,T} + \kappa\varepsilon_{i,T}B_{i,T} \\ &\leq -k_{i,T}\gamma_{i,T}^{2} + \tilde{\tau}_{i,T}^{T}\left(\Lambda_{i,T}^{-1}\dot{\hat{\tau}}_{i,T} - \gamma_{i,T}\alpha_{i,T}\xi_{i,T}\right) + \kappa\varepsilon_{i,T}B_{i,T} \\ &\quad + \tilde{B}_{i,T}\left(\frac{1}{l_{i,T}}\dot{\hat{B}}_{i,T} - \gamma_{i,T}\alpha_{i,T}\tanh\left(\frac{\gamma_{i,T}\alpha_{i,T}}{\varepsilon_{i,T}}\right)\right)\end{aligned} \tag{65}$$

Substituting the adaptive laws (60), (61) and by completion of squares, we have

$$\begin{aligned}\dot{V}_{i,T}^{*} &\leq -k_{i,T}\gamma_{i,T}^{2} - \sigma_{i,T1}\tilde{\tau}_{i,T}^{T}\hat{\tau}_{i,T} - \sigma_{i,T2}\tilde{B}_{i,T}\hat{B}_{i,T} + \kappa\varepsilon_{i,T}B_{i,T} \\ &\leq -k_{i,T}\gamma_{i,T}^{2} - \frac{\sigma_{i,T1}\|\tilde{\tau}_{i,T}\|^{2}}{2} - \frac{\sigma_{i,T2}\tilde{B}_{i,T}^{2}}{2} + \Delta_{i,T}\end{aligned} \tag{66}$$

with $\Delta_{i,T} = \kappa\varepsilon_{i,T}B_{i,T} + \sigma_{i,T1}\|\tau_{i,T}\|^{2}/2 + \sigma_{i,T2}B_{i,T}^{2}/2$. Hence, the $\dot{V}_{i,T}^{*}$ will become negative as long as

$$\left|\gamma_{i,T}\right| > \sqrt{\frac{\Delta_{i,T}}{k_{i,T}}} \tag{67}$$

or

$$\left\|\tilde{\tau}_{i,T}\right\| > \sqrt{\frac{2\Delta_{i,T}}{\sigma_{i,T1}}} \tag{68}$$

or

$$\left|\tilde{B}_{i,T}\right| > \sqrt{\frac{2\Delta_{i,T}}{\sigma_{i,T2}}} \tag{69}$$

Based on the standard Lyapunov theorem extension [23], $\gamma_{i,T}$, $\tilde{\tau}_{i,T}$ and $\tilde{B}_{i,T}$ are bounded.

According to the properties of error transformation, the boundedness of $\gamma_{i,T}$ concludes that the guaranteed performance described by (45) is achieved, and thus $e_{i,T}$ is bounded. The reference T^{*} generated by the MPPT algorithm is bounded. It thus follows that T_{em} is bounded. Since $\hat{\tau}_{i,T} = \tilde{\tau}_{i,T} + \tau_{i,T}$, and $\tau_{i,T}$ is bounded from definition, we have that $\hat{\tau}_{i,T}$ is bounded as well. Since $\hat{B}_{i,T} = \tilde{B}_{i,T} + B_{i,T}$, and $B_{i,T}$ is bounded from definition, we have that $\hat{B}_{i,T}$ is also bounded.

The value of desired reactive power Q^* is determined by grid needs, e.g., a specified amount of reactive power compensated in grid can improve the role of the grid power factor, lower power transformer and transmission line losses. In this chapter, for analysis convenience, the desired reactive power Q^* is assumed to be a known signal, and the control objective is to track a desired Q^* with guaranteed performance, i.e., to obtain the tracking error $e_{i,Q}$ satisfying (46). By recalling (19), the time derivative of $\gamma_{i,Q}$ can be obtained as

$$\begin{aligned}\dot{\gamma}_{i,Q} &= \frac{\partial \gamma_{i,Q}}{\partial \ell_{i,Q}}\left(-\frac{3U_s L_m}{2L_s}\dot{I}_{d,r} - \dot{Q}^*\right) + \frac{\partial \gamma_{i,Q}}{\partial \bar{\ell}_{i,Q}(t)}\dot{\bar{\ell}}_{i,Q}(t) + \frac{\partial \gamma_{i,Q}}{\partial \underline{\ell}_{i,Q}(t)}\dot{\underline{\ell}}_{i,Q}(t) \\ &= \alpha_{i,Q}\left(-\frac{3U_s L_m}{2L_s}\dot{I}_{d,r} - \dot{Q}^*\right) + \beta_{i,Q} \end{aligned} \tag{70}$$

where $\alpha_{i,Q} = \frac{\partial \gamma_{i,Q}}{\partial \ell_{i,Q}}$ and $\beta_{i,Q} = \frac{\partial \gamma_{i,Q}}{\partial \bar{\ell}_{i,Q}(t)}\dot{\bar{\ell}}_{i,Q}(t) + \frac{\partial \gamma_{i,Q}}{\partial \underline{\ell}_{i,Q}(t)}\dot{\underline{\ell}}_{i,Q}(t)$, which are available as feedback signals.

By recalling (15), (70) and considering the modeling error, we have

$$\dot{\gamma}_{i,Q} = \alpha_{i,Q}\left(f_{i,Q}(Z_{i,Q}) + g_{i,Q}U_{d,r} + d_{i,Q}(t)\right) \tag{71}$$

with

$$\begin{aligned} f_{i,Q}(Z_{i,Q}) &= \frac{3U_s L_m R_r}{2L_e} I_{d,r} - \frac{3U_s L_m}{2L_s}\left(\Omega_s - p\Omega_g\right) I_{q,r} - \dot{Q}^* + \frac{\gamma_{i,Q}}{\alpha_{i,Q}} \\ Z_{i,Q} &= [I_{q,r}, I_{d,r}, \Omega_g, \dot{Q}^*, \gamma_{i,Q}, \alpha_{i,Q}]^T \\ g_{i,Q} &= -\frac{3U_s L_m}{2L_e} \end{aligned} \tag{72}$$

with $d_{i,Q}(t) = -\frac{3U_s L_m \varpi_d(t)}{2L_s}$ being the bounded disturbance term embodying modeling errors in the dynamics of $\ell_{i,Q}$. To facilitate the reactive power control design, it can be observed that

$$-\frac{f_{i,Q}(Z_{i,Q})}{g_{i,Q}} = {\tau_{i,Q}}^T \xi_{i,Q} \tag{73}$$

where $\tau_{i,Q} = -[1.5U_s L_m R_r/L_e, -1.5U_s L_m \Omega_s/L_s, 1.5U_s L_m \ p/L_s, -1, 1]^T/g_{i,Q}$, and $\xi_{i,Q} = [I_{d,r}, I_{q,r}, \Omega_g I_{q,r}, \dot{Q}^*, \gamma_{i,Q}/\alpha_{i,Q}]^T$. Moreover, assume that the disturbance term $d_{i,Q}(t)$ is bounded such that

$$\left|\frac{d_{i,Q}(t)}{g_{i,Q}}\right| \le B_{i,Q} \tag{74}$$

Since both $\tau_{i,Q}$ and $B_{i,Q}$ are unknown, let their estimates to be $\hat{\tau}_{i,Q}$ and $\hat{B}_{i,Q}$, we propose the following input $U_{d,r}$

$$U_{d,r} = k_{i,Q}\frac{\gamma_{i,Q}}{\alpha_{i,Q}} + \hat{\tau}_{i,Q}^T\xi_{i,Q} + \tanh\left(\frac{\alpha_{i,Q}\gamma_{i,Q}}{\varepsilon_{i,Q}}\right)\hat{B}_{i,Q} \tag{75}$$

with $k_{i,Q}$ being a user-defined positive constant. The adaptive laws for $\hat{\tau}_{i,Q}$ and $\hat{B}_{i,Q}$ are given by

$$\begin{aligned} \dot{\hat{\tau}}_{i,Q} &= \Lambda_{i,Q}\left(\gamma_{i,Q}\alpha_{i,Q}\xi_{i,Q} - \sigma_{i,Q1}\hat{\alpha_o}\right) \\ \dot{\hat{B}}_{i,Q} &= l_{i,Q}\left(\gamma_{i,Q}\alpha_{i,Q}\tanh\left(\frac{\alpha_{i,Q}\gamma_{i,Q}}{\varepsilon_{i,Q}}\right) - \sigma_{i,Q2}\hat{B}_{i,Q}\right) \end{aligned} \tag{76}$$

where the learning rate $\Lambda_{i,Q} \in \Re^{5\times 5}$ is a positive definite matrix, and $l_{i,Q}$ is a positive constant.

The stability and control performance of the reactive power closed-loop system is given in the following theorem.

Theorem 3 *Consider the inner loop control characterized by (13) and (14). If the control voltage $U_{d,r}$ is designed by (75) with adaptive laws (76), the reactive power Q can track its desired value Q^* with guaranteed performance in terms of tracking error $\ell_{i,Q}$ satisfying (46).*

Proof The proof is similar to Theorem 2 and thus omitted here.

6 Validation Results

To validate the proposed inner loop control and outer loop control, we have conducted numerical analysis using NREL FAST code [24] on the NREL WP 1.5 MW wind turbine, which has three blades on a horizontal axis [25, 26]. The parameters of the wind turbine are given in Table 1.

We use the FAST module in the Simulink environment as shown in Fig. 6. We choose the following system parameters in our validation: air density $\rho =$

Table 1 Parameters of wind turbine

Number of blades	3
Rotor radius	35 m
Hub height	84.3 m
Rated power	1.5 MW
Turbine total inertia	4.4532×10^5 kg m^2

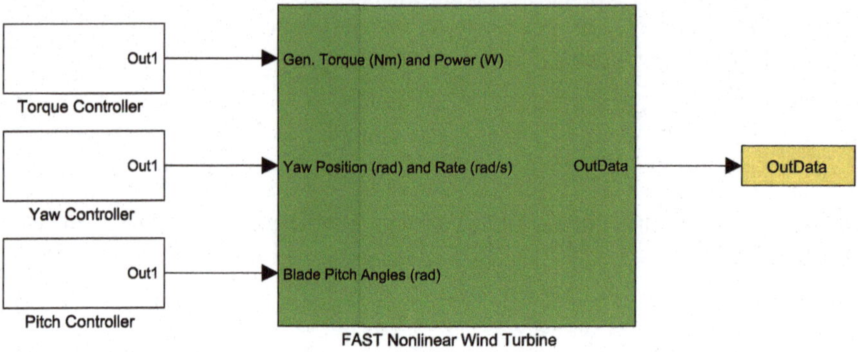

Fig. 6 FAST simulator block

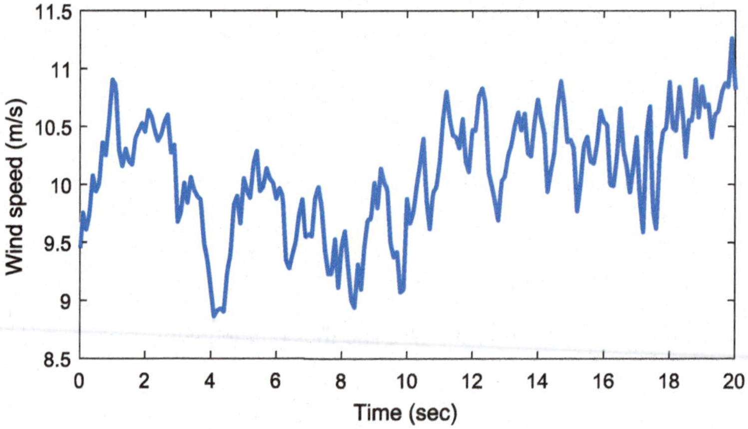

Fig. 7 Wind profile for outer loop control

$1.225\,\mathrm{kg/m^3}$, ratio $n_p = 0.9$, maximum power ratio $C_p^{\max} = 0.412$. The controller parameters are listed as follows: $k_o = 5$, $\Lambda_o = \mathrm{diag}(10^{-25}, 10^{-15}, 10^{-3})$, $\sigma_o = 10$, $\varepsilon_o = 10$.

The wind speed used in this test is given in Fig. 7. It is generated by the TurbSim [27] with the mean wind speed as $9.5\,\mathrm{m/s}$ and turbulence intensity as 15%.

The tracking error performance is depicted in Fig. 8, which can be observed that our proposed outer loop controller can ensure the imposed performance requirements. Figure 9 shows the output power trajectory of the generator. It can be observed that the power output can follow the maximum available power from wind. Finally, Fig. 10 depicts the generator torque input.

For the inner loop control, in order to consider the external noises, two Gaussian distribution noises with standard deviations 0.1 and 0.5 are added in the dynamics of $I_{d,r}$ and $I_{q,r}$. In this case study, the generated wind speed is shown in Fig. 11, which is also created using the Kaimal turbulence model with a mean value of $6\,\mathrm{m/s}$

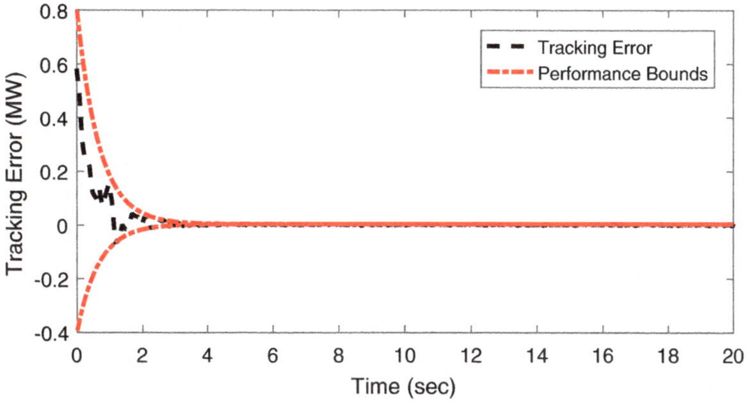

Fig. 8 Tracking error ℓ_o

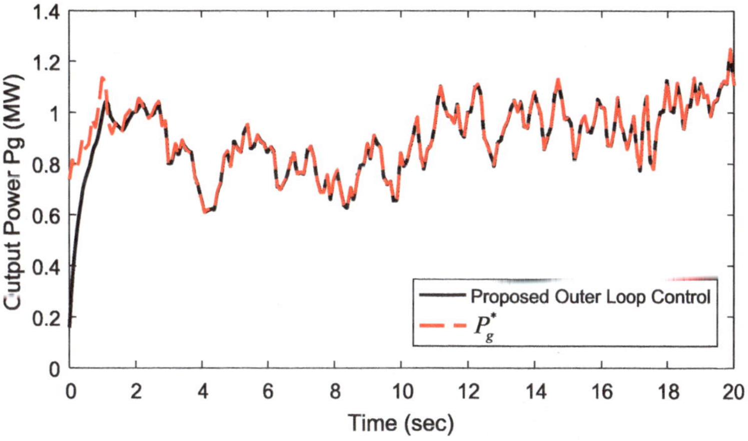

Fig. 9 Generator output power

and turbulence intensity of 10%. To be more realistic, the temporal evolution of the electrical parameters (resistances, inductances), varying from their nominal values is considered as shown in Fig. 12.

One of the control goal is to drive electromagnetic torque T_{em} to track T^* by setting $k_{opt} = 0.2357$, with tracking error $\ell_{i,T}$ satisfying predefined constraints. The corresponding upper bound $\bar{\ell}_T(t)$ is determined as $9 \times \exp(-2t) + 1$ along with the lower bound as $-149 \times \exp(-3t) - 1$. Moreover, control parameters in control of electromagnetic torque are listed as follows: $k_{i,T} = 3 \times 10^{-4}$, $\Lambda_{i,T} = \text{diag}\{10^{-6}, 2 \times 10^{-6}, 1, 2 \times 10^{-10}, 10^{-5}, 3 \times 10^{-7}, 3 \times 10^{-7}\}$, $\sigma_{i,T11} = 2 \times 10^{-5}$, $l_{i,T} = 0.01$, $\sigma_{i,T2} = 3$, $\varepsilon_{i,T} = 2$.

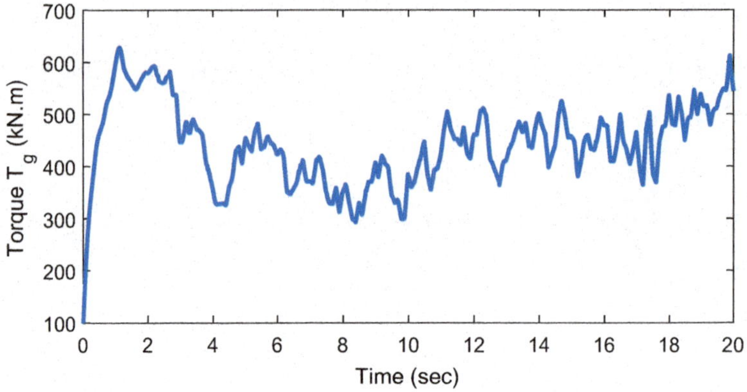

Fig. 10 Generator torque

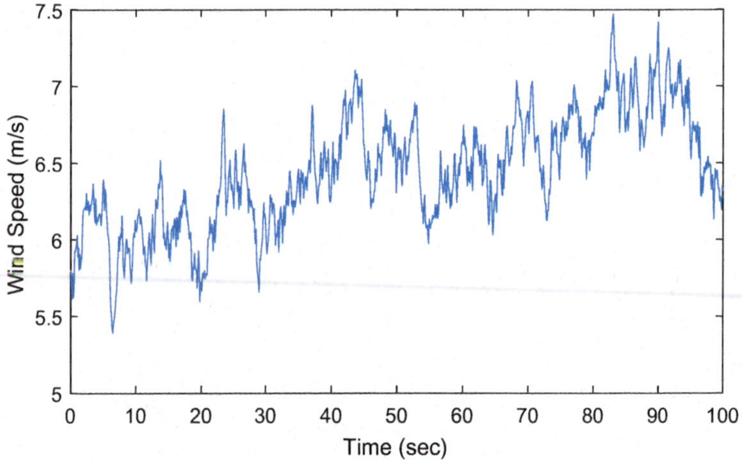

Fig. 11 Wind speed profile

Figure 13 depicts the electromagnetic torque and its desired value T^*, which shows good tracking performance. The tracking error $\ell_{i,T}$ with its performance bounds is given in Fig. 14. It can be observed that the prescribed performance is achieved.

The desired reactive power is given by $Q^* = 1000 + 30\sin(0.1t)$. One of the control goals is to drive the the reactive power to follow this desired power with tracking error $\ell_{i,Q}$ satisfying predefined constraints. The corresponding upper bound $\bar{\ell}_Q(t)$ is determined as $3.5 \times \exp(-2t) + 1$ along with the lower bound as $-118.5 \times \exp(-5t) - 1$. Moreover, control parameters in control of reactive power are listed as follows: $k_{i,Q} = 4 \times 10^{-7}$, $\Lambda_{i,Q} = \text{diag}\{10^{-7}, 9 \times 10^{-8}, 4 \times 10^{-12}, 0.001, 10^{-5}\}$, $\sigma_{Q1} = 2, l_{i,Q} = 5 \times 10^{-7}, \sigma_{Q2} = 1, \varepsilon_Q = 5$.

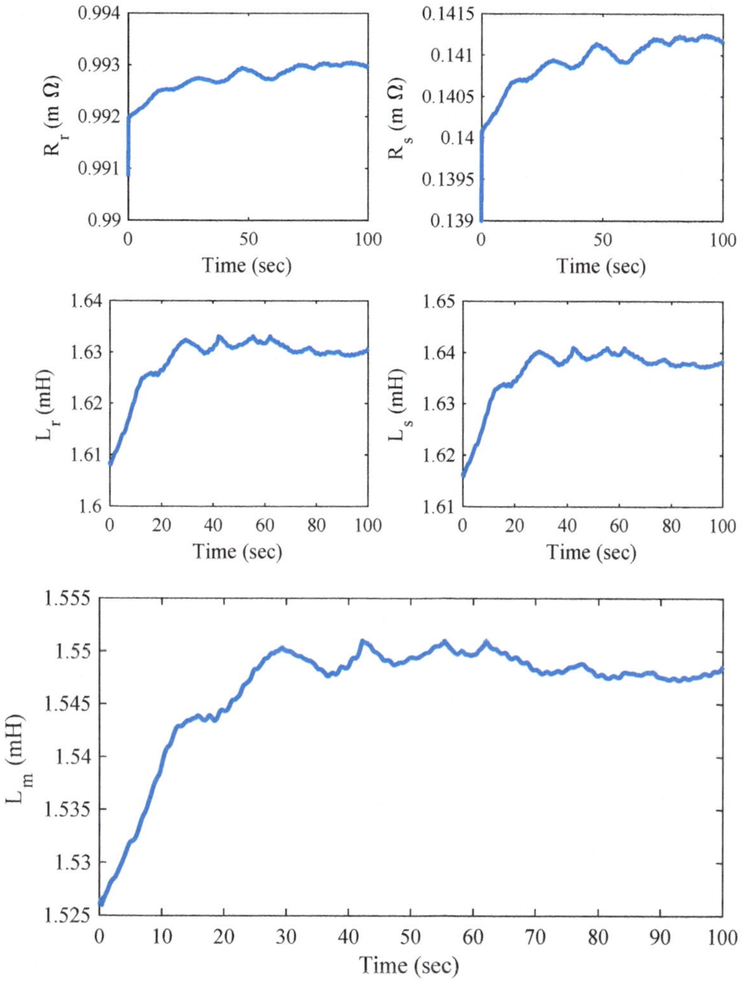

Fig. 12 Temporal evolution of the electromagnetic parameters

Figure 15 shows the reactive power and its desired value Q^*, and the corresponding tracking error $\ell_{i,Q}$ along with its performance bounds is plotted in Fig. 16. It can be observed that the corresponding prescribed performance can also be ensured.

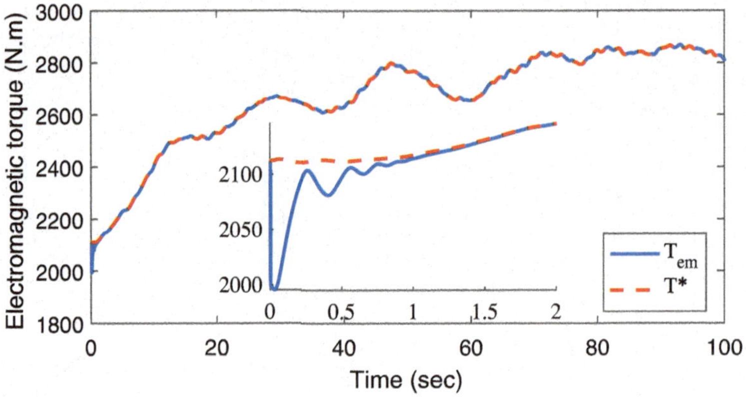

Fig. 13 Electromagnetic torque and its reference

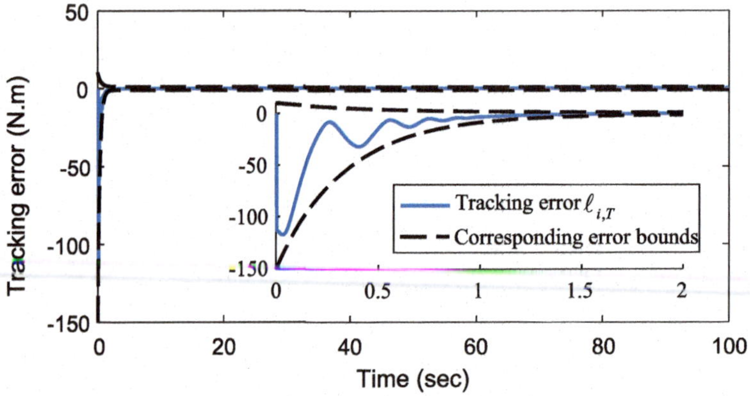

Fig. 14 Tracking error $\ell_{i,T}$ along with its bounds

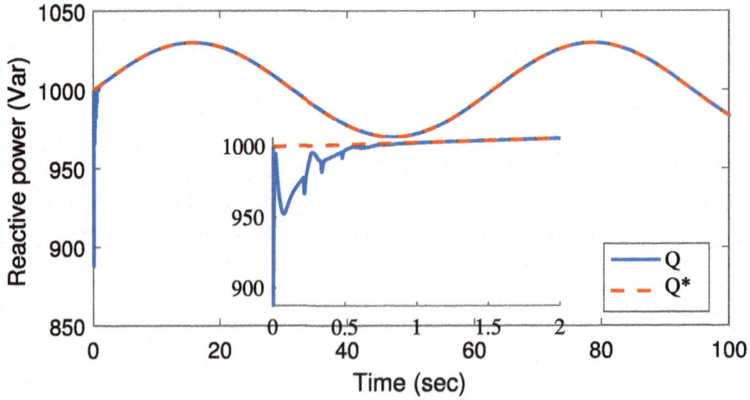

Fig. 15 Reactive power and its reference

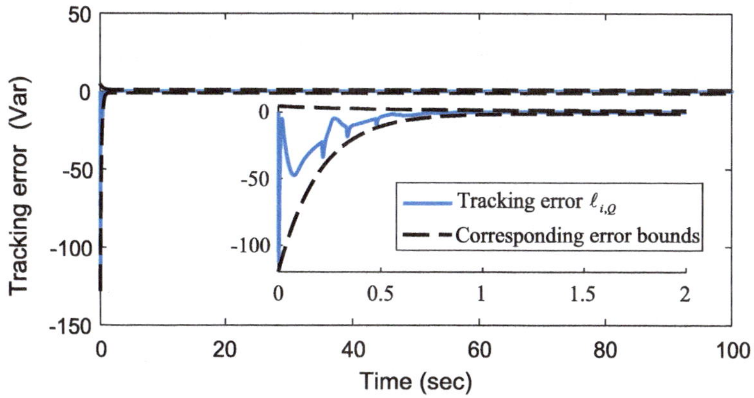

Fig. 16 Tracking error $\ell_{i,Q}$ along with its bounds

7 Conclusion

We have proposed an adaptive guaranteed performance controller for WECS equipped with DFIG. The WECS comprises the outer loop control concerning the aeroturbine mechanical subsystem, and the inner loop control concerning the electrical subsystem. Because the aeroturbine runs much slower than electrical subsystem, the stability analysis in the outer control loop and inner control loop is addressed separately. With the help of error transformation, our proposed method is capable of quantifying and further guaranteeing the system performance on both transient and steady state stages.

References

1. Bose B (2010) Global warming: energy, environmental pollution, and the impact of power electronics. IEEE Ind Electron Mag 4(1):6–17
2. Kaldellis JK (2008) The wind potential impact on the maximum wind energy penetration in autonomous electrical grids. Renew Energy 33(7):1665–1677
3. Joselin Herbert GM, Iniyan S, Sreevalsan E, Rajapandian S (2007) A review of wind energy technologies. Renew Sustain Energy Rev 11(6):1117–1145
4. Todeschini G, Emanuel AE (2011) Transient response of a wind energy conversion system used as active filter. IEEE Trans Energy Convers 26(2):522–531
5. Munteanu I, Cutululis NA, Bratcu AI, Ceanga E (2005) Optimization of variable speed wind power systems based on a LQG approach. Control Eng Pract 13(7):903–912
6. Bououden S, Chadli M, Filali S, Hajjaji AE (2012) Fuzzy model based multivariable predictive control of a variable speed wind turbine: LMI approach. Renew Energy 37(1):434–439
7. Boukhezzar B, Siguerdidjane H (2009) Nonlinear control with wind estimation of a DFIG variable speed wind turbine for power capture optimization. Energy Convers Manag 50(4):885–892

8. Benbouzid M, Beltran B, Amirat Y, Yao G, Han J, Mangel H (2014) Second-order sliding mode control for DFIG-based wind turbines fault ride-through capability enhancement. ISA Trans 53(3):827–833
9. Meng W, Yang Q, Ying Y, Sun Y, Yang Z, Sun Y (2013) Adaptive power capture control of variable-speed wind energy conversion systems with guaranteed transient and steady-state performance. IEEE Trans Energy Convers 28(3):716–725
10. Meng W, Yang Q, Sun Y (2016) Guaranteed performance control of DFIG variable-speed wind turbines. IEEE Trans Control Syst Technol 24(6):2215–2223
11. Beltran B, Ahmed-Ali T, Benbouzid M (2009) High-order sliding-mode control of variable-speed wind turbines. IEEE Trans Ind Electron 56(9):3314–3321
12. Evangelista C, Valenciaga F, Puleston P (2013) Active and reactive power control for wind turbine based on a MIMO 2-sliding mode algorithm with variable gains. IEEE Trans Energy Convers 28(3):682–689
13. Valenciaga F, Puleston PF, Spurgeon SK (2009) A geometric approach for the design of MIMO sliding controllers. Application to a wind-driven doubly fed induction generator. Int J Robust Nonlinear Control 19(1):22–39
14. Boukhezzar B, Lupu L, Siguerdidjane H, Hand M (2007) Multivariable control strategy for variable speed, variable pitch wind turbines. Renew Energy 32(8):1273–1287
15. Kumar A, Stol K (2010) Simulating feedback linearization control of wind turbines using high-order models. Wind Energy 13(5):419–432
16. Beltran B, El Hachemi Benbouzid M, Ahmed-Ali T (2012) Second-order sliding mode control of a doubly fed induction generator driven wind turbine. IEEE Trans Energy Convers 27(2):261–269
17. Bechlioulis CP, Rovithakis GA (2008) Robust adaptive control of feedback linearizable mimo nonlinear systems with prescribed performance. IEEE Trans Autom Control 53(9):2090–2099
18. Slotine JJE, Li W et al (1991) Applied nonlinear control. Prentice-Hall Englewood Cliffs, NJ
19. Chen M, Ge SS, Ren B (2011) Adaptive tracking control of uncertain mimo nonlinear systems with input constraints. Automatica 47(3):452–465
20. Narendra K, Annaswamy A (1987) A new adaptive law for robust adaptation without persistent excitation. IEEE Trans Autom Control 32(2):134–145
21. Yang Q, Jagannathan S (2012) Reinforcement learning controller design for affine nonlinear discrete-time systems using online approximators. IEEE Trans Syst Man Cybern Part B Cybern 42(2):377–390
22. Bechlioulis CP, Rovithakis GA (2014) A low-complexity global approximation-free control scheme with prescribed performance for unknown pure feedback systems. Automatica 50(4):1217–1226
23. Lewis FL, Jagannathan S, Yesildirak A (1999) Neural network control of robot manipulators and non-linear systems. Taylor & Francis, Philadelphia, PA
24. National Renewable Energy Laboratory, Golden, CO. (2007, Feb). http://wind.nrel.gov/designcodes/simulators/fast/
25. Beltran B, Ahmed-Ali T, El Hachemi Benbouzid M (2008) Sliding mode power control of variable-speed wind energy conversion systems. IEEE Trans Energy Convers 23(2):551–558
26. Buhl Jr ML, Manjock A (2006) A comparison of wind turbine aeroelastic codes used for certification. Natl Renew Energy Lab, Golden, CO, NREL/CP-500-39113
27. Jonkman BJ, Buhl Jr ML, Turbsim user's guide. Technical Report NREL/TP-500-41136. National Renewable Energy Laboratory (NREL), Golden, CO. http://wind.nrel.gov/designcodes/simulators/fast/

Machine Learning and Meta-heuristic Algorithms for Renewable Energy: A Systematic Review

Essam H. Houssein

Abstract The demand for energy is become essential due to industrial activities and increasing agricultural of any nation. According to the aforementioned, the renewable energy resources available are very suitable to meet the ever-growing requirement of energy by the humanity rather causing any harmful effects to nature. Therefore, several research studies have been introduced in the renewable energy field such as solar, wind, biomass, and biogas due to the clean and sustainability. To better scheme and utilize this energy resource, good forecasting and optimization are necessary and intrinsic. So, this review introduces an overview of the renewable energy forecasting techniques that have been utilized in this field based on meta-heuristic optimization algorithms and machine learning (ML). In addition, several challenges have been addressed, recommendations for future research are provided, and a comprehensive bibliography is conducted. Eventually, in general speaking, this comprehensive review of renewable energy resources may help the researchers, energy planners, and policymakers.

1 Introduction

Sustainable development and global economic growth lead to increased demand for energy. Approximately 80% of global energy requirements were supplied by fossil fuels [1]. Regarding to the rapid rate and exhausting of the fossil fuels which simulates the supplement of renewable energy resources such as biomass, solar, hydropower, biogas, biodiesel, wind power, etc. A rapid increase in global energy demand is due. To the rate of population growth, high rate of industrialization, and socio-economic development [2]. The world population growth, energy consumption, and manufacturing activities effect on the human activities around all nations. So, the renewable energy resources are regrading as the most effective and effectiveness solutions. Therefore, the traditional energy resources in the most of the growing countries has

E. H. Houssein (✉)
Faculty of Computers and Information, Minia University, Minya, Egypt
e-mail: essam.halim@mu.edu.eg

R.-E. Precup et al. (eds.), *Advanced Control and Optimization Paradigms for Wind Energy Systems*, Power Systems, https://doi.org/10.1007/978-981-13-5995-8_7

been provided extra than 90% of total energy used to set off rapid deforestation, reducing soil fertility, etc. [3].

The rest of this review is organized in the following sections. Preliminaries for the meta-heuristic optimization algorithms and machine learning techniques are presented in Sect. 2. Several types of renewable energy resources also are introduced in Sect. 3. Then, the existing studies of meta-heuristic optimization algorithms and machine learning techniques applied in renewable energy are discussed in Sect. 4. Section 5 address some challenges and open problems. Finally, Sect. 6 concludes this systematic review and provide some future directions.

2 Preliminaries

2.1 Machine Learning Techniques

2.1.1 Artificial Neural Networks

In a wide range of applications, ANN models have been widely used such as simulation, optimization, pattern recognition, classification, forecasting (prediction), etc. Also, a large number of cite technologies was introduced [4]. In the last decade, ANN excels on classical and traditional statistical methods for classification and prediction. ANN is a powerful tool in several application fields and the advantages of ANN are; (1) no need for any assumptions; (2) extrapolating from historical data to generate expectations; and (3) solving complex nonlinear problems respectively. On the other hand, ANN chooses the important weights during the training phase and estimates the output retained for the following use [5]. The ability to design a multivariate problem is regrading the big advantage that merits the ANN models. In general speaking, ANN consists of many nodes and consists of an input layer, some hidden layers, and an output layer. Therefore, ANNs are parallel arithmetic models widely used for information processing and data representation. In dealing with most simulation and prediction problems, NN have characteristics that make them potentially successful because they are capable of learning the complex nonlinear relationships and relationships of a large set of data due to their adaptability, intrinsic nonlinearity, circularity, and robustness [6].

2.1.2 Support Vector Machines

SVM is regarding as one of the most popular and powerful tool under classification problems suggested in biomedical science, pattern recognition and classification, etc. [7–9]. The advantages of SVMs are; (1) classify two different sets in their relevant class; (2) deal with nonlinear and high-dimensional data. To improve the learning ability and accuracy of classification for SVM, specifying features and improving the parameter for SVM is a critical problem. SVM parameters and kernel parameters effect strongly on the final results defined as follows:

$$\varphi(X) : X \subset R^k \rightarrow R^m, k \leq m \tag{1}$$

Hyperplane selection is done with the help of a set of functions in the form of universal approximation for the linearly separable hyperplanes as shown by the following Equation:

$$f(x) = \sum_{i=1}^{n} w_i x_i + b \tag{2}$$

The following Equations represents the hyperplane targeted by SVMs in order to obtain linear separability in feature space.

$$f(x) = \sum_{i=1}^{n} w_i \varphi(x_i) + b \geq 1 \forall i : y_i = (+1) \tag{3}$$

$$f(x) = \sum_{i=1}^{n} w_i \varphi(x_i) + b \leq -1 \forall i : y_i = (-1) \tag{4}$$

Eventually, linear kernel, polynomial kernel, Redial Basis Function (RBF) and Sigmoid is regrading as the most popular four kernels functions used with SVM.

2.2 Meta-heuristic Optimization Algorithms

In computer science and mathematical optimization, "Meta-heuristics" is regarding as a higher-level procedure applied to search, create, define a search algorithm in order to provide a an optimal solution to an specific problem, especially with incomplete complete information or ability to account is limited. Four important categories in the meta-heuristics optimization algorithms are; Swarm Intelligence (SI) and bio-inspired optimization algorithms, chemical/physics and other algorithms. The first and second type may termed as natural inspired algorithms. In this chapter, we will focus on SI algorithms that related to the behavior of many interacting clients who follow the same rules. Although each agent may be considered unwise, the whole system shows some self-organization behavior and is seen as collective intelligence. Depending on the inspiration of the SI systems in nature, several algorithms have been suggested. Therefore, SI is a set of nature-inspired algorithms under the large umbrella of Evolutionary Computation (EC). SI is a population-based algorithm that interacts with each other and is statistically better across generations and ultimately finding good enough solutions. All swarm intelligence algorithms uses the concept

Table 1 Brief list for existing meta-heuristic optimization algorithms

Ref.	Algorithm	Year	Ref.	Algorithm	Year
[10]	Chaotic bird swarm optimization	2018	[11]	Water evaporation optimization	2016
[12]	Sperm whale algorithm	2016	[13]	Sine cosine algorithm	2016
[14]	Water wave optimization	2015	[15]	Dragonfly algorithm	2016
[16]	Elephant herding optimization	2015	[17]	Grey wolf optimizer	2014
[18, 19]	Whale optimization algorithm	2019	[20]	Salps algorithm	2017
[21]	Grasshopper optimization algorithm	2017	[22]	Opposition learning algorithm	2018
[23]	Elephant herding optimization	2018	[24]	Swarm optimizer	2018
[25]	Spotted hyena optimizer	2017	[26]	Chemotherapy science algorithm	2017
[27]	Artificial flora	2018	[28]	Emperor penguin optimizer	2018
[29]	Intelligent water drops algorithm	2009	[30]	Ions motion algorithm	2015
[31]	Water wave optimization	2017	[32]	Binary whale optimization algorithm	2017
[33]	Whale optimization algorithm	2017	[34]	Vortex search algorithm	2017

of multiple factors that mimics the behavior of social insects like bees, ants, termites, wasps, birds, fish, etc. A brief overview of analytical meta-optimization algorithms is presented in Table 1.

3 Overview of Renewable Energy Resources

Despite the availability improvement of power in the worldwide, but demand has orderly overridden the supply. Therefore, non-traditional energy resources such as renewable energy have become vital desirable. Wind energy system is regrading as the most recognized resource for renewable energy and has achieved a rapid growth recently [35]. Also, by 2020, it is estimated about 12% of the worlds electricity will be available through wind generation [36]. In the following subsections, we will introduce a brief overview of the different renewable energy resources.

3.1 Wind Energy

Wind data for analysis was introduced by the National Renewable Energy Laboratory (NREL). Also, a Western wind resources data collection is a large regional study on solar and wind energy in the United States [37]. Data extracted from the Western wind resources data collection were generated in part with the help of numerical weather forecasts and sampling every 10 min every 2 km and 1.2 million network points were aggregated to 32,043 locations. Also, the graphical user interface for the west wind data permits to download the corresponding string data times and identify network points. For more details, Potter et al. [38] shows how NREL data were sampled. The two vertical (row) and horizontal (line) components of the training package are as follows; (1) Rows means the vertical axis demonstrates the grid points and the number of grid points. (2) Lines addresses the horizontal axis determines the window time.

3.2 Biomass Energy

Biomass is regarding as a one of the renewable sources of energy and is produced from plant crops. Biomass measurement is directly destructive and costly. The most recent estimates are based on remote sensing data, such as Vegetation Indicators (VIs) [39]. Also, Gnyp et al. [40] shows that the soil-adjusted vegetation index has stronger relationships with rice biomass at the conjugation stage than at the time of takeoff. In addition, Gao et al. [41] prove that the accuracy of wheat biomass is higher when using Chinese environmental satellite (HJ) images and Radar Polarimetric Parameters (RPPs) rather than using VIs or RPPs individually [42].

3.3 Solar Energy

Solar energy is now a negligible contribution to the world of electricity production mix. Accordingly to the level of solar radiation is high and stable this make the solar energy has been grown in many regions [43, 44].

3.4 Other Energy Resources

In isolated islands, the inspired energy from fossil fuels is regarding very expensive due to the cost of transportation. Therefore, to avoid this economic cost of fuel electricity, wind and solar are considered the proper solution as mentioned in [45, 46]. Also, greenhouse gases are introduced as a new source of renewable energy.

Nowadays, a considerable attention has gotten to the biodiesel production and became a hopeful alternative fuel to conventional diesel [47, 48]. Nevertheless, biomass exchange to methane in biogas plants has achieved an increasing the worldwide energy production in large-scale government aid. According to the complexity of anaerobic digestion processes, the biogas plant operation faces various difficulties [49, 50]. Various techniques have been developed for biogas prediction as presented in [51, 52].

4 Applications of Machine Learning and Meta-heuristic Algorithms

4.1 Machine Learning in Wind Energy

Various researchers have been applied the ANNs and SVMs tools to develop many techniques which are useful to solve complicated systems, various types of control, fault diagnosis and prediction/estimation specifically in the wind speed.

4.1.1 Artificial Neural Networks

In [53], a soft computing methodology of wind speed has been figured out the best fitting circulation. In order to foresee the wind speed, the authors have been utilized Adaptive Neuro-Fuzzy Inference System (ANFIS). ANFIS is utilized to optimize wind speed parameters. In [54], a hybrid forecasting model has been presented by combining GA and ANN and it is called GANN. ANN is utilized to determine the amount of power output with good accuracy. the proposed GANN model has been applied on Taiwan wind power and the results revealed that GANN has predicted the power output with a maximum difference 6.52% less than the real output values. Also, Peng et al. in [55] a short-term wind prediction for a wind farm have proposed two different prediction methods. To train the two models, wind direction, wind speed, and temperature from 40 wind turbines in Mongolia has been used for data collected. In addition, in [56], for wind speed prediction, a control algorithm using ANN is introduced. Moreover, the dynamic performance in all evaluation tests revealed that the proposed model demonstrated superior results. Also, a short-term prediction method based on Differential Evolution (DE) and Particle Swarm Optimization (PSO) has proposed using two well-known prediction models such as NN and the K-Nearest Neighbor (KNN) by Jursa and Rohrig [57]. Ten wind farms in Germany in the period 2004–2007 was used to evaluate the proposed method. The best result obtained is about 5% for RMSE measure for wind farms 9 and 42. Further, Chen et al. [58], in order to predict the average hourly wind speed, a novel technique based on Orthogonal Least-Squares (OLS) algorithm using Radial Basis Function

(RBF) has been proposed. The results showed that a good performance especially with the RBF better than Multi-Layer Perceptron (MLP).

In the same regard, Fonte et al. in [59], in order to estimate the average wind speed per hour, ANN was applied. For the proposed technical evaluation, the data collected was used in the period from 2003 to 2004 and was divided into training, validation and test data. The proposed technique was also trained using the back-propagation algorithm. In addition, Li et al. in [60], an approach based on three NNs and an Bayesian model to improve the performance of wind speed prediction has presented. The data set used is compiled from two sites in North Dakota, USA. The Bayesian model achieved better results than the NN model in terms of absolute error rate. Further, Zhao et al. in [61] introduced new models based on ANN and Numerical Weather Prediction (NWP) to evaluate the performance of wind energy predictions for China's wind farm. The proposed performance error was evaluated in the terms of the Normalized Mean Square Error(NRMSE) and the results obtained were 16.47% for NRMSE. Also, Rasit Ata in [62] provided an overview of several ANN-based applications used in wind power systems.

To find wind speed profile, a clustering algorithm using neuro-fuzzy has proposed by Mohandes et al., in [63]. Also, an approach based on fuzzy logic and ANN has proposed by Monfared et al., in [64]. Furthermore, a three-layered feed-forward ANN for short-term wind power forecasting in Portugal has presented in [65]. Gopi and Palanisamy in [66], to model the class-conditional Probability Density Function (pdf), a NN approach was proposed. In order to evaluate the proposed approach, two techniques called Gaussian based Kernel and fisher's Linear Discriminant Analysis (LDA) are used. Also, fuzzy rough sets to predict wind speed based on a Fuzzy Rough Regression Prediction Algorithm (FRRPA) has been proposed in [67]. In addition, a hybrid computing model based on multilayer perceptron network to predict wind speed has presented and tested on real-time wind data was presented by Sheela et al., in [68]. Also, in order to classify raw wind data, a Probabilistic Neural Network (PNN) model was introduced by Liu et al., in [69]. Further, a short-term prediction approach based on ANFIS for wind velocity forecasts has introduced by Ernesto et al., in [70].

4.1.2 Support Vector Machines

Recently, a lot of forecasting techniques have been presented. SVM classifier is widely known and has been applied in the renewable energy domain. This section will review some studies were performed to detect wind speed based on SVM.

Prediction techniques based on the Support Vector Regression (SVR) and Recurrence Plot (RP) have proposed by Wang et al., in [71]. In order to analyzed the wind speed series, RP is used also, SVM using kernel function RBF is applied to forecast wind speed. To optimize gamma and penalties factor of the kernel function RBF, three meta-heuristic algorithms such as Cuckoo Search (CS), PSO and Genetic Algorithm (GA) are utilized. Also, a hybrid approach for wind speed forecasting is proposed based on SVR in [72]. An autoregressive model using time delay coordinates

Table 2 Brief list for existing machine learning techniques

Ref.	Technique	Year	Ref.	Technique	Year
[76]	ANN	2009	[83]	SVM & ANN	2014
[86]	SVM	2009	[85]	ANN	2014
[88]	ANN	2009	[77]	NN	2015
[80]	SVM	2010	[84]	ANN	2015
[81]	NN	2010	[75]	Autoregressive	2016
[79]	NN	2012	[87]	SVR	2017
[78]	NN	2013	[89]	SVR	2017
[82]	SVR	2013	[90]	TSVR	2017

is applied for feature selection and hence univariate wind speed time series is used to train SVR model. To optimize the SVR parameters, GA is applied. Further, in [73], PSO and Programming Algorithm (EP) have been applied to optimize the hyper-parameters estimation problem in SVR and tested on real wind speed data collected from wind turbines of a Spanish wind farm. In addition, CS, PSO and GA have been used to proposes an ensemble approach based on SVR and NN model by Chen et al., in [74]. The experimental results revealed that the CS has achieved a superior results compared with both PSO and GA in terms of global searching capacity and convergence.

As a summary, according to all the aforementioned studies, it can be concluded a brief overview of machine learning techniques in Table 2.

4.2 *Meta-heuristic Optimization Algorithms in Renewable Energy*

a hybrid model based on PSO and SVM called PSO-SVM to improve the classification accuracy is proposed in [94]. Also, a hybrid forecasting technique based on SVR and RP to forecast short-term wind speed series is proposed by Chen et al., in [74]. Further, PSO, GA and CS to optimize both of SVR and the kernel function RBF parameters. Also, a forecasting technique using SVR to estimate the short wind speed has been proposed in [95, 96]. A forecasting model based on Whale Optimization Algorithm (WOA) was presented in [97].

In the same regard, several studies have been proposed to deal with the wind power. In [98], a long term wind speed forecasting based on hybrid SVR and WOA. Ren et al., in [99], have proposed an approach called PSO-BP that combines Back Propagation Neural Network (BPNN) based on Particle Swam Optimization (PSO-BP) with Input parameter Selection (IS). Based on Bayesian theory and structural break for short-term wind speed forecasting and tested with actual wind speed data, a time series forecasting model has been proposed by Jiang et al., in [100]. Sheng-wei and Yong in [101], a hybrid model of Artificial Bee Colony algorithm-based Relevance Vector Machine (ABCRVM) has been proposed for wind speed estimation to improve the estimation ability of Relevance Vector Machine (RVM) for wind speed. In addition, in [102] a hybrid forecasting model has proposed based on Gaussian and SVM (g-SVM) and Chaotic Particle Swarm Optimization to investigate its feasibility in forecasting regional electricity in China. Carneiro et al. [103] proposed a novel approach for estimation of Weibull parameters based on PSO. Data are collected form 80 tower are distributed in a different area in Brazilian. Also based on SVR, a long-term wind speed prediction has been presented in [87]. Finally, a wind speed forecasting approach based on PSO combined with Twin Support Vector Regression (TSVR) has been introduced in [104]. Also, [105, 106] have been introduced some meta-heuristic algorithm related to renewable energy.

Regarding to photovoltaic (PV), Soon and Low in [107] have proposed a method for a single diode PV model based on PSO to identify the unknown parameters. Moreover, various maximum power point tracking (MPPT) based on PSO have been presented in [108]. Also, Kashif et al. have used the PSO to track the MPP of PV system. The results have revealed that the proposed method regarding tracking speed and steady state oscillations superior the traditional hill climbing method as presented in [109]. Further, MPPT based on PSO have proposed by Tumbelaka and Miyatake [110]. Fu and Tong in [111] have presented several PV prediction based on PSO. Also, a new control method is proposed using PSO to extract MPPT from an PV panel has presented in [112]. A hybrid approach consists of a resistive load based on PSO has introduced in [113].

4.3 Hybrid Meta-heuristic with Machine Learning in Renewable Energy

4.3.1 Wind Energy

Table 3 summarize some previous studies in the filed of wind speed forecasting based on hybrid machine learning and meta-heuristic optimization algorithms.

4.3.2 Biomass Energy

Wang et al., [114], Random Forest (RF) regression have used for remotely estimating wheat biomass and the performance evaluation of the proposed model is compared with SVR and ANN. Also, Lpez et al., in [115], have proposed binary PSO approach for an electric system using biomass to determine the supply area and best location. Also, the strategic planning of the biomass supply chain was investigated using PSO by Izquierdo et al., in [116], and the model has been applied to the Mountain, Italy.

Table 3 Summary of existing studies for wind energy prediction

Studies	Model	Measure metric	Results
[64]	Fuzzy and ANN	Root mean square errora (RMSE)	NA
[65]	Bees Algorithm and PSO	Objective function	NA
[66]	PSO-BP	Mean absolute error (MAE)	15.51%
		Mean square error (MSE)	
		MAPE	
[67]	ANN	False positive rate (FPR)	0.08
[68]	Time series method	MAE	1.385
		MSE	2.996
		RMSE	2.996
[69]	NN and Bayes classifier	Accuracy	89.24%
[70]	Artificial Bee colony (ABC)	MAPE	19.83
[91]	ANN	Absolute relative error (ARE)	35%
[92]	ANN	Mean absolute percentage error (MAPE)	19.05%
[93]	Fuzzy rough sets	Correlation coefficients	0.8945
		RMSE	0.0318
[98]	Hybrid NN	RMSE	0.0828
[99]	PNN	MAE	14.691
		RMSE	18.584
		MAPE	29.192
[100]	ANFIS	MSE	1.569
[101]	Kernel ridge regression (KRR)	RMSE	1.7835
		MAE	1.3559
		Normalized mean square error (NMSE)	0.8775
[102]	Gaussian-SVM	MAE	18.75
		MAPE	0.135
		MSE	1.486
[103]	PSO	RMSE	NA
		Relative bias (RB)	

In addition, Sedighizadeh et al., in [117] have combined GA and PSO to determine the optimal power plant size and biomass supply area hat offer the best profitability.

Table 4 summarize some previous studies in the filed of biomass based on hybrid machine learning and meta-heuristic optimization algorithms.

4.3.3 Solar Energy

In 2016, according to [118], the solar PV has reached nearly 50%, per hour, more than 31,000 solar panels installed. A long-term analysis of four concentrating solar

Table 4 Summary of previous studies on Biomass

Studies	Model	Measure metric	Results
[114]	Emerging random forest (RF)	RMSE	45%
[115]	Binary PSO	Fitness function	76%
[116]	PSO	NA	NA
[117]	GA and PSO	NA	NA

Table 5 Summary of previous studies on solar systems

Studies	Technique	Results
[119]	CSP and SAM	As the result of SAM, the lowest for the Greece site and the highest values in Egypt
[120]	Fuzzy-neural	Obtained performance of the hybrid fuzzy-neural is 99%
[121]	SVM and FFA	RMSE is 0.6988, R^2 is 0.8024, and MAPE is 6.1768
[122]	Bat algorithms	On average improved by 14.71% over K-means

power plants using SAM software have been presented by Polo et al., in [119]. ANN is applied with system advisor model for the plant output. Also, Alireza and Majid in [120] have proposed a novel fuzzy-neural combined with MPPT method to obtain the maximum power for intelligent PV system. Also, the experimental results revealed that the proposed method superior by 2% points in comparison with the conventional methods. In [121], a hybrid approach has introduced for solar radiation prediction based on Firefly Algorithm (FA) and SVMs called SVM-FFA approach. In [122], Munshi et al., have applied the Bat algorithm to select the efficient pattern in grouping PV power data with different objective functions in clustering PV power patterns.

Table 5 summarize some previous studies in the filed of solar based on hybrid machine learning and meta-heuristic optimization algorithms.

In the past decade, the research into solar energy has gained momentum. Therefore, there are a great challenge in choosing the mathematical parameters model "extraction/estimation" of PV cells and modules for finding suitable PV parameter. Two common approaches for the PV parameter estimation techniques have been categorized; (1) analytical and (2) numerical extraction method.

As a summary, according to all the aforementioned, it can be concluded a brief overview of meta-heuristic algorithms for PV system as shown in Table 6.

Eventually, the challenge of power losses for PV ability have addressed in several research works and all have proved that the use of a power optimizer known as MPPT. Most of the MPPT techniques has been inspired to receive the same irradiance. In last few decades, several research studies on MPPT have been proposed concluded in Table 7.

Table 6 Brief list for PV systems

Ref.	Technique	Year	Ref.	Technique	Year
[123]	Parameter estimation	2017	[124]	Parameter extraction	2016
[125]	Parameter extraction	2017	[126]	Parameter extraction	2017
[127]	Parameter estimation	2016	[128]	Parameter extraction	2018
[129]	Parameter extraction	2016	[130]	Parameter identification	2016
[131]	Parameters identification	2016	[132]	Parameters extraction	2016
[133]	Parameter estimation	2016	[134]	Parameter estimation	2017
[135]	Parameter extraction	2016	[136]	Predict the performance	2017
[137]	Parameter estimation	2016	[138]	Parameter estimation	2018
[139]	Parameter estimation	2017	[140]	Parameters extraction	2017
[141]	Parameter estimation	2017	[142]	Performance evaluation	2018
[143]	Parameter estimation	2018	[144]	Parameter estimation	2018
[123]	Parameter estimation	2017	[145]	Parameters extraction	2017
[146]	Parameters estimation	2017	[147]	Parameters optimization	2016
[148]	Parameters estimation	2016	[149]	Monitoring platform for solar	2017
[150]	Determine the dynamic parameters	2016	[151]	Parameter estimation	2016
[152]	Parameters extraction	2016	[153]	Parameters identification	2017
[154]	Parameter estimation	2015	[155]	Parameter estimation	2017
[156]	Parameters identification	2017	[157]	Optimal power flow solution	2018

Table 7 Brief list for maximum power point tracking (MPPT) technology

Ref.	Technique	Year	Ref.	Technique	Year
[158]	Partial shading conditions	2017	[159]	Ant colony optimization algorithm	2017
[160]	Partial shaded conditions	2018	[161]	Particle swarm optimization	2015
[162]	Artificial bee colony algorithm	2015	[163]	Flower pollination algorithm	2017
[164]	Particle swarm optimization	2016	[165]	Particle swarm optimization	2015
[166]	Particle swarm optimization	2015	[167]	Cuckoo search	2017
[168]	Cuckoo search	2014	[169]	Bird mating optimizer	2013

4.3.4 Other Energy Resources

One of the main points to reduce the energy consumption is to decrease the energy demand. Otherwise the aforementioned discussing for the renewable energy in the previous section, several energy resources existing such as; (1) Geothermal energy [170]. (2) Hydro plants (HP) [171]. (3) the CO_2 processes of energy production

[172]. (4) Biodiesel [173]. Also, (5) Biogas which considered an important renewable energy resource based on producing of a group of gases such as methane gas and carbon dioxide as introduced in [174].

5 Challenges and Open Problems

Recently, one of the most promising renewable sources of energy, wind power has attracted increasing global attention. According to the World Wind Energy Association [175] by the end of 2010, global wind power reached 2.5% of world electricity consumption. The major challenge for wind power production is determined by the wind speed of the wind farm. Due to their dependence on temperature, atmospheric pressure, altitude and terrain, wind is irregular and evaporates. As well as, solar energy became one of the biggest resources for the renewable energy in the world and still an open problem that need more effort and researches.

The price of the wind established descend by more than 20 over the last 30 years. Therefore, Table 8 show the average wind power marketing signed in 2012 [176].

In the worldwide, wind power is now the cheapest one of renewable energy resources. In generally speaking, Figure 1 demonstrates the amount of wind power that has gained by a 10x around over the last ten years [176].

Due to the volatility of diesel prices and government pressure to reduce energy costs, hence relying on a renewable energy became an urgent demand. Regarding to Navigant Research report [177] predict the following:

Table 8 Wind cost per kWh (US) from the year 1980–2012

Year	US cents per kWh	Year	US cents per kWh	Year	US cents per kWh
1980	55	1981	51	1982	45
1983	42	1984	38	1985	32
1986	28	1987	25	1988	22
1989	18	1990	17	1991	15
1992	13	1993	12	1994	11
1995	9	1996	8	1997	7
1998	6	1999	5	2000	4
2001	3	2002	2	2003	2.5
2004	3	2005	3.5	2006	4
2007	5	2008	7	2009	8
2010	7	2011	5	2012	4

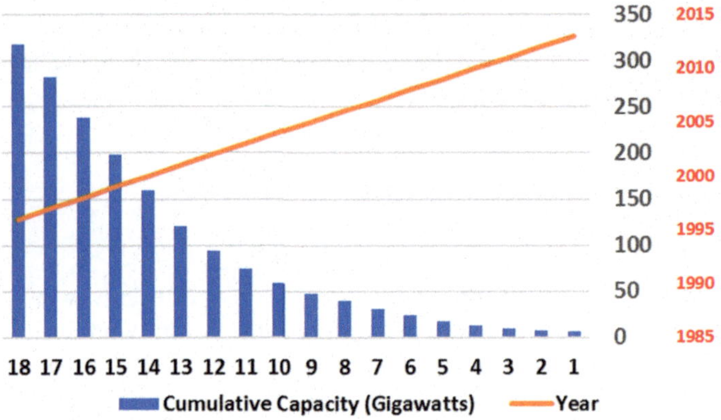

Fig. 1 Global wind power cumulative capacity

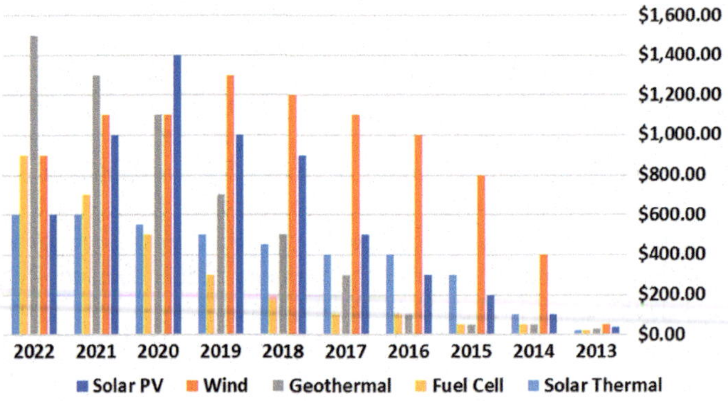

Fig. 2 Renewable energy for the mining industry revenue by technology

- Renewable energy will participate by the end of 2022 around between 5 and 8% of the worlds industry power consumption as shown in Fig. 2.
- Due to the increasing the energy resources and its low cost, the natural gas generators will be utilized widespread.
- In order to encourage widespread adoption by mining companies, energy costs must come down enough and this limited through 2022.

In addition, due to the great evolution of renewable energy prediction, therefore, almost trends and prospects can be drawn as follows:

- Study methods that improve the training algorithm such as ANN that aim at more accurate results.
- In the long and short term forecasting, the need to combine different physical and statistical models for good results.
- More researches on the practical application, not only in theory based on new mathematical methods.

- Hybridization of machine learning techniques can be developed with other meta-heuristic optimization algorithms may be vital research point.
- Improving the stability of wind generators using machine learning and meta-heuristic algorithms may be vital research point, which needs more detailed attention.

Eventually, with regard to the above discussion, this review attempts to give a clear idea of renewable energy, especially wind speed and solar system. It is clear that some of the topics that have not been identified are an urgent need for the feasibility of operational tools, such as:

- Three-dimensional measurements of wind speed and wind turbines are an urgent need.
- Machine learning techniques such as ANN can be used to predict energy for energy allocation of the country as well as improvement the performance accuracy of existing models.
- Integration between the mathematical/statistical and physical/meteorological models have been becoming a desirable task.
- The goal of a MPPT is to satisfy the precision, accuracy, speed, and robustness. Therefore, meta-heuristic optimization algorithms may regarded a superior alternative over traditional methods to achieve this goal.

6 Conclusions

Our main focus in this review was on renewable energy resources. In general, wind energy is regarding as one of the significant sources of sustainable renewable energy. To support wind energy it is important to estimate the generation of wind power. According to, the continuous inconstancy of wind resources represents a great challenge for accurate and reliable short-term forecasts of wind speed. The current state of the biogas and biomass market shows the growing need for new developments in the areas of measurement, analysis, control and optimization. All these areas are still a hot topics and need more effort and research. In addition, there are increasing demands for manufacturers of solar systems to create a sustainable energy resource with greater efficiency. In this regard, we summarize the results of many renewable energy resource technologies. This review shows that machine learning as well as meta-heuristic optimization algorithms plays important roles in renewable energy and produces high resolution results. The main findings of this review concluded as follows; (1) ANN is regarding as one of the most famous machine learning tools that has been used to forecast/estimate renewable energy systems. (2) GA and PSO are regrading the most popular meta-heuristic algorithms applied in this field. (3) Hybrid meta-heuristic algorithms with machine learning tools seems to be a more efficient.

References

1. Asif M, Muneer T (2007) Energy supply, its demand and security issues for developed and emerging economies. Renew Sustain Energy Rev 11(7):1388–1413
2. Farooq M, Ramli A, Naeem A (2015) Biodiesel production from low FFA waste cooking oil using heterogeneous catalyst derived from chicken bones. Renew Energy 76:362–368
3. Jebaraj S, Iniyan S (2006) A review of energy models. Renew Sustain Energy Rev 10(4):281–311
4. Esen H et al (2008) Performance prediction of a ground-coupled heat pump system using artificial neural networks. Expert Syst Appl 35(4):1940–1948
5. Kalogirou Soteris A (2006) Artificial neural networks in energy applications in buildings. Int J Low-Carbon Technol 1(3):201–216
6. Kalogirou Soteris A (2003) Artificial intelligence for the modeling and control of combustion processes: a review. Prog Energy Combust Sci 29(6):515–566
7. Vapnik V (2013) The nature of statistical learning theory. Springer science & business media
8. Hamad A, Houssein EH, Hassanien AE, Fahmy AA, Bhattacharyya S (2018) A hybrid gray wolf optimization and support vector machines for detection of epileptic seizure. Series in machine perception and artificial intelligence, hybrid metaheuristics, pp 197–225
9. Hassanien AE, Kilany M, Houssein EH, AlQaheri H (2018) Intelligent human emotion recognition based on elephant herding optimization tuned support vector regression. Biomed Signal Process Control 45:182–191
10. Ismail FH, Houssein EH, Hassanien AE (2018) Chaotic bird swarm optimization algorithm. In: International conference on advanced intelligent systems and informatics. Springer, Cham
11. Kaveh A, Bakhshpoori T (2016) Water evaporation optimization: a novel physically inspired optimization algorithm. Comput Struct 167:69–85
12. Ebrahimi A, Khamehchi E (2016) Sperm whale algorithm: an effective metaheuristic algorithm for production optimization problems. J Nat Gas Sci Eng 29:211e222
13. Seyedali Mirjalili SCA (2016) A sine cosine algorithm for solving optimization problems. Knowl Syst 96:120–133
14. Zheng Yu-Jun (2015) Water wave optimization: a new nature-inspired metaheuristic. Comput Op Res 55:1–11
15. Seyedali M (2016) Dragonfly algorithm: a new meta-heuristic optimization technique for solving single-objective, discrete, and multi-objective problems. Neural Comput Appl 27(4):1053–1073
16. Wang G-G, Deb S, Coelho LDS Elephant herding optimization. In: 2015 3rd international symposium on computational and business intelligence
17. Seyedali M, Mirjalili SM, Lewis A, Wolf G (2014) Optimizer. Adv Eng Softw 69:46–61
18. Sauber AM, Nasef MM, Houssein EH, Hassanien AE (2018) Parallel whale optimization algorithm for solving constrained and unconstrained optimization problems. arXiv:1807.09217
19. Hussien AG et al (2019) S-shaped binary whale optimization algorithm for feature selection. Recent Trends Signal Image Process. Springer, Singapore, pp 79–87
20. Hussien AG, Hassanien AE, Houssein EH (2017) Swarming behaviour of salps algorithm for predicting chemical compound activities. In: 2017 eighth international conference on intelligent computing and information systems (ICICIS), IEEE
21. Tharwat A et al (2017) MOGOA algorithm for constrained and unconstrained multi-objective optimization problems. Appl Intell 1–16
22. Ewees AA, Elaziz MA, Houssein EH (2018) Improved grasshopper optimization algorithm using opposition-based learning. Exp Syst Appl
23. Hassanien AE et al (2018) Intelligent human emotion recognition based on elephant herding optimization tuned support vector regression. Biomed Signal Process Control 45:182–191
24. Houssein EH, Ewees AA, ElAziz MA (2018) Improving twin support vector machine based on hybrid swarm optimizer for heartbeat classification. Pattern Recognit Image Anal 28(2):243–253

25. Gaurav D, Vijay K (2017) Spotted hyena optimizer: a novel bio-inspired based metaheuristic technique for engineering applications. Adv Eng Softw 114:48–70
26. Salmani MH, Eshghi K (2017) A metaheuristic algorithm based on chemotherapy science: CSA. J Opt
27. Long C, Wu X, Yan W (2018) Artificial flora (AF) optimization algorithm. Appl Sci 8(3):329
28. Gaurav D, Vijay K (2018) Emperor penguin optimizer: a bio-inspired algorithm for engineering problems. Knowl Syst
29. Hamed SH (2009) The intelligent water drops algorithm: a nature-inspired swarm-based optimization algorithm. Int J Bio-Inspired Comput 1(1–2):71–79
30. Behzad J, Abdolreza H, Seyedali M (2015) Ions motion algorithm for solving optimization problems. Appl Soft Comput 32:72–79
31. Kilany M, Houssein EH, Hassanien AE, Badr A (2017) Hybrid water wave optimization and support vector machine to improve EMG signal classification for neurogenic disorders. In: 2017 12th international conference on computer engineering and systems (ICCES). IEEE, pp 686–691
32. Hussien AG, Houssein EH, Hassanien AE (2017) A binary whale optimization algorithm with hyperbolic tangent fitness function for feature selection. In: 2017 eighth international conference on intelligent computing and information systems (ICICIS), IEEE
33. Ahmed MM et al (2017) Maximizing lifetime of wireless sensor networks based on whale optimization algorithm. In: International Conference on Advanced Intelligent Systems and Informatics. Springer, Cham
34. Houssein EH, Wazery YM (2017) Vortex search topology control algorithm for wireless sensor networks. Int J Intell Eng Syst 10(6):87–97
35. Bhaskar K, Singh SN (2012) AWNN-assisted wind power forecasting using feed-forward neural network. IEEE Trans Sustain Energy 3(2):306–315
36. Shu F, Liao James R, Ryuichi Y, Luonan C, Wei-Jen L (2009) Forecasting the wind generation using a two-stage network based on meteorological information. IEEE Trans Energy Conv 24(2):474–482
37. Lew D, Milligan M, Jordan G, Freeman L, Miller N, Clark K, Piwko R (2009) How do wind and solar power affect grid operations: the western wind and solar integration study. In: 8th international workshop on large scale integration of wind power and on transmission networks for offshore wind farms, pp 14–15
38. Cameron P, Debra L, Jim M, Sam C, Scott E, Eric G (2008) Creating the dataset for the western wind and solar integration study (USA). Wind Eng 32(4):325–338
39. Yuanyuan F, Guijun Y, Jihua W, Xiaoyu S, Haikuan F (2014) Winter wheat biomass estimation based on spectral indices, band depth analysis and partial least squares regression using hyperspectral measurements. Comput Electron Agric 100:51–59
40. Gnyp Martin L, Yuxin M, Fei Y, Ustin Susan L, Yu K, Yinkun Y, Shanyu H, Georg B (2014) Hyperspectral canopy sensing of paddy rice aboveground biomass at different growth stages. Field Crops Res 155:42–55
41. Shuai G, Zheng N, Ni H, Xuehui H (2013) Estimating the leaf area index, height and biomass of maize using HJ-1 and RADARSAT-2. Int J Appl Earth Observ Geoinf 24:1–8
42. Xiuliang J, Guijun Y, Xu X, Hao Y, Haikuan F, Zhenhai L, Jiaxiao S, Yubin L, Chunjiang Z (2015) Combined multi-temporal optical and radar parameters for estimating LAI and biomass in winter wheat using HJ and RADARSAR-2 data. Remote Sens 7(10):13251–13272
43. Valentina B, Michela C, Giulia F, Elena V (2012) The future prospect of PV and CSP solar technologies: an expert elicitation survey. Energy Policy 49:308–317
44. Klaus B, Philipp G, Luise H (2015) Comparing the incomparable: Lessons to be learned from models evaluating the feasibility of Desertec. Energy 82:905–913
45. Iglesias G, Carballo R (2011) Wave resource in El Hierroan island towards energy self-sufficiency. Renew Energy 36(2):689–698
46. Merino J, Veganzones C, Sanchez JA, Martinez S, Platero Carlos A (2012) Power system stability of a small sized isolated network supplied by a combined wind-pumped storage generation system: A case study in the Canary Islands. Energies 5(7):2351–2369

47. Ashnani MH, Mohammadi AJ, Hashim H, Hasani E (2014) A source of renewable energy in Malaysia, why biodiesel? Renew Sustain Energy Rev 35:244–257
48. Lee HV, Juan JC, Taufiq-Yap YH (2015) Preparation and application of binary acidbase CaOLa2 O3 catalyst for biodiesel production. Renew Energy 74:124–132
49. BDO Deutsche Warentreuhand AG, Der Biogas market nach der EEG-Novelle (The biogas market after the Renewable Energy Sources Act amendment), p 89 (2008)
50. Koppe P, Stozek A (1993) Municipal wastewaterits ingredients according to origin, composition and reactions in waste water treatment processes including biosolid. Auflage, Essen
51. Batstone Damien J, Keller J, Angelidaki I, Kalyuzhnyi SV, Pavlostathis SG, Rozzi A, Sanders WTM, Siegrist H, Vavilin VA (2002) The IWA anaerobic digestion model no 1 (ADM1). Water Sci Technol 45(10):65–73
52. Strik David PBTB, Domnanovich AM, Zani L, Braun R, Holubar P (2005) Prediction of trace compounds in biogas from anaerobic digestion using the MATLAB neural network toolbox. Environ Model Softw 20(6):803–810
53. Shamshirband S et al (2014) Survey of four models of probability density functions of wind speed and directions by adaptive neuro-fuzzy methodology. Adv Eng Softw (76):148–153
54. Kolhe M, Lin TC, Maunuksela J (2011) GA-ANN for short-term wind energy prediction. In: Power and energy engineering conference (APPEEC), 2011 Asia-Pacific. IEEE
55. Peng H, Liu F, Yang X (2013) A hybrid strategy of short term wind power prediction. Renew Energy 50:590–595
56. Kalogirou Soteris A (2000) Applications of artificial neural-networks for energy systems. Appl Energy 67(1):17–35
57. Jursa R, Rohrig K (2008) Short-term wind power forecasting using evolutionary algorithms for the automated specification of artificial intelligence models. Int J Forecast 24(4):694–709
58. Chen B et al (2009) Wind speed prediction using OLS algorithm based on RBF neural network. In: 2009 Asia-Pacific power and energy engineering conference, IEEE
59. Fonte PM, Silva GX, Quadrado JC (2005) Wind speed prediction using artificial neural networks. WSEAS Trans Syst 4(4):379–384
60. Li G, Shi J, Zhou J (2011) Bayesian adaptive combination of short-term wind speed forecasts from neural network models. Renew Energy 36(1):352–359
61. Zhao P et al (2012) Performance evaluation and accuracy enhancement of a day-ahead wind power forecasting system in China. Renew Energy 43:234–241
62. Ata R (2015) Artificial neural networks applications in wind energy systems: a review. Renew Sustain Energy Rev 49:534–562
63. Mohandes M, Rehman S, Rahman SM (2011) Estimation of wind speed profile using adaptive neuro-fuzzy inference system (ANFIS). Appl Energy 88(11):4024–4032
64. Monfared M, Rastegar H, Kojabadi HM (2009) A new strategy for wind speed forecasting using artificial intelligent methods. Renew Energy 34(3):845–848
65. Catalo JPS, Pousinho HMI, Mendes VMF (2009) An artificial neural network approach for short-term wind power forecasting in Portugal. In: 15th international conference on in intelligent system applications to power systems, ISAP '09, pp 1–5
66. Gopi ES, Palanisamy P (2015) Neural network based class-conditional probability density function using kernel trick for supervised classifier. Neurocomputing 154:225–229
67. An S, Shi H, Qinghua H, Li X, Dang Jianwu (2014) Fuzzy rough regression with application to wind speed prediction. Inf Sci 282:388–400
68. Sheela KG, Deepa SN (2013) Neural network based hybrid computing model for wind speed prediction. Neurocomputing 122:425–429
69. Liu Z, Gao W, Wan Y-H, Muljadi E (2012) Wind power plant prediction by using neural networks. In: 2012 IEEE energy conversion congress and exposition (ECCE), pp 3154–3160, IEEE
70. Prez EC, Algredo-Badillo I, Rodrguez VHG (2012) Performance analysis of ANFIS in short term wind speed prediction. IJCSI Int J Comput Sci Iss 9(3)
71. Wang J et al (2015) Short-term wind speed forecasting using support vector regression optimized by cuckoo optimization algorithm. Math Probl Eng

72. Santamaría-Bonfil G, Reyes-Ballesteros A, Gershenson C (2016) Wind speed forecasting for wind farms: a method based on support vector regression. Renew Energy 85:790–809
73. Salcedo-Sanz S et al (2011) Short term wind speed prediction based on evolutionary support vector regression algorithms. Exp Syst Appl 38(4):4052–4057
74. Chen X et al (2015) Short-term wind speed forecasting study and its application using a hybrid model optimized by cuckoo search. Math Probl Eng
75. Koivisto M, Seppnen J, Mellin I, Ekstrm J, Millar John, Mammarella Ivan, Komppula Mika, Lehtonen Matti (2016) Wind speed modeling using a vector autoregressive process with a time-dependent intercept term. Int J Electr Power Energy Syst 77:91–99
76. Catalo JPS, Pousinho HMI, Mendes VMF (2009) An artificial neural network approach for short-term wind power forecasting in Portugal. In: 15th international conference on intelligent system applications to power systems, 2009. ISAP '09, IEEE
77. Gopi ES, Palanisamy P (2015) Neural network based class-conditional probability density function using kernel trick for supervised classifier. Neurocomputing 154:225–229
78. Sheela KG, Deepa SN (2013) Neural network based hybrid computing model for wind speed prediction. Neurocomputing 122:425–429
79. Liu Z et al (2012) Wind power plant prediction by using neural networks. In: Energy conversion congress and exposition (ECCE), IEEE
80. Wu Q (2010) A hybrid-forecasting model based on Gaussian support vector machine and chaotic particle swarm optimization. Exp Syst Appl 37(3):2388–2394
81. Ren C et al (2014) Optimal parameters selection for BP neural network based on particle swarm optimization: a case study of wind speed forecasting. Knowl Syst 56:226–239
82. Douak F, Melgani F, Benoudjit N (2013) Kernel ridge regression with active learning for wind speed prediction. Appl energy 103:328–340
83. De Giorgi M et al (2014) Comparison between wind power prediction models based on wavelet decomposition with least-squares support vector machine (LS-SVM) and artificial neural network (ANN). Energies 7(8):5251–5272
84. Potonik P, Strmnik E, Govekar E (2015) Linear and neural network-based models for short-term heat load forecasting. Strojniki vestnik-J Mech Eng 61(9):543–550
85. Azad HB, Mekhilef S, Ganapathy VG (2014) Long-term wind speed forecasting and general pattern recognition using neural networks. IEEE Trans Sustain Energy 5(2):546–553
86. Kusiak A, Zheng H, Song Z (2009) Short-term prediction of wind farm power: a data mining approach. IEEE Trans Energy Conv 24(1):125–136
87. Osama S, Darwish A, Houssein EH, Hassanien AE, Fahmy AA, Mahrous A (2017) Long-term wind speed prediction based on optimized support vector regression. In: 2017 eighth international conference on intelligent computing and information systems (ICICIS), pp 191–196, IEEE
88. Monfared M, Rastegar H, Kojabadi HM (2009) A new strategy for wind speed forecasting using artificial intelligent methods. Renew energy 34(3):845–848
89. Osama S, Houssein EH, Hassanien AE, Fahmy AY (2017) Forecast of wind speed based on whale optimization algorithm. In: Proceedings of the 1st international conference on internet of things and machine learning, ACM, p 62
90. Hamad A, Houssein EH, Hassanien AE, Fahmy AA (2016) Feature extraction of epilepsy EEG using discrete wavelet transform. In: Computer engineering conference (ICENCO), 2016 12th international, IEEE, pp 190–195
91. Guo Z, Chi D, Jie W, Zhang W (2014) A new wind speed forecasting strategy based on the chaotic time series modelling technique and the Apriori algorithm. Energy Conv Manag 84:140–151
92. Sallis PJ, Claster W, Hernndez S (2011) A machine-learning algorithm for wind gust prediction. Comput Geosci 37(9):1337–1344
93. Douak F, Melgani F, Benoudjit N (2013) Kernel ridge regression with active learning for wind speed prediction. Appl Energy 103:328–340
94. Huang C-L, Dun J-F (2008) A distributed PSOSVM hybrid system with feature selection and parameter optimization. Appl Soft Comput 8(4):1381–1391

95. Hamad A, Houssein EH, Hassanien AE, Fahmy AA (2017) A hybrid EEG signals classification approach based on grey wolf optimizer enhanced SVMs for epileptic detection. In: International conference on advanced intelligent systems and informatics. Springer, Cham, pp 108–117
96. Hamad A, Houssein EH, Hassanien AE, Fahmy AA (2018) Hybrid grasshopper optimization algorithm and support vector machines for automatic seizure detection in EEG signals. In: International conference on advanced machine learning technologies and applications. Springer, Cham, pp 82–91
97. Osama S, Houssein EH, Hassanien AE, Fahmy AA (2017) Forecast of wind speed based on whale optimization algorithm. In: International conference on internet of things and machine learning (IML, 2017) Liverpool city. ACM, United Kingdom
98. Osama S, Houssein EH, Darwish A, Hassanien AE, Fahmy AA (2018) An optimized support vector regression using whale optimization for long term wind speed forecasting. Series in machine perception and artificial intelligence, hybrid metaheuristics, pp 171–196
99. Ren C, An N, Wang J, Li L, Bin Hu, Shang Duo (2014) Optimal parameters selection for BP neural network based on particle swarm optimization: a case study of wind speed forecasting. Knowl Syst 56:226–239
100. Jiang Y, Song Z, Kusiak A (2013) Very short-term wind speed forecasting with Bayesian structural break model. Renew Energy 50:637–647
101. Fei S, He Y (2015) Wind speed prediction using the hybrid model of wavelet decomposition and artificial bee colony algorithm-based relevance vector machine. Int J Electr Power Energy Syst 73:625–631
102. Wu Q (2010) A hybrid-forecasting model based on Gaussian support vector machine and chaotic particle swarm optimization. Exp Syst Appl 37(3):2388–2394
103. Carneiro TC, Melo SP, Carvalho PCM, Braga APS (2016) Particle swarm optimization method for estimation of Weibull parameters: a case study for the Brazilian northeast region. Renew Energy 86:751–759
104. Houssein EH (2017) Particle swarm optimization-enhanced twin support vector regression for wind speed forecasting. J Intell Syst De Gruyter
105. Fei S, He Y (2015) Wind speed prediction using the hybrid model of wavelet decomposition and artificial bee colony algorithm-based relevance vector machine. Int J Electr Power Energy Syst 73:625–631
106. Mahto T, Mukherjee V (2016) Evolutionary optimization technique for comparative analysis of different classical controllers for an isolated winddiesel hybrid power system. Swarm Evolut Comput 26:120–136
107. Soon JJ, Low K-S (2012) Optimizing photovoltaic model parameters for simulation. In: IEEE international industrial electronics (ISIE)
108. Azab M (2012) Optimal power point tracking for stand-alone PV system using particle swarm optimization. Int J Renew Energy Technol
109. Ishaque K, Salam Z, Amjad M, Mekhilef S (2012) An improved particle swarm optimization (PSO) based MPPT for PV with reduced steady-state oscillation. IEEE Trans Power Electron 27
110. Tumbelaka HH, Miyatake M (2010) A grid current-controlled inverter with particle swarm optimization MPPT for PV generators. World Acad Sci Eng Technol 43
111. Fu Q, Tong N (2010) A new PSO algorithm based on adaptive grouping for photovoltaic MPP prediction. In: International workshop on intelligent systems and applications, China
112. Boutasseta N (2012) PSO-PI based control of photovoltaic arrays. Int J Comput Appl
113. Ngan MS, Tan CW (2011) Multiple peaks tracking algorithm using particle swarm optimization incorporated with artificial neural network. World Acad Sci Eng Technol 58
114. Wang L, Zhou X, Zhu X, Dong Z, Guo Wenshan (2016) Estimation of biomass in wheat using random forest regression algorithm and remote sensing data. Crop J 4(3):212–219
115. Lpez PR, Jurado F, Ruiz Reyes N, Garca Galn S, Gmez M (2008) Particle swarm optimization for biomass-fuelled systems with technical constraints. Eng Appl Artif Intell 21(8):1389–1396

116. Izquierdo J, Minciardi R, Montalvo I, Robba M, Tavera M (2008) Particle swarm optimization for the biomass supply chain strategic planning. In: Proceedings of the international congress on environmental modelling and software, pp 1272–1280
117. Sedighizadeh M, Rafiei M, Hakimi A (2013) Optimizing a typical biomass fueled power plant using genetic algorithm and binary particle swarm optimization. Int J Tech Phys Probl Eng 5:15–21
118. Chen X, Bin X, Mei C, Ding Yuhan, Li Kangji (2018) Teaching learning based artificial bee colony for solar photovoltaic parameter estimation. Appl Energy 212:1578–1588
119. Polo J, Tllez FM, Tapia C (2016) Comparative analysis of long-term solar resource and CSP production for bankability. Renew Energy 90:38–45
120. Rezvani A, Gandomkar M (2017) Simulation and control of intelligent photovoltaic system using new hybrid fuzzy-neural method. Neural Comput Appl 28(9):2501–2518
121. Olatomiwa L et al (2015) A support vector machine firefly algorithm-based model for global solar radiation prediction. Solar Energy 115:632–644
122. Munshi AA, Mohamed Yasser A-RI (2017) Comparisons among Bat algorithms with various objective functions on grouping photovoltaic power patterns. Solar Energy 144:254–266
123. Et-torabi K et al (2017) Parameters estimation of the single and double diode photovoltaic models using a GaussSeidel algorithm and analytical method: a comparative study. Energy Conv Manage 148:1041–1054
124. Tong NT, Pora W (2016) A parameter extraction technique exploiting intrinsic properties of solar cells. Appl Energy 176:104–115
125. Patel SJ, Panchal AK, Kheraj V (2014) Extraction of solar cell parameters from a single currentvoltage characteristic using teaching learning based optimization algorithm. Appl Energy 119:384–393
126. Prasanth RJ, Sudhakar Babu T, Rajasekar N (2017) A comprehensive review on solar PV maximum power point tracking techniques. Renew Sustain Energy Rev 67:826–847
127. Nassar-Eddine I et al (2016) Parameter estimation of photovoltaic modules using iterative method and the Lambert W function: a comparative study. Energy Conv Manage 119:37–48
128. Gao X et al (2018) Parameter extraction of solar cell models using improved shuffled complex evolution algorithm. Energy Conv Manage 157:460–479
129. Babu TS et al (2016) Parameter extraction of two diode solar PV model using fireworks algorithm. Solar Energy 140:265–276
130. Guo L et al (2016) Parameter identification and sensitivity analysis of solar cell models with cat swarm optimization algorithm. Energy Conv Manage 108:520–528
131. Chen X et al (2016) Parameters identification of solar cell models using generalized oppositional teaching learning based optimization. Energy 99:170–180
132. Allam D, Yousri DA, Eteiba MB (2016) Parameters extraction of the three diode model for the multi-crystalline solar cell/module using Moth-Flame optimization algorithm. Energy Conv Manage 123:535–548
133. Saad NH, El-Sattar AA, Mansour AE-AM (2016) Improved particle swarm optimization for photovoltaic system connected to the grid with low voltage ride through capability. Renew Energy 85:181–194
134. Fathy A, Rezk H (2017) Parameter estimation of photovoltaic system using imperialist competitive algorithm. Renew Energy 111:307–320
135. Ali EE et al (2016) Parameter extraction of photovoltaic generating units using multi-verse optimizer. Sustain Energy Technol Assess 17:68–76
136. Li W et al (2017) A coupled optical-thermal-electrical model to predict the performance of hybrid PV/T-CCPC roof-top systems. Renew Energy 112:166–186
137. Chin VJ, Salam Z, Ishaque K (2016) An accurate modelling of the two-diode model of PV module using a hybrid solution based on differential evolution. Energy Conv Manage 124:42–50
138. Louzazni M et al (2018) Metaheuristic algorithm for photovoltaic parameters: comparative study and prediction with a firefly algorithm. Appl Sci 8(3):339

139. Derick M et al (2017) An improved optimization technique for estimation of solar photovoltaic parameters. Solar Energy 157:116–124
140. Abbassi A et al (2017) An improved single-diode model parameters extraction at different operating conditions with a view to modeling a photovoltaic generator: a comparative study. Solar Energy 155:478–489
141. Elena CL, Saul PO, Hernandez M, Bandarra FEP (2017) Comparison of four methods for parameter estimation of mono-and multi-junction photovoltaic devices using experimental data. Renew Sustain Energy Rev
142. Baig H et al (2018) Conceptual design and performance evaluation of a hybrid concentrating photovoltaic system in preparation for energy. Energy 147:547–560
143. Jordehi AR (2018) Enhanced leader particle swarm optimisation (ELPSO): an efficient algorithm for parameter estimation of photovoltaic (PV) cells and modules. Solar Energy 159:78–87
144. Xu S, Wang Y (2017) Parameter estimation of photovoltaic modules using a hybrid flower pollination algorithm. Energy Conv Manage 144:53–68
145. Lin P et al (2017) Parameters extraction of solar cell models using a modified simplified swarm optimization algorithm. Solar Energy 144:594–603
146. Kler D et al (2017) PV cell and module efficient parameters estimation using evaporation rate based water cycle algorithm. Swarm Evolut Comput 35:93–110
147. Barth N et al (2016) PV panel single and double diode models: optimization of the parameters and temperature dependence. Solar Energy Mater Solar Cells 148:87–98
148. Li W et al (2016) Six-parameter electrical model for photovoltaic cell/module with compound parabolic concentrator. Solar Energy 137:551–563
149. Rohit AK et al (2017) Virtual lab based real-time data acquisition, measurement and monitoring platform for solar photovoltaic module. Res Eff Technol 3(4):446–451
150. Cotfas DT, Cotfas PA, Kaplanis S (2016) Methods and techniques to determine the dynamic parameters of solar cells. Renew Sustain Energy Rev 61:213–221
151. Jordehi AR (2016) Parameter estimation of solar photovoltaic (PV) cells: a review. Renew Sustain Energy Rev 61:354–371
152. Humada AM et al (2016) Solar cell parameters extraction based on single and double-diode models: a review. Renew Sustain Energy Rev 56:494–509
153. Yu K, Chen X, Wang X, Wang Z (2017) Parameters identification of photovoltaic models using self-adaptive teaching-learning-based optimization. Energy Conv Manage 145:233–246
154. Alam DF, Yousri DA, Eteiba MB (2015) Flower pollination algorithm based solar PV parameter estimation. Energy Conv Manage 101:410–422
155. Oliva D, El Aziz MA, Hassanien AE (2017) Parameter estimation of photovoltaic cells using an improved chaotic whale optimization algorithm. Appl Energy 200:141–154
156. Yu K et al (2017) Parameters identification of photovoltaic models using self-adaptive teaching-learning-based optimization. Energy Conv Manage 145:233–246
157. Attia A-F, El Sehiemy RA, Hasanien HM (2018) Optimal power flow solution in power systems using a novel Sine-Cosine algorithm. Int J Electric Power Energy Syst 99:331–343
158. Prasanth RJ, Rajasekar N (2017) A new global maximum power point tracking technique for solar photovoltaic (PV) system under partial shading conditions (PSC). Energy 118:512–525
159. Titri S et al (2017) A new MPPT controller based on the Ant colony optimization algorithm for photovoltaic systems under partial shading conditions. Appl Soft Comput 58:465–479
160. Chaieb H, Sakly Anis (2018) A novel MPPT method for photovoltaic application under partial shaded conditions. Solar Energy 159:291–299
161. Mirhassani SM et al (2015) An improved particle swarm optimization based maximum power point tracking strategy with variable sampling time. Int J Electric Power Energy Syst 64:761–770
162. Soufyane Benyoucef A et al (2015) Artificial bee colony based algorithm for maximum power point tracking (MPPT) for PV systems operating under partial shaded conditions. Appl Soft Comput 32:38–48

163. Diab AAZ, Rezk H (2017) Global MPPT based on flower pollination and differential evolution algorithms to mitigate partial shading in building integrated PV system. Solar Energy 157:171–186
164. Chao K-H, Lin Y-S, Lai U-D (2015) Improved particle swarm optimization for maximum power point tracking in photovoltaic module arrays. Appl Energy 158:609–618
165. Babu TS, Rajasekar N, Sangeetha K (2015) Modified particle swarm optimization technique based maximum power point tracking for uniform and under partial shading condition. Appl Soft Comput 34:613–624
166. Shi J, Zhang W, Zhang Y, Xue F, Yang T (2015) MPPT for PV systems based on a dormant PSO algorithm. Electric Power Syst Res 123:100–107
167. Rezk H, Fathy A, Abdelaziz AY (2017) A comparison of different global MPPT techniques based on meta-heuristic algorithms for photovoltaic system subjected to partial shading conditions. Renew Sustain Energy Rev 74:377–386
168. Ahmed J, Salam Z (2014) A Maximum Power Point Tracking (MPPT) for PV system using Cuckoo Search with partial shading capability. Appl Energy 119:118–130
169. Askarzadeh A, Rezazadeh A (2013) Extraction of maximum power point in solar cells using bird mating optimizer-based parameters identification approach. Solar Energy 90:123–133
170. Jiang P, Li X, Ruina X, Zhang F (2016) Heat extraction of novel underground well pattern systems for geothermal energy exploitation. Renew Energy 90:83–94
171. Martnez-Lucas G, Sarasa JI, Snchez-Fernndez J, Wilhelmi JR (2016) Frequency control support of a wind-solar isolated system by a hydropower plant with long tail-race tunnel. Renew Energy 90:362–376
172. Boubaker K, Colantoni A, Marucci A, Longo L, Gambella Filippo, Cividino Sirio, Gallucci Francesco, Monarca Danilo, Cecchini Massimo (2016) Perspective and potential of CO_2: a focus on potentials for renewable energy conversion in the Mediterranean basin. Renew Energy 90:248–256
173. Alves JCL, Henriques CB, Poppi RJ (2012) Determination of diesel quality parameters using support vector regression and near infrared spectroscopy for an in-line blending optimizer system. Fuel 97:710–717
174. Wolf C, McLoone S, Bongards M (2008) Biogas plant optimization using genetic algorithms and particle swarm optimization, pp 244–249
175. Gsanger S, Pitteloud Jean-Daniel (2012) World wind energy report 2011. World Wind Energy Association, Bonn, Germany
176. Naam R (2013) The infinite resource: the power of ideas on a finite planet, UPNE
177. Hardcastle JL Mining industry slow to adopt renewable energy tech, 28 Oct 2013. http://www.energymanagertoday.com/renewables-in-the-mining-industry-draft-096453/

Design of a Supervisory Control System Based on Fuzzy Logic for a Hybrid System Comprising Wind Power, Battery and Ultracapacitor Energy Storage System

Raúl Sarrias-Mena, Luis M. Fernández-Ramírez, Carlos Andrés García-Vázquez and Francisco Jurado

Abstract Hybrid configurations involving renewable energies and storage devices pose certain challenges regarding their energy management strategies, such as the intermittent and fluctuating power generation from renewable sources, the time-varying available energy in the storage systems, or their maximum charge and discharge limitations. Observing these aspects is mandatory in order to develop a smart energy management strategy within the hybrid system. This chapter presents a control strategy for the coordinated operation of a wind power generator and two different energy storage devices. The proposed control scheme is based on fuzzy logic to monitor the state of charge of the storage systems, while defining their power references to comply with an imposed grid demand. The control strategy has been evaluated through simulation under different operating conditions, proving a satisfactory regulation of the monitored parameters and an adequate supply of the grid requirements.

R. Sarrias-Mena
Department of Engineering in Automation, Electronics and Computer Architecture & Networks, Research Group in Electrical Technologies for Sustainable and Renewable Energy (PAIDI-TEP-023), University of Cádiz, 11202 EPS Algeciras, Algeciras (Cádiz), Spain
e-mail: raul.sarrias@uca.es

L. M. Fernández-Ramírez (✉) · C. A. García-Vázquez
Department of Electrical Engineering, Research Group in Electrical Technologies for Sustainable and Renewable Energy (PAIDI-TEP-023), University of Cádiz, 11202 EPS Algeciras, Algeciras (Cádiz), Spain
e-mail: luis.fernandez@uca.es

C. A. García-Vázquez
e-mail: carlosandres.garcia@uca.es

F. Jurado
Department of Electrical Engineering, Research Group in Research and Electrical Technology (PAIDI-TEP-152), University of Jaen, 23700 EPS Linares, Linares (Jaen), Spain
e-mail: fjurado@ujaen.es

R.-E. Precup et al. (eds.), *Advanced Control and Optimization Paradigms for Wind Energy Systems*, Power Systems, https://doi.org/10.1007/978-981-13-5995-8_8

1 Introduction

The increasing penetration of renewable sources poses certain inconveniences. For instance, the intermittency and variability of wind speed may lead to undesired fluctuations of power, voltage or frequency in power systems [2, 4, 15]. Large-scale ESSs are often presented as an adequate alternative to mitigate the risks of a growing participation of renewable energy generation [4, 38]. Detailed reviews of the most interesting ESSs potentially applicable for such tasks have been carried out [8, 11, 13]. It is frequently highlighted a clearly differentiated performance among the storage technologies, what makes them more or less valid for a specific application. In this sense, the ability to respond quickly to fluctuating charge/discharge cycles has been attributed to UCs [10, 16]. Regarding electrochemical batteries, a slower response compared with the UCs has been noted in the literature [10]. Nevertheless, longer charge or discharge cycles can be achieved with BESSs [12]. The main qualities of both ESSs have been regarded to design a control strategy that enhances the operation of the WECS in the hybrid scheme proposed in this chapter.

An additional option to mitigate the uncertainty of wind power generation is the use of wind speed forecasting techniques to estimate the future power production of a wind farm [40]. Such studies can be carried out with different time spans to pursue different objectives, such as daily scheduling [14, 22], or hourly forecasting for day-ahead predictions [26]. The concept presented in [26] can be seen as a valid option for the wind speed prediction considered in this configuration. Nonetheless, the aim of this work is to develop a control strategy that uses this prediction as a known parameter. Therefore, a forecasting algorithm is not implemented herein.

The feasible alternatives regarding the design of a SCS in a hybrid system comprising several power sources are numerous and varied. In this sense, intelligent control techniques have a key role to play. A hybrid system using hydrogen fuel cells as the main power source to supply domestic loads is presented in [18]. The hybrid scheme includes a UC as an auxiliary source to cover the quick changes in demand that the hydrogen system cannot deliver due to its slower dynamics. The authors designed a power control algorithm to manage the different power flows within the hybrid system. A similar approach is presented in [24], where a gas-fuelled microturbine is added to ensure continuous power supply. Wind power generation is added to this hybrid configuration as the main power source in [19]. Again, implementing an adequate control management system is crucial to achieve a satisfactory coordinated operation of several power sources. A DC microgrid with wind power and ESSs is evaluated in [17]. Power from the different sources is managed in [17] using fuzzy logic to provide an optimal operation of the system and fulfil load demand. Furthermore, the SOC of the battery is also considered as an input to the fuzzy controller. A supervisory controller based on fuzzy logic for a hybrid system with wind power and ESS is also used in [23]. The ESS improves the power quality of the wind power generation. Nonetheless, the hybrid system proposed does not have the ability to follow load variations. An advanced droop control with fuzzy rules for an isolated microgrid is presented in [30]. The authors claim a more accurate, adaptive and robust control of power sharing among several distributed generators in the microgrid due to the inclusion of the fuzzy method. An isolated microgrid with renewable energy, diesel generator and energy storage is also studied in [5]. The frequency of the AC power supply is regulated against changes in load or generation through a fuzzy gain scheduling of a PID controller, achieving a better transient response compared to the classic PID controller structure. However, the SOC of the storage devices was not monitored in the control strategy proposed. In the literature, fuzzy controllers have also been applied to WECS to regulate parameters such as DC voltage [31], pitch angle [20, 25, 35], rotor speed [7], or active and reactive power [27, 32]. Nevertheless, in such cases the focus was particularly put on the operation of the WECS, rather than using the fuzzy controller as a smart decision-maker in a hybrid system with various power sources.

In this chapter, a hybrid system comprising a WECS and two different ESSs is evaluated. The main objective of this hybrid scheme is to provide a controlled active power to the grid. Due to the fluctuating and uncontrolled power generation of the WECS, the ESSs will be responsible for absorbing or providing the power imbalances between generation and demand. Therefore, it becomes crucial to monitor the SOC of both ESSs. For this purpose, an intelligent SCS has been designed. The SCS is based on fuzzy logic to generate a compensating active power term that is exchanged between both ESSs to regulate their SOC according to the instantaneous operating conditions. This hybrid system has been tested under simulation and compared to other alternatives.

2 Configuration of the Hybrid System

A wind turbine coupled to a DFIG is the primary power source in this configuration. Additionally, two different storage technologies are considered. BESS and UC have been chosen for their distinctive characteristics to develop two complementary duties in the hybrid system. Due to their fast response and large rated power, UCs can achieve a satisfactory performance when operating under quick and frequent charge/discharge cycles with potentially large power boosts or sags. Nonetheless, they are less adequate to store or provide a large amount of power for a long time. On the other hand, such tasks can be successfully carried out by electrochemical batteries, whereas their response to large and fast power peaks is poorer than that of the UCs. The qualities of each of the storage technologies considered have been taken into account to define their role in the hybrid system. The decision between the use of BESS or UC is made by the active power SCS specifically implemented for this purpose. Furthermore, the availability of these two ESSs has also been used to develop a smart control of their SOC.

Power conversion stages are needed to accommodate the BESS and the UC in the hybrid system. A bidirectional DC/DC converter allows regulating their active power exchange. Then, AC/DC converters must be used to connect the bidirectional

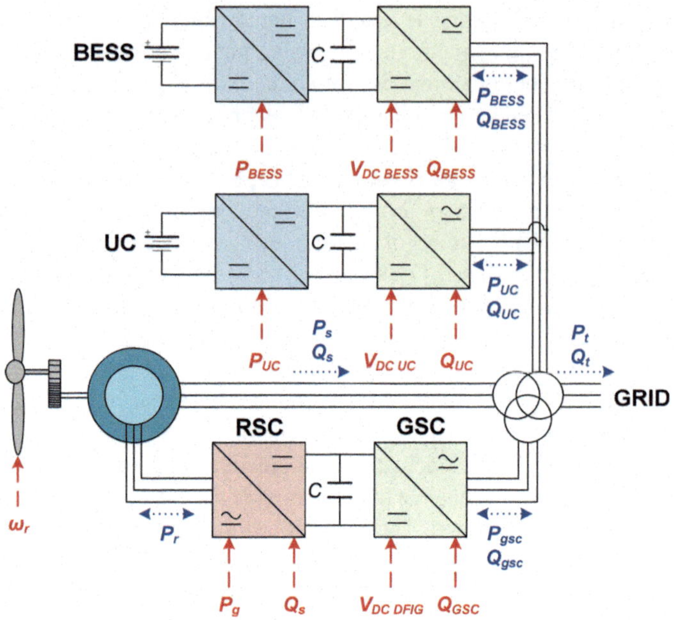

Fig. 1 Schematic of the hybrid system configuration

DC/DC converter to the AC output of the DFIG. The DC link voltage and their reactive power exchange is controlled in these converters.

A complete scheme of the hybrid configuration is shown in Fig. 1, where the control parameters are also illustrated.

3 Modelling of the WECS

The WECS is the prime power source in the hybrid scheme. The power harnessed in the wind turbine is typically given by (1) [21].

$$P_{WT} = \frac{\rho}{2} \cdot A_r \cdot u^3 \cdot C_p \tag{1}$$

The factor C_p is a dimensionless parameter that defines the fraction of the incoming wind energy that can be collected by the wind turbine. It depends on the aerodynamics of the wind turbine, and it can be expressed in a set of curves as a function of the blade pitch angle and the tip-speed ratio.

The torque of the wind turbine can be calculated dividing the wind turbine power by its angular speed. This torque is transmitted along the powertrain to the rotor of the electric generator. This mechanical system can be modelled by two masses linked through a flexible coupling characterized by its stiffness and damping. This is the well-known two-masses model [36], which is given by (2)–(4).

$$T_{WT} - T_{pt} = J_{WT} \cdot \frac{d\omega_{WT}}{dt} \tag{2}$$

$$T_{pt} = D_{pt}\left(\omega_{WT} - \omega_r\right) + K_{pt} \int \left(\omega_{WT} - \omega_r\right) dt \tag{3}$$

$$T_e - T_{pt} - \omega_r \cdot F = J_r \cdot \frac{d\omega_r}{dt} \tag{4}$$

The wind turbine drives a DFIG that has been modelled through the equations of the stator and the rotor voltages expressed in a reference frame where the d and the q axes rotate at synchronous speed. These voltages can be expressed as in (5)–(6) [28]:

$$\begin{aligned} u_{ds} &= R_s \cdot i_{ds} + \frac{d}{dt}\varphi_{ds} - \omega \cdot \varphi_{qs} \\ u_{qs} &= R_s \cdot i_{qs} + \frac{d}{dt}\varphi_{qs} + \omega \cdot \varphi_{ds} \end{aligned} \tag{5}$$

$$\begin{aligned} u_{dr} &= R_r \cdot i_{dr} + \frac{d}{dt}\varphi_{dr} - (\omega - \omega_r) \cdot \varphi_{qr} \\ u_{qr} &= R_r \cdot i_{qr} + \frac{d}{dt}\varphi_{qr} + (\omega - \omega_r) \cdot \varphi_{dr} \end{aligned} \tag{6}$$

The electromagnetic torque used as an input in the powertrain model can be derived from (7).

$$T_e = 1.5 \cdot p\left(\varphi_{ds} \cdot i_{qs} - \varphi_{qs} \cdot i_{ds}\right) \tag{7}$$

The rotor windings of the DFIG are fed through a back-to-back power converter based on VSC. It provides the capacity to adjust the frequency of the rotor currents by decoupling the mechanical speed at the rotor shaft and the electrical frequency of the grid [3], thus allowing the regulation of the wind turbine rotating speed. A switched model has been used for the two-level VSCs, where the three-phase AC voltage can be derived from (8) [42]:

$$V_a = \frac{V_{DC}}{2}\left(s_1 - s_4\right),\ V_b = \frac{V_{DC}}{2}\left(s_2 - s_5\right),\ V_c = \frac{V_{DC}}{2}\left(s_3 - s_6\right) \tag{8}$$

Based on the power balance that must be maintained in the lossless converter, the performance of the converter at the DC side can be obtained from (9):

$$V_{DC} \cdot i_{DC} = V_a \cdot i_a + V_b \cdot i_b + V_c \cdot i_c \tag{9}$$

Three control systems are implemented in the WECS. First, the pitch angle of the blades is responsible for limiting the angular speed of the wind turbine to a reference value. For above rated wind speeds, the pitch angle increases to reduce the power capture. Subsequently, the angular speed of the rotor remains regulated at the reference value. On the other hand, the pitch angle is set at 0° for below rated wind speeds, thus optimizing the power extracted from the wind. This performance can be achieved with a PI on a feedback control loop, as depicted in Fig. 2, where the mechanical limitations of the actuators, such as maximum speed of variation and maximum/minimum angle, have been included.

The control strategy implemented on the RSC allows an independent regulation of the active and reactive power generation through the dq components of the rotor voltage. Hence, P_g is regulated using a cascaded PI control loop. The reference is set using the optimal speed vs. power curve of the WECS, receiving the angular speed as an input. The reactive power exchange through the stator is also controlled in the RSC, and it can be expressed as a function of the stator operating conditions. This structure is illustrated in Fig. 3. As seen, the voltage references are obtained from the controller of the rotor currents in the inner loops, while the outer loops are responsible for setting the current references. Finally, a PWM generator provides the trigger signals to the VSC model.

In the GSC, the d component of voltage controls the DC voltage, whereas the q component regulates the reactive power exchange through the converter. The config-

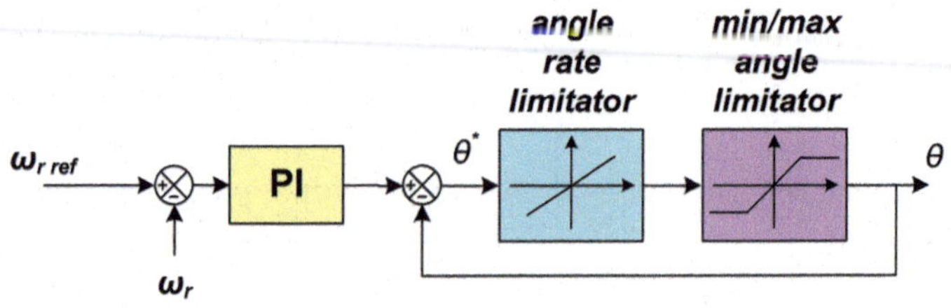

Fig. 2 Pitch angle control scheme

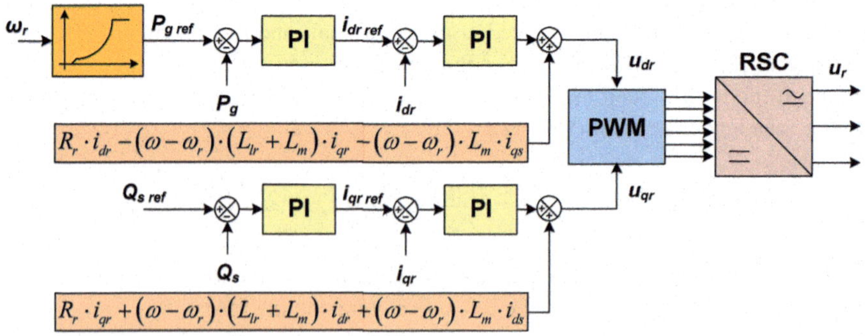

Fig. 3 RSC control scheme

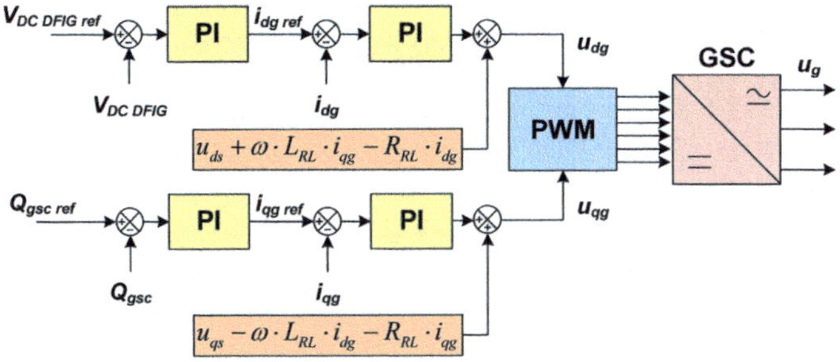

Fig. 4 GSC control scheme

uration of this controller is depicted in Fig. 4. As seen, the control strategy consists of two cascaded PI control loops, one for each control parameter. The GSC is responsible for controlling the DC bus voltage close to a constant reference.

4 Modelling of the ESSs

The BESS model consists of the updated version of the Shepherd model proposed in [39], which can reproduce the dynamic performance of different battery types. Hence, E_{batt} is a function of the nominal parameters of the device and the operating conditions, as calculated in Fig. 5. Then, U_{batt} is computed with a series resistor that models the internal losses of the device [39].

Another key aspect of the battery model is the calculation of its SOC. This parameter is continuously calculated by (10) [39].

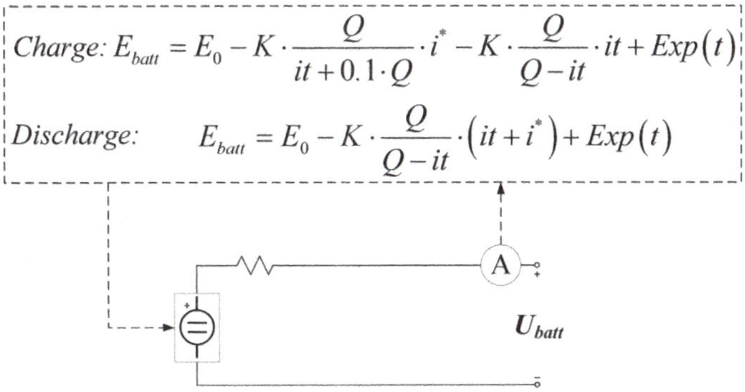

Fig. 5 Scheme of the BESS model

$$SOC_{BESS}\,(\%) = 100\left(1 - \frac{\int i_{batt}\,(t)\,dt}{Q}\right) \tag{10}$$

The parameters of the model have been obtained from a commercial reference. The discharge curves of product D121000BD from *Discover Battery* [9] have been used to extract the necessary parameters of the battery model. In the proposed configuration, 288 cells are connected in series to reach a nominal voltage of 576 V. Three of these branches are connected in parallel for an energy rating of 100 kWh. These values show the actual capacity of the BESS considered in this hybrid configuration.

Regarding the UC, a reduced UC model has been found a valid alternative in previous studies of the electrical performance of hybrid systems with wind energy [1, 41]. Therefore, the UC model implemented consists of an ideal capacitor in series with a resistor that models the internal losses of the UC, and SOC_{UC} is given by (11).

$$SOC_{UC}\,(\%) = 100 \cdot \frac{v_{UC}\,(t)}{V_{UC\ rated}} \tag{11}$$

The parameters of the UC model have been obtained for validation purposes from *Maxwell Technologies'* product BMOD0063 P125 [29]. Parallel and series association of several UC modules is necessary to achieve a 2.5 kWh capacity, with a fully-charged voltage of 625 V.

For both ESSs, a DC/DC converter is needed to connect the storage device at the PCC through the corresponding inverters. The configuration chosen for this purpose is the classical bidirectional buck-boost converter widely used in renewable energy applications [33, 37]. This converter presents two electronic switches based on IGBT and diodes, as well as an inductor and capacitors to reduce current and voltage ripple respectively. An averaged model of this DC/DC converter has been implemented on the hybrid configuration.

The DC/DC converters are controlled to track the active power reference that the SCS defines for each ESS. In this context, PI controllers are able to provide a satisfactory performance. The control scheme developed for the ESS DC/DC converters consists of a single loop based on a PI controller that generates the duty cycle for the converter model, as shown in Fig. 6.

Finally, a DC/AC converter is needed to complete the conversion of the power exchanged between each ESS and the PCC. The GSC converter model and control strategy have been used also for these DC/AC converters, and they have been described in Sect. 3.

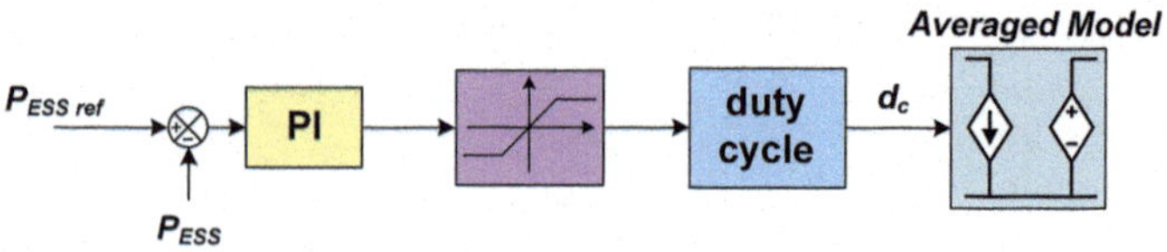

Fig. 6 DC/DC converter control scheme

5 Design of the SCS

A proper coordination between the three power sources in the hybrid system is crucial in order to achieve a satisfactory operation. The WECS is considered the main power source. The DFIG is controlled under a MPPT strategy, which can be developed autonomously by the DFIG controllers without the participation of the SCS, since the power reference is calculated through the speed vs. power characteristic curve of the WECS. The measured rotating speed of the wind turbine is used as an input to the optimal power curve. Then, the maximum power that can be generated at that speed is obtained from the optimal curve and used as a reference in the RSC controller, as shown in Fig. 3. This controller is responsible for regulating the total power generation of the DFIG to this optimal reference, thus achieving the MPPT operation of the WECS.

The secondary power sources are the energy storage devices selected. Two different technologies contribute as an active power backup for the system. This offers a wide range of possibilities regarding their operation and use. In this case, the BESS supports the WECS with active power exchanges (either storage or supply) maintained for longer intervals compared to the UC. On the other hand, the UC covers fast peaks of active power that switch frequently from generation to consumption and vice versa. The active power SCS has been designed under these premises.

According to the previous description, the active power SCS is structured into two stages. The first stage sets the primary references for the BESS and the UC. The distinctive performance and characteristics of both ESSs are observed here. A second stage of the SCS includes a fuzzy-logic controller that supervises the SOC of the ESSs and modifies the primary active power references accordingly. The aim is to use the BESS to avoid SOC_{UC} exceeding certain operating boundaries. Similarly, the UC can support SOC_{BESS} when necessary. Figure 7 shows a schematic of the described two-stages SCS for the active power references and SOC regulation. As seen, three parameters are used as inputs on the first stage. The total active power provided by the hybrid system must equal the active power demanded by the grid. In traditional power plants, this demand is typically set according to the requirements of the TSO or to commercial agreements on a day-ahead market. Thanks to the ESSs, the hybrid system can operate similarly to traditional power plants responding to changes in the power demanded by the TSO. Obviously, the limited capacity of the ESSs poses certain boundaries to the flexible generation of the hybrid system, but with these storage devices, the WECS undoubtedly gains the capacity to manage energy with higher efficiency. Since P_{dem} corresponds to the orders of the TSO, and could be agreed in a day-ahead market, this parameter can be expected to remain stable for relatively long periods, experience ramps, and be set either above or below the rated power of the WECS. Additionally, it is not expected to show quick fluctuations.

Another input signal of the SCS is $P_{prediction}$. The power generation of a wind turbine can be estimated using a wind speed forecast and the speed versus power curve of the generator [26]. Wind speed forecasting techniques for different time spans have been addressed in the literature and are not within the scope of this

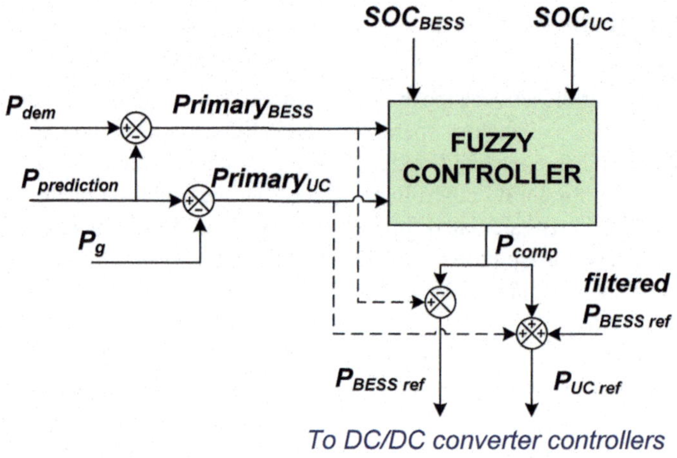

Fig. 7 Structure of the proposed SCS

study. Therefore, $P_{prediction}$ is considered a known value and used as an input in the SCS. With this power forecast available, the wind farm operator can provide a future power production estimation and participate in electric markets similarly to traditional power plants. Certainly, this power prediction is subject to inaccuracies due to forecasting errors on the wind speed. Nonetheless, when kept within certain SOC and capacity limits, the ESSs can participate to guarantee that the total generation of the hybrid power systems equals the time-ahead power prediction. The result is not only an instantly controlled active power output, but also a power generation that matches the commitment agreed in time-ahead electric markets. In this sense, the hybrid configuration of WECS with ESS reproduces the operation of traditional power generators. The power prediction signal can remain stable for long periods. For instance, a wind speed forecast with hourly average values is presented in [26]. Under such conditions, the active power prediction can be considered constant for intervals of one-hour duration, and that is the time variation used herein for this parameter.

The third signal inputted to the SCS is P_g, calculated as the sum of the stator and rotor active power. This parameter will vary with the wind speed fluctuations for winds below the rated value. Hence, P_g can present relatively fast and frequent fluctuations due to instantaneous wind sag and gusts.

The first stage of the active power SCS sets $Primary_{BESS}$ and $Primary_{UC}$ according to the characteristics of the input signals. As described in the preceding paragraphs, the grid demand could be set by the TSO either higher or lower than P_g, and remain constant for long periods. Since it is not necessarily linked to the wind speed conditions, this demand can be notably different from the active power prediction, which shows an hourly variation pattern. This means that a significant amount of energy must be delivered or absorbed by an ESS during time intervals ranging from minutes to hours in order to supply the grid with the demanded active power. Such performance is suitable for the BESS, which presents a higher capacity than UCs. Subsequently, $Primary_{BESS}$ is defined as the difference between P_{dem} and $P_{prediction}$. Therefore, the BESS gives the hybrid system the ability to adapt its active power generation to an external requirement. Nevertheless, the instantaneous SOC and maximum capacity of the BESS are boundaries to this improved performance.

The primary active power reference for the BESS relies on a power generation prediction. Within the time span considered for the power prediction there will be deviations of the actual generation above and below this mean value because of wind speed gusts and sags. These deviations can show a rapidly changing profile, due to the inherent characteristics of wind. The fast response to power peaks of the UC is adequate for this duty, and they can be used to compensate these rapid fluctuations. Therefore, $Primary_{UC}$ is calculated in the SCS as the difference between $P_{prediction}$ and P_g. Subsequently, the power generation resulting from the WECS plus the UC equals the power prediction forecasted time ahead, and it can be considered that the power prediction used to calculate the BESS primary reference is always true, subject to the limitations of the UC capacity and SOC.

Ideally, the operation of the hybrid system given the primary active power references calculated as described is analogous to that of a traditional power plant, where the hourly active power production can be anticipated to participate in day-ahead electric markets. The role of the two ESSs is clearly differentiated here. The UC ensures that the power prediction is accurately provided the day after, whereas the BESS allows time shifting the WECS power generation. Nonetheless, this ideal performance is constrained by the limited capacity of the ESSs and their instantaneous SOC variation. The first limitation can be addressed by evaluating the size of the ESS. This has been studied in the literature and it is not within the scope of this work. Alternatively, a smart control of the SOC of the BESS and the UC is proposed here as a strategy to enhance the contribution of both ESSs in the hybrid system and provide a controlled power supply.

BESSs and UCs present maximum and minimum SOC boundaries that should not be passed in normal operation in order to avoid damaging these storage devices. In the control scheme proposed herein, the SOC of both ESSs is supervised in the second stage of the active power SCS by means of a fuzzy logic-based controller. The smart regulation of the SOC is performed through the calculation of an additional compensating active power term that modifies the primary active power references. This compensating term is conceived as an active power that virtually flows from one ESS to the other. Since the active power balance between the whole hybrid system and the grid must be ensured at the PCC, all the additional power injected by the UC that breaks this power balance must be absorbed by the BESS to recover the power balance. The same applies to an additional power injection by the BESS that breaks the power balance. Hence, P_{comp} can be considered a direct active power exchange between the UC and the BESS. The main task of the second stage of the SCS is to compute the magnitude and direction of this power exchange to compensate the SOC variations of both ESSs and reduce the risk of exceeding their maximum and minimum recommended values.

The compensating power term P_{comp} is calculated by a controller based on fuzzy logic as illustrated in Fig. 7. The inputs for this fuzzy logic controller are the SOC and primary active power references of both ESSs, and P_{comp} is obtained as output. A positive value of P_{comp} indicates an additional power injection of the UC to support the SOC recovery of the BESS. Analogously, a negative value indicates that the BESS supports the UC. Implementing a fuzzy controller on the SCS requires the definition of the MFs for inputs and outputs, as well as the rules that describe the input/output relations. Usually, these parameters are derived based on the expertise of the designer and on empirical experiences. Under this perspective, the configuration of the fuzzy controller has been designed as described below:

- The MFs for the SOC of the ESSs have been divided into three possible levels, namely 'High', 'Normal' or 'Low'. The 'Normal' SOC range is considered when the ESS operates with a SOC between its maximum and minimum boundaries. Consequently, the 'High' and 'Low' SOC are defined for SOC values above or below the corresponding limits. Preferably, the ESSs would work on the 'Normal' SOC region. The maximum and minimum SOC limits depend on the storage technology, and they have been set at different values for the BESS and the UC in Fig. 8.
- The primary active power references for UC and BESS are classified into one of the following four situations: Deep charge ('DC'), soft charge ('SC'), soft discharge ('SD') and deep discharge ('DD'). Under a soft charge/discharge cycle, the ESS is receiving/releasing a low or medium amount of active power and the SOC is expected to vary slowly. Nonetheless, a deep charge/discharge implies a faster variation of SOC that can drive the ESS to a limiting situation more rapidly.

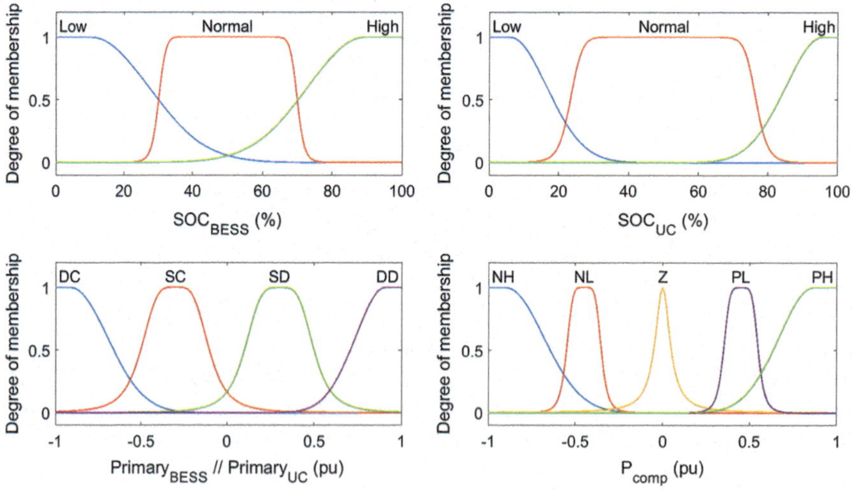

Fig. 8 MFs for the inputs and output of the fuzzy controller

- The compensating active power outputted from the fuzzy controller will be included in one of the following five possibilities: Negative high ('NH'), negative low ('NL'), zero ('Z'), positive low ('PL') and positive high ('PH'). The positive and negative ranges are distributed symmetrically around the zero output, when no compensation is needed between the ESSs. The positive and negative values of P_{comp} determine which storage device receives support from the other to avoid limiting SOC situations, whereas the high and low alternatives define how strong this support needs to be, which is reflected on the amount of active power exchanged between them.

The MFs for the inputs and the output of the fuzzy controller are shown in Fig. 8, and have been derived through an experience-based design procedure. It can be observed that the same MFs are used for $Primary_{BESS}$ and $Primary_{UC}$. Nonetheless, the MFs of their SOC differ, which is justified by the fact that both devices typically show a different recommended depth of discharge.

The SOC inputs vary in the range of [0, 100]%, and normalization was not considered necessary for these parameters. Regarding the active power inputs and output, they are expressed in p.u., using the rated power of the WECS as the base power. Therefore, their variation is already in the range of $[-1, 1]$ p.u., and no additional normalization gains were needed in the fuzzy controller.

Some initial parameters of the MFs, such as their number, shape and their range of variation, were chosen based on the knowledge and the desired performance of the system. A second step consisted on the fine adjustment of the initial set of MFs. This was carried out through simulation of the control system under various scenarios, and observation of the responses obtained (i.e.: SOC variation of the ESSs). This stage led to the complete definition of the MFs finally implemented on the fuzzy controller. Nevertheless, other alternatives were evaluated and rejected during this process. For instance, a control system with less MFs was tested. This configuration showed a limited controllability of the SOC of both ESSs, difficulty to discriminate between deep/soft charge/discharge cycles, or steep and sudden changes on the compensating power term among other undesired effects, and it was not considered valid for this application. Opposite, a configuration with more MFs than those finally selected was also studied, observing a higher precision on the compensating power outputted from the fuzzy controller. However, the complexity of the system (number of fuzzy rules needed) and the computational time increased notably. Hence, it was considered that the accuracy gained in the SOC controllability did not justify the inconveniences derived from the increased number of MFs. In conclusion, the set of MFs presented in this work provided satisfactory results, and it was considered an adequate configuration for the purpose pursued herein.

Together with the MFs, the fuzzy rules are also responsible for defining the output of the fuzzy controller depending on the value of the inputs. In this configuration, 144 fuzzy rules are implemented on the fuzzy controller. The list of fuzzy rules is presented in Table 1. With this configuration, the compensating active power is adequately defined to support the SOC regulation between both ESSs. Certainly, the computation of inputs and outputs in a system with 144 fuzzy rules can pose a

Table 1 Tables of the fuzzy rules (Note that $Primary_{BESS}$ and $Primary_{UC}$ have been shortened to $Prim_{BESS}$ and $Prim_{UC}$ respectively for the sake of conciseness)

SOC_{BESS}: High, SOC_{UC}: High

$Prim_{UC}$ / $Prim_{BESS}$	DC	SC	SD	DD
DC	Z	Z	NL	NH
SC	PL	PL	NL	NH
SD	PH	PL	Z	NL
DD	PH	PL	Z	Z

(a)

SOC_{BESS}: Normal, SOC_{UC}: High

$Prim_{UC}$ / $Prim_{BESS}$	DC	SC	SD	DD
DC	PL	PL	Z	NL
SC	PL	PL	Z	NL
SD	PH	PL	PL	Z
DD	PH	PH	PL	Z

(b)

SOC_{BESS}: Low, SOC_{UC}: High

$Prim_{UC}$ / $Prim_{BESS}$	DC	SC	SD	DD
DC	PH	PL	Z	Z
SC	PH	PL	PL	PL
SD	PH	PH	PH	PL
DD	PH	PH	PH	PH

(c)

SOC_{UC}: Normal

$Prim_{UC}$ / $Prim_{BESS}$	DC	SC	SD	DD
DC	NL	NL	NH	NH
SC	NL	NL	NH	NH
SD	Z	Z	NL	NH
DD	Z	Z	NL	NL

(d)

SOC_{UC}: Normal

$Prim_{UC}$ / $Prim_{BESS}$	DC	SC	SD	DD
DC	Z	Z	NL	NH
SC	Z	Z	NL	NL
SD	PL	Z	Z	NL
DD	PL	PL	Z	Z

(e)

SOC_{UC}: Normal

$Prim_{UC}$ / $Prim_{BESS}$	DC	SC	SD	DD
DC	Z	Z	Z	Z
SC	PL	PL	Z	Z
SD	PH	PH	PL	PL
DD	PH	PH	PH	PH

(f)

SOC_{UC}: Low

$Prim_{UC}$ / $Prim_{BESS}$	DC	SC	SD	DD
DC	NH	NH	NH	NH
SC	NH	NH	NH	NH
SD	NL	NL	NH	NH
DD	Z	NL	NL	NH

(g)

SOC_{UC}: Low

$Prim_{UC}$ / $Prim_{BESS}$	DC	SC	SD	DD
DC	NH	NH	NH	NH
SC	NL	NH	NH	NH
SD	NL	NL	NH	NH
DD	Z	Z	NL	NH

(h)

SOC_{UC}: Low

$Prim_{UC}$ / $Prim_{BESS}$	DC	SC	SD	DD
DC	Z	Z	NL	NL
SC	Z	Z	NL	NL
SD	PL	PL	Z	Z
DD	PL	PL	Z	Z

(i)

challenge. Nevertheless, these rules were necessary to achieve sufficient accuracy in the controller and to avoid abrupt fluctuations of the output. The system was tested with a lower number of MFs obtaining a significantly poorer performance. Furthermore, other studies have used fuzzy controllers with a similar number of rules in applications with renewable energies and hybrid systems [5, 6, 17, 27], thus supporting its adequacy for such purposes. A flowchart of the decision-making process in the SCS is shown in Fig. 9.

As a final step in the SCS, another active power term is added to the UC power reference. It consists of the high frequency components of the power reference for the BESS. Due to its slower dynamic response, the BESS will not be able to react quickly to steep changes on its power reference. Therefore, the UC will be demanded to provide a fast response to compensate this deficiency of the BESS. This operation increases the coordination between both ESSs and enhances the benefits of

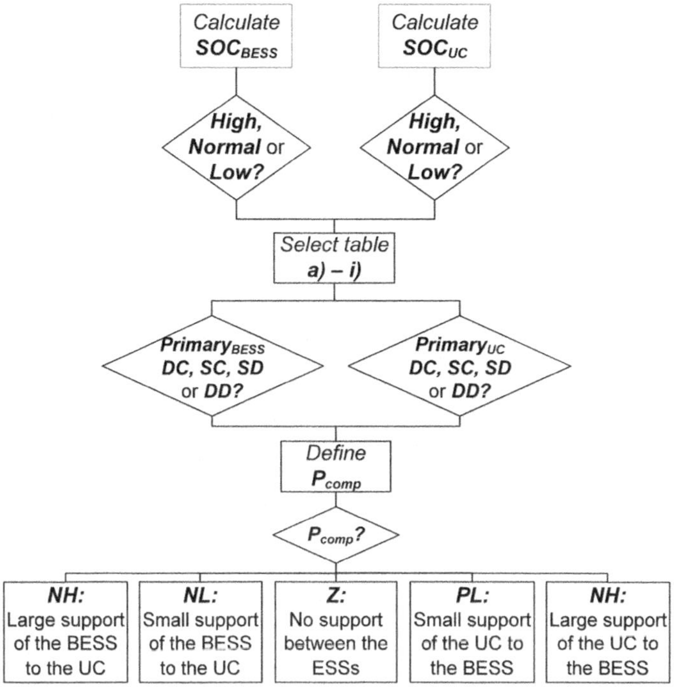

Fig. 9 Flowchart of the proposed SCS

implementing a configuration with two storage technologies with differentiated characteristics.

The fuzzy controller consists of four inputs and one output. Asymmetric Gaussian MFs have been chosen for the extreme SOC values, whereas a generalized bell distribution is used for the 'Normal' range. The SOC is considered 'Low' below 10% for the UC and 20% for the BESS, 'Normal' between 25% and 75% for the UC, and between 30% and 70% for the BESS, and 'High' above 90% and 85% for UC and BESS respectively. These values have been considered to define the corresponding MFs. For the primary active power references, the MFs for UC and BESS show the same characteristics. These input variables are expected to vary between ±1 p.u. The extreme ranges (i.e. DC and DD) have been modelled as asymmetric Gaussian MFs, reaching the maximum point of the curve at ±0.9 p.u. On the other hand, the intermediate values (i.e. SC and SD) are represented by generalized bell MFs, with a peak at ±0.3 p.u., in order to achieve a symmetrical performance. For primary active power references near 0 p.u., no specific MFs have been defined. Regarding the output, a similar approach was used. Asymmetric Gaussian MFs are used for the extreme values of the output (i.e. ±0.8 p.u.), whereas generalized bell MFs are considered for the intermediate values. The NL and PL curves are defined symmetrically with the maximums at ±0.45 p.u., and the Z output is defined as a sharp peak at 0 p.u.

The output MF is defined, depending on the values of the inputs, through a set of `if-then` rules. For instance, in a particular situation, the output could be defined by the following sentence: `if` SOC_{UC} is **Low** & SOC_{BESS} is **High** & $Primary_{UC}$ is **SD** & $Primary_{BESS}$ is **DC** `then` P_{comp} is **NH**, that is, a large amount of power flows from the BESS to the UC to help increase its SOC.

The fuzzy system implemented is based on the Mamdani-type inference, according to which each rule is computed as in (12), and the output is then generated through a min-max algorithm described by (13)–(14) [27]:

$$f_i = \min \left\{ \begin{array}{l} \mu\left(SOC_{UC}\right), \mu\left(SOC_{BESS}\right), \\ \mu\left(\text{Primary}_{UC}\right), \mu\left(\text{Primary}_{BESS}\right) \end{array} \right\} \tag{12}$$

$$D_i = \min\left\{f_i, MF\left(P_{comp}\right)\right\} \tag{13}$$

$$MF\left(P_{comp}\right) = \max\left\{D_i\right\} \quad \text{for } i = 1, 2, 3, \ldots \tag{14}$$

Hence, through (14), the rule with maximum membership value is selected to deliver the output with a maximum possibility distribution. Once the adequate MF of the output is chosen, it is delivered out of the fuzzy system as P_{comp}.

6 Results and Discussion

Three alternative options for the SCS are compared in this section. The first one corresponds to the SCS including the fuzzy controller proposed herein. A second alternative is the proposed SCS without the fuzzy controller, where $Primary_{BESS}$ and $Primary_{UC}$ are directly delivered to the respective ESS. Therefore, the SOC of both devices is not monitored in this second SCS. Finally, the third benchmark is a SCS based on a state machine that monitors the SOC of the ESSs and their primary active power references, and defines the active power references for both storage devices accordingly. The state machine SCS is based on our previous work for a hybrid system with a single ESS [34].

6.1 Case Study 1: $SOC_{UC} = 25\%$–$SOC_{BESS} = 50\%$

This first simulation addresses the case when the UC starts with a 'Low' SOC situation, and the BESS is in a 'Normal' state. Additionally, the grid demand is defined in three levels throughout the simulation. An initial level as 1 p.u. during the first 20 s of the simulation. Then, it decreases to 0.7 p.u. from 20 to 40 s. Finally, it is set at 1.2 p.u. until the end of the simulation. Furthermore, the wind speed varies as shown in Fig. 10a, and $P_{prediction}$ is set at a constant value of 0.92 p.u.

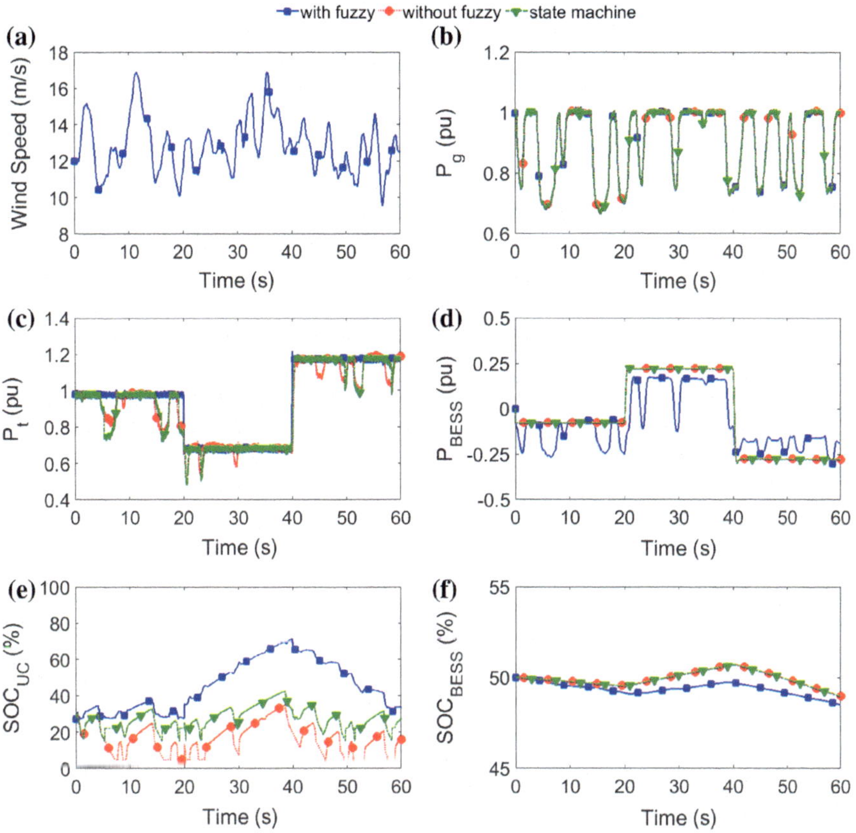

Fig. 10 Case study 1. **a** Wind speed; **b** total WECS active power; **c** total active power output of the hybrid system; **d** BESS active power; **e** SOC of the UC; **f** SOC of the BESS

Under the conditions described, P_g is illustrated in Fig. 10b. As seen, the generation of the WECS varies with the incoming wind speed. Furthermore, the three configurations evaluated show the same response for this parameter. This is an expected result, since the performance of the WECS should not be affected by the power exchange of the ESSs. On the other hand, the contribution of the ESSs in each configuration is clearly distinguished when P_t is represented. As seen in Fig. 10c, the proposed SCS is able to supply P_{dem} throughout the whole simulation period, even though the UC starts with a low SOC condition. This can be achieved due to the support of the BESS, which provides an additional active power injection to avoid an increased discharge of the UC. For instance, in the interval 4.6–7.8 s the demand is not fully satisfied by the 'without fuzzy' and the 'state machine' configurations. During this same period, a deeper discharge can be observed in the BESS active power (Fig. 10d) of the 'with fuzzy' SCS compared to the other two alternatives. This additional discharge of the BESS compensates the difference between the forecasted and the actual generation, and relieves the UC from having to provide this power. Subsequently, SOC_{UC} (Fig. 10e) does not decrease as much as for the 'without fuzzy' and the 'state machine' SCSs in this interval. A better regulation of SOC_{UC} for the 'with fuzzy' SCS compared to the other two alternatives is observed, as it remains clearly higher all along the simulation. As a counterpart, SOC_{BESS} is slightly reduced for the 'with fuzzy' SCS (Fig. 10f). This consequence could be expected from the previous results. Nevertheless, this parameter never falls in a low SOC_{BESS} situation, and the slight differences observed are not a potentially risky situation for the BESS.

It must be highlighted that, as seen in Fig. 10e, the 'without fuzzy' SCS causes the UC to reach a minimum value of 5% for SOC_{UC}, while the 'state machine' maintains this parameter around 25% (limiting region between 'Low' and 'Normal' SOC) for most of the simulation. On the other hand, the 'with fuzzy' configuration proves a better regulation of this parameter.

6.2 Case Study 2: $SOC_{UC} = 50\%$–$SOC_{BESS} = 50\%$

The second simulation analyses 'Normal' SOC conditions for both ESSs under the same wind speed and grid demand as in the previous case study. Regarding P_t (Fig. 11a), it can be seen that, although the UC starts with a 50% SOC, the configurations 'without fuzzy' and with 'state machine' SCS are not able to fulfil P_{dem} in two short periods reaching the end of the simulation. On the other hand, when the fuzzy controller is included, P_{dem} is adequately supplied along the whole simulation. These two unsatisfied load intervals are due to low SOC conditions for the UC in the 'without fuzzy' and 'state machine' configurations, as observed in Fig. 11b. In this figure, it can be also appreciated that the 'with fuzzy' SCS accomplishes a better regulation of SOC_{UC}, while being able to comply with the grid requirements at all times. By the end of the simulation, a 41% SOC_{UC} has been registered for the 'with fuzzy' SCS, whereas 26% is measured for the 'state machine' option, which borders the low SOC operation. In the case of the 'without fuzzy' SCS, SOC_{UC} falls below 20% at time 60 s, and records two minimum sags around 5%, which is an undesired situation for the device.

Regarding SOC_{BESS}, no relevant differences can be observed in Fig. 11c. A maximum difference of around 0.4% (lower for the 'with fuzzy' SCS) is registered at 40 s, but this is not a significant value. Furthermore, almost the same SOC_{BESS} is registered at the end of the simulation, despite the fact that a notable improvement of the 'with fuzzy' SCS was observed for SOC_{UC}.

The active power exchange between both ESSs is illustrated in Fig. 11d. As seen, this parameter varies between positive and negative values. For a negative P_{comp}, the BESS supports the UC to increase its SOC. Hence, a certain energy that otherwise would remain unused in the BESS is delivered to the UC, thus increasing SOC_{UC}. This can be seen in Fig. 11c, d from 20 to 40 s. With a higher SOC, the UC has enough capacity to cover the difference between the power prediction and the generation when necessary, thus avoiding unmet grid demand situations. If P_{comp} is not generated

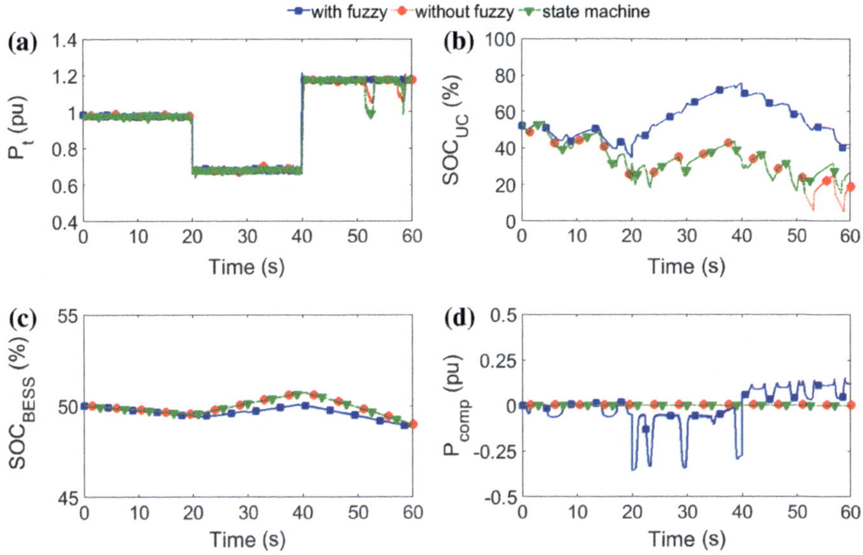

Fig. 11 Case study 2. **a** Total active power output of the hybrid system; **b** SOC of the UC; **c** SOC of the BESS; **d** compensating active power term

(i.e. 'without fuzzy' and 'state machine' SCSs), SOC_{UC} decreases throughout the simulation and P_{dem} cannot be addressed at some intervals. On the other hand, positive values of P_{comp} indicate that the UC backs the recovery of SOC_{BESS} when the combination of primary active powers and SOCs allows it.

6.3 *Case Study 3:* $SOC_{UC} = 85\%–SOC_{BESS} = 30\%$

A third case study has been developed to present the situation with a 'Low' SOC_{BESS} and 'High' SOC_{UC}. The incoming wind speed and grid demand were maintained as in the previous experiences. In this case, the aim is to illustrate that, with the proposed fuzzy controller, the UC can support the operation of the BESS when it has a 'Low' SOC by assuming a part of the primary active power reference requested to the battery. Therefore, SOC_{BESS} can be monitored while the hybrid system supplies P_{dem} for longer. This can be clearly noticed in Fig. 12a. With a 'Low' SOC_{BESS} at the beginning of the simulation, the 'state machine' SCS prevents the battery from discharging. Subsequently, P_{dem}, which is above the power prediction, cannot be met during the first 20 s for this SCS. For the last 20 s of the simulation, SOC_{BESS} has not recovered to a 'Normal' SOC region yet, the BESS cannot release any active power, and the load is unmet from 40 to 60 s again. On the other hand, SOC_{BESS} (Fig. 12b) does not decrease during these intervals for this SCS. Regarding the 'without fuzzy' SCS, it can be seen that P_{dem} is addressed during the whole simulation (Fig. 12a).

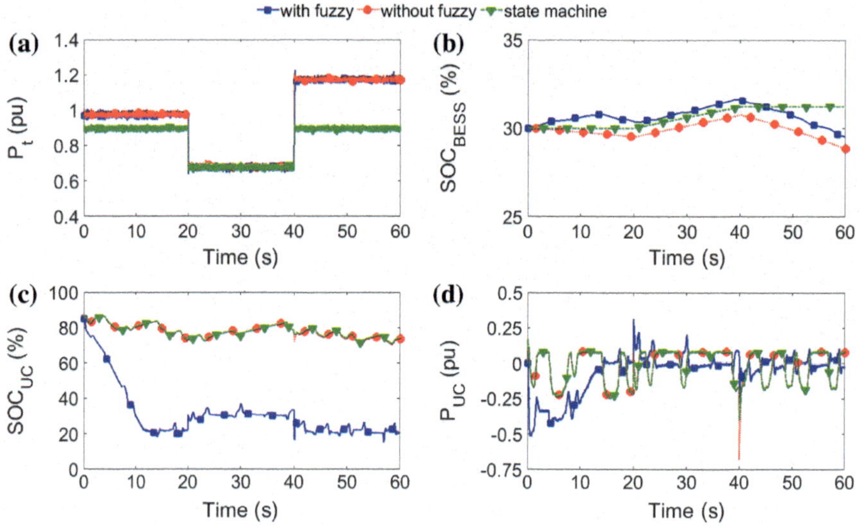

Fig. 12 Case study 3. **a** Total active power output of the hybrid system; **b** SOC of the BESS; **c** SOC of the UC; **d** UC active power

However, this can only be made at the cost of reducing SOC_{BESS} to a lower value, which might eventually damage the device if this situation was maintained. On the contrary, the 'with fuzzy' SCS allows addressing P_{dem} throughout the simulation while increasing SOC_{BESS} for some periods. The result is that this SCS complies with the two main objectives of meeting the grid requirements and avoiding excessively low charge/discharge of the ESSs.

Due to the design of the 'with fuzzy' SCS, the energy needed to fulfil P_{dem} with a 'Low' SOC_{BESS} must be obtained from the UC. Hence, a large reduction in SOC_{UC} can be observed for this SCS compared to the 'without fuzzy' and the 'state machine' alternatives (Fig. 12c). By the end of the simulation, the hybrid system faces a situation of both ESSs approaching a 'Low' SOC. If the external conditions (i.e. wind or grid demand) do not change, the system might have to leave some demand unmet in order to avoid a potentially risky over discharge of the ESSs. On the other hand, the 'without fuzzy' and 'state machine' SCSs have maintained a higher SOC_{UC} in this period. However, the energy management has not been adequate since either the SOC_{BESS} has been decreased excessively, or the load has not been met for long periods. Hence, the UC has been storing energy that was more necessary in other elements of the hybrid system in order to support the BESS or to be supplied to the grid. Therefore, it can be stated that the 'with fuzzy' SCS has performed a smarter use of the available energy.

Figure 12d shows the active power exchanged by the UC. As seen, the UC releases more power in the 'with fuzzy' SCS compared to the other two alternatives, which could be expected from Fig. 12c. This additional power supplied by the UC is useful to meet the grid demand and support SOC_{BESS}. Moreover, it can be appreciated that

P_{UC} changes rapidly and moves frequently from charge to discharge cycles. This was consciously pursued when defining $Primary_{UC}$ and $Primary_{BESS}$, as this type of behaviour is better suited for an UC than for a BESS.

7 Conclusions

This chapter presents a control strategy designed for the coupled operation of several power sources in a hybrid system. The proposed SCS is in charge of calculating the references for the active power to be stored in or delivered by the ESSs. These references are defined on a primary stage according to a time-ahead wind power generation prediction, the instantaneous grid demand, and the actual wind power generation. On a second stage, a compensating power term is defined to supervise excessively low or high SOC conditions on the ESSs that may jeopardize their performance. A fuzzy logic controller has been used to compute this power compensating term. The fuzzy controller decides the amount of power exchanged between the ESSs to support their SOC mutually, while attempting to fulfil the demanded active power as long as possible. Therefore, the ESSs included in the hybrid system are used with a double purpose. Firstly, they add flexibility and capacity to control and regulate the active power generation of the hybrid system, adapting to changes on the grid demand. Secondly, they offer mutual support for their SOC regulation to avoid or reduce the risk of exceeding the maximum and minimum recommended boundaries. Such performance improves the exploitation of the ESSs in the hybrid system.

In the simulations carried out, it could be observed that the inclusion of the ESSs in the hybrid system allowed delivering a controlled active power generation from a fluctuating and intermittent resource such as wind power. This was feasible in most cases, only interrupted in those situations when the ESSs were depleted and no additional energy could be released. This inconvenience remarks the importance of regulating adequately the SOC of the ESSs in a hybrid system.

Furthermore, the results showed that the configuration with the fuzzy controller can comply with the grid specifications longer than the other two alternatives considered. This can be achieved due to a smarter use of the available energy between the ESSs. The storage devices support each other on their SOC regulation. As a consequence of this improved energy management, a better regulation of the SOC of the ESSs was observed in the simulations for the SCS with the fuzzy controller. In general, the BESS finds it easier to support the UC due to the larger capacity of the former. Nonetheless, the UC also proved its capacity to take charge of the BESS power demand when it experiences low SOC, thus preventing the battery from a deeper discharge.

Apart from the better SOC monitoring accomplished with the fuzzy controller, the proposed SCS also takes advantage of the main qualities of the ESSs used. The primary active power references are computed in the SCS considering the quick response and ability to change rapidly between charge and discharge cycles of the UC, and the larger capacity and slower dynamics of the BESS. Subsequently, from different perspectives, an improved exploitation of the ESSs available has been achieved with the proposed SCS based on a fuzzy controller.

References

1. Abbey C, Joos G (2007) Supercapacitor energy storage for wind energy applications. IEEE Trans Ind Appl 43(3):769–776
2. Akhmatov V, Knudsen H (2007) Large penetration of wind and dispersed generation into Danish power grid. Electr Power Syst Res 77(9):1228–1238
3. Anaya-Lara O, Jenkins N, Ekanayake J, Cartwright P, Hughes M (2009) Wind energy generation modelling and control. Wiley, West Sussex, UK
4. Beaudin M, Zareipour H, Schellenberglabe A, Rosehart W (2010) Energy storage for mitigating the variability of renewable electricity sources: an updated review. Energy Sustain Dev 14(4):302–314
5. Bisht MS, Sathans (2014) Fuzzy based intelligent frequency control strategy in standalone hybrid AC microgrid. In: Proceedings of the IEEE conference on control applications
6. Capizzi G, Tina G (2007) Long-term operation optimization of integrated generation systems by fuzzy logic-based management. Energy 32(7):1047–1054
7. de Almeida RG, Peças Lopes JA, Barreiros JAL (2004) Improving power system dynamic behavior through doubly fed induction machines controlled by static converter using fuzzy control. IEEE Trans Power Syst 19(4):1942–1950
8. Díaz-González F, Sumper A, Gomis-Bellmunt O, Villafáfila-Robles R (2012) A review of energy storage technologies for wind power applications. Renew Sustain Energy Rev 16(4):2154–2171
9. Discover Battery (2018) D121000BD AGM series features & benefits. http://discoverbattery.com/assets/dropbox/Datasheets/en/D121000BD.pdf. Cited 15 Jun 2018
10. Evans A, Strezov V, Evans TJ (2012) Assessment of utility energy storage options for increased renewable energy penetration. Renew Sustain Energy Rev 16(6):4141–4147
11. Ferreira HL, Garde R, Fulli G, Kling W, Peças Lopes J (2013) Characterisation of electrical energy storage technologies. Energy 53:288–298
12. Gallo AB, Simões-Moreira JR, Costa HKM, Santos MM, Moutinho dos Santos E (2016) Energy storage in the energy transition context: a technology review. Renew Sustain Energy Rev 65:800–822
13. Guney MS, Tepe Y (2017) Classification and assessment of energy storage systems. Renew Sustain Energy Rev 75:1187–1197
14. Guo Z, Zhao J, Zhang W, Wang J (2011) A corrected hybrid approach for wind speed prediction in Hexi Corridor of China. Energy 36(3):1668–1679
15. Gutiérrez-Martín F, Da Silva-Álvarez RA, Montoro-Pintado P (2013) Effects of wind intermittency on reduction of CO_2 emissions: the case of the Spanish power system. Energy 61:108–117
16. Hadjipaschalis I, Poullikkas A, Efthimiou V (2009) Overview of current and future energy storage technologies for electric power applications. Renew Sustain Energy Rev 13:1513–1522
17. Hajizadeh A, Soltani M, Norum LE (2017) Intelligent power control of DC microgrid. In: Proceedings of the IEEE international conference on ubiquitous wireless broadband
18. Hassan SZ, Li H, Kamal T, Mumtaz S, Khan L (2016) Fuel cell/electrolyzer/ultra-capacitor hybrid power system: focus on integration, power control and grid synchronization. In: Proceedings of the international Bhurban conference on applied sciences and technology
19. Hassan SZ, Li H, Cagriyener S, Kamal T, Mufti GM, Waseem MH, Gohar H (2017) Integration and simulation of wind with hydrogen/supercapacitor storage hybrid system. In: Proceedings of the international conference on electrical engineering

20. Hassan SZ, Li H, Kamal T, Abbas MO, Khan MA, Mufti GM (2017) An intelligent pitch angle control of wind turbine. In: Proceedings of the international symposium on recent advances in electrical engineering
21. Heier S (1998) Grid integration of wind energy conversion systems. Wiley, West Sussex, UK
22. Hong YY, Chang HL, Chiu CS (2010) Hour-ahead wind power and speed forecasting using simultaneous perturbation stochastic approximation (SPSA) algorithm and neural network with fuzzy inputs. Energy 35(9):3870–3876
23. Jerbi L, Krichen L, Ouali A (2009) A fuzzy logic supervisor for active and reactive power control of a variable speed wind energy conversion system associated to a flywheel storage system. Electr Power Syst Res 79(6):919–925
24. Kamal T, Hassan SZ, Hui L, Awais M (2015) Design and power control of fuel cell/electrolyzer/microturbine/ultra-capacitor hybrid power plant. In: Proceedings of the international conference on emerging technologies
25. Kamel RM, Chaouachi A, Nagasaka K (2010) Wind power smoothing using fuzzy logic pitch controller and energy capacitor system for improvement micro-grid performance in islanding mode. Energy 35(5):2119–2129
26. Kavasseri RG, Seetharaman K (2009) Day-ahead wind speed forecasting using f-ARIMA models. Renew Energy 34(5):1388–1393
27. Kazemi MV, Moradi M, Kazemi RV (2012) Minimization of powers ripple of direct power controlled DFIG by fuzzy controller and improved discrete space vector modulation. Electr Power Syst Res 89:23–30
28. Krause PC, Wasynczuk O, Sudhoff SD (2002) Analysis of electric machinery and drive systems. Wiley, New York, USA
29. Maxwell Technologies Products catalog (2016) BMOD0063 P125 User Manual. http://www.maxwell.com/images/documents/MANUAL_HTM125_20BMOD0063_1014343_8.pdf. Cited 15 Jun 2018
30. Mi Y, Zhang H, Fu Y, Wang C, Loh PC, Wang P (2018) Intelligent power sharing of DC isolated microgrid based on fuzzy sliding mode droop control. IEEE Trans Smart Grid. https://doi.org/10.1109/TSG.2018.2797127
31. Pandey K, Bharath KVS (2017) Intelligent approach for active and reactive power control in doubly fed induction generator wind turbine system. In: Proceedings of the India international conference on power electronics
32. Pichan M, Rastegar H, Monfared M (2013) Two fuzzy-based direct power control strategies for doubly-fed induction generators in wind energy conversion systems. Energy 51:154–162
33. Rahim AHMA, Nowicki EP (2012) Supercapacitor energy storage system for fault ride-through of a DFIG wind generation system. Energy Convers Manag 59:96–102
34. Sarrias R, Fernández LM, García CA, Jurado F (2012) Coordinate operation of power sources in a doubly-fed induction generator wind turbine/battery hybrid power system. J Power Sources 205:354–366
35. Shrinath K, Paramasivam S, Palanisamy K (2017) An intelligent self-tuning fuzzy logic controller for pitch angle control for a wind turbine fed induction generator. In: Proceedings of the innovations in power and advanced computing technologies
36. Slootweg JG, Kling WL (2003) Wind power: modelling and impact on power system dynamics. TU Delft, Netherlands PhD thesis
37. Tamalouzt S, Benyahia N, Rekioua T, Rekioua D, Abdessemed R (2016) Performances analysis of WT-DFIG with PV and fuel cell hybrid power sources system associated with hydrogen storage hybrid energy system. Int J Hydrog Energy 41(45):21006–21021
38. Teleke S, Baran ME, Huang AQ, Bhattacharya S, Anderson L (2009) Control strategies for battery energy storage for wind farm dispatching. IEEE Trans Energy Convers 24(3):725–732
39. Tremblay O, Dessaint LA (2009) Experimental validation of a battery dynamic model for EV applications. World Electr Veh J 3(2):289–298
40. Varanasi J, Tripathi MM (2016) Artificial neural network based wind speed & power forecasting in US wind energy farms. In: Proceedings of the IEEE international conference on power electronics, intelligent control and energy systems

41. Wei L, Joos G, Belanger J (2010) Real-time simulation of a wind turbine generator coupled with a battery supercapacitor energy storage system. IEEE Trans Ind Electron 57(4):1137–1145
42. Yazdani A, Iravani R (2010) Voltage-sourced converters in power systems: modeling, control and applications. Wiley, New Jersey, USA

Neural-Based P-Q Decoupled Control for Doubly Fed Induction Generator in Wind Generation System

Moulay Rachid Douiri

Abstract This chapter describes an Artificial Neural Network (ANN) approach for active and reactive decoupled control based Direct Power Control (DPC) in Doubly Fed Induction Generator (DFIG) for Wind Generation System (WGS) by using the suitable voltage vectors on the rotor side. To avoid the computational complexity of DPC, we develop a neuronal approach using an individual training technique with fixed weight and supervised networks. For this, the neural system is split into 5 sub-networks namely: reactive and real power measurement sub-networks with dynamic neurons and fixed-weight; reactive calculation and reference real sub-networks with square neurons and fixed-weight; reference stator current computation sub-network with logarithm of sigmoid, tangent sigmoid neurons and supervised weight, reference rotor current computation sub-network with recurrent neurons and fixed-weight; and reference rotor voltage calculation sub-networks with dynamic neurons and fixed-weight. Under transient conditions, and for step changes of the real and the reactive power references, the DFIG is capable of tracking the references with a response time of less than 1 s. This is fast enough for changes made by the power system operator, and for tracking wind speed variations. Thus, the sensorless measurement of the position is effective in controlling P and Q.

1 Introduction

The Doubly Fed Induction Generator (DFIG) is a very promising candidate for Wind Turbine (WT) application. The direct connection of DFIG to the grid forces the rotation speed to remain constant so that the machine remains close to synchronism [1]. This restriction results in reduced efficiency for wind turbine applications at high

M. R. Douiri (✉)
Department of Electrical Engineering, Higher School of Technology,
Cadi Ayyad University, BP. 383 Essaouira, Morocco
e-mail: douirirachid@hotmail.com

R.-E. Precup et al. (eds.), *Advanced Control and Optimization Paradigms for Wind Energy Systems*, Power Systems, https://doi.org/10.1007/978-981-13-5995-8_9

wind speeds. The introduction of the power converter between the stator and the grid gives rise to a decoupling between the frequency of the electrical grid and the speed of rotation of the machine which makes it possible to operate at variable speed [2]. However, this converter must be sized to pass all the power generated by the machine. The Voltage Source Converter (VSC) connected to the slip-ring terminals can control at the rotor-side the active and reactive power from the stator-side [3]. A second VSC (Front End Converter) is connected back-to-back on its dc side to a rotor-side VSC which transmits the rotor slip frequency power to 60 Hz power of the distribution grid [2–4]. A power transformer is usually necessary to step up the low voltages of the VSC to ac grid voltages.

In fact, several studies related to decoupled control of P-Q power algorithms for DFIG in Wind Generation System(WGS)such as sliding mode control [5], non-linear control [6] and model-based predictive control [7]; which have afterwards been proven its instability in handling various changes in weather conditions.

In Ref. [8] the author suggests a distributed coordinated control of P-Q power scheme for WGS using the model predictive control (MPC), wherein the distributed reactive/voltage control of WGS is coordinated and realized with the real power control.

A H^{∞} control algorithm was adopted in [9] to regulate and decouple both active and reactive powers of wind energy conversion system considering the uncertainties in the permanent magnet synchronous generator dynamics.

Others [10] present a decoupling control approach for the P-Q power of DFIG to improve the performance of wind turbines, which relies on proportional-integral (PI) controllers. This method has the disadvantage of being less precise and less stable as the coordination between these regulators is based on classical trial and error parameter tuning techniques a very difficult and challenging task.

Conventional techniques used for P-Q power decoupling control are generally less robust, also, these techniques face a major drawback with functions that include noisy experimental data. Therefore, researchers thought of finding new different algorithm techniques for solving these problems, such as Metaheuristic Optimization Algorithms, Fuzzy Logic (FL), and Artificial Neural Networks (ANNs), which can easily integrate human intelligence into complex control system based on human knowledge and experience. They can also be used for curve fitting with experimental data as they can find the global minimum even if many local minima exist [11, 12].

The authors in [13] develop an energy conversion structure integrate a combined control mode with three different methods: a conventional PI regulators, a FL controller and a matrix inverter. This technique ensures the system transient stability and the accuracy, but it does not improve the voltage quality supplied on the DFIG level.

A technique has been proposed by [14] using a Flywheel Energy Storage System control based on a FL supervisor capable of reducing the active power oscillation caused by the random wind speed fluctuations and generate to the grid a cleaner power. The command law ensures reactive power by providing a smooth reactive power to the load supplied by the DFIG in WGS.

Others [15] present an artificial neural network controllers replacing the conventional regulators in the vector control of the DFIG systems. Also, the authors in [16] trained an ANN-based power controllers applying back-propagation approach for decoupled control of DFIG wind turbine system in different wind speed conditions, machine parameters and faults.

An approach metaheuristic reactive power planning active power dispatch is presented in [17] to maximize the profit of the utility while satisfying the operating constraints of the power system in the presence of DFIG wind turbines and various uncertainties loads.

In this situation, the principal contribution of this chapter is a controlled the P-Q power of a wind-driven DFIG using ANN under fault conditions and varying wind speed conditions. The power estimators based on Neural networks are trained by back-propagation method and is divided into five sub-nets namely: real and reactive power measurement; reference P-Q computation; reference stator current computation; reference rotor current computation; and reference rotor voltage computation.

The rest of this chapter is organized as follows. Section 2 describes an overview of the developed control algorithm of WGS and DFIG Mathematical model. In Sect. 3, the ANN-based wind speed estimation is presented. In Sect. 3, the optimal active and reactive control using ANN controller of DFIG system is proposed. The numerical results are presented and discussed in Sect. 4, followed by conclusions.

2 WGS and DFIG Mathematical Model

The wind speed is always fluctuating, and thus the energy content of the wind is always changing. The $C_p - \lambda - \beta$ curves characterize the aerodynamic model of a WT, where C_p is the performance factor, which is a non-linear function of both tip-speed-ratio λ and the blade pitch angle β. It can be calculated through the following expression [18, 19]:

$$C_p(\beta, \lambda) = \sum_{i=0}^{4} \sum_{j=0}^{4} \alpha_{ij} \beta^i \lambda^j \ , \tag{1}$$

The model is found to be accurate for the range $2 < \lambda < 13$, λ is defined as the ratio of rotor tip speed to free wind speed [20]:

$$\lambda = \frac{\omega_t R_t}{\nu_\omega} \ , \tag{2}$$

where R_t and ν_ω is the rotor radius and the incoming wind speed respectively. Together $\omega_t R_t$ make up the blades linear speed at the outer tip. Taking into account the performance coefficient, the mechanical power (P_m) that the WT extracts from the wind is given by [20]:

$$P_m = f(\nu_\omega, \omega_t, \beta) = \frac{1}{2} \rho A_r \nu_\omega^3 C_p(\lambda, \beta) \ , \tag{3}$$

where P_m, A_r and ρ is the mechanical power in the moving mass of air, swept area by the rotor and air density respectively. Power factor characteristic of a WT can be calculated through analytical models using non-linear functions if actual measurements from a wind turbine manufacturer are not available. One such model is given [18, 20] as follows:

$$C_p(\lambda, \beta) = 0.22\left(\frac{116}{\lambda_i} - 0.4\beta - 5\right) exp\left(\frac{-12.5}{\lambda_i}\right) , \tag{4}$$

with

$$\frac{1}{\lambda_i} = \frac{1}{\lambda + 0.08\beta} - \frac{0.035}{\beta^3 + 1} , \tag{5}$$

Equations used to model DFIG are similar to those used for modelling a squirrel cage induction generator with just one exception, the rotor windings are not shorted, hence rotor voltages (v_{dr}, v_{qr}) are not equal to zero. The induction generator used in doubly-fed configuration can be modelled through a full 5th order stator and rotor voltage equations in d-q reference frame using generator convention as below [21, 22]:

$$v_{ds} = -R_s i_{ds} - \omega_s \psi_{qs} + \frac{d\psi_{ds}}{dt} , \tag{6}$$

$$v_{qs} = -R_s i_{qs} + \omega_s \psi_{ds} + \frac{d\psi_{qs}}{dt} , \tag{7}$$

$$v_{dr} = -R_r i_{dr} - s_l \omega_s \psi_{qr} + \frac{d\psi_{dr}}{dt} , \tag{8}$$

$$v_{qr} = -R_r i_{qr} - s_l \omega_s \psi_{dr} + \frac{d\psi_{qr}}{dt} , \tag{9}$$

where s_l is the slip, defined as:

$$s_l = 1 - \frac{p\omega_m}{2\omega_s} , \tag{10}$$

Stator and rotor flux linkages are given as:

$$\psi_{ds} = -(L_{s\sigma} + L_m)I_{ds} - L_m I_{dr} , \tag{11}$$

$$\psi_{qs} = -(L_{s\sigma} + L_m)i_{qs} - L_m i_{qr} , \tag{12}$$

$$\psi_{dr} = -(L_{r\sigma} + L_m)i_{dr} - L_m i_{ds} , \tag{13}$$

$$\psi_{qr} = -(L_{r\sigma} + L_m)i_{qr} - L_m i_{qs} , \tag{14}$$

The difference between mechanical and electrical torque results in change of generator speed that can be calculated from the following expression [18]:

$$\frac{d\omega_m}{dt} = \frac{1}{2H_m}(T_m - T_e) \,, \tag{15}$$

$$T_e = \psi_{ds} i_{qs} - \psi_{qs} i_{ds} = \psi_{qr} i_{dr} - \psi_{dr} i_{qr} \,, \tag{16}$$

where T_e is the electric torque of the generator. The equations for active and reactive power exchange with the grid are similar to that of a squirrel cage induction generator, except the rotor windings can also be accessed in a DFIG hence the rotor component in Eqs. (21) and (22). Converters can consume or produce reactive power but they cannot produce or consume active power, thus total active power fed into the grid by a DFIG can be expressed by P_{total}. However, reactive power fed into the grid is not the same as Q_{total} in Eq. (22) because it is affected by the converter [18].

$$P_s = v_{ds} i_{ds} + v_{qs} i_{qs} \,, \tag{17}$$

$$Q_s = v_{qs} I_{ds} - v_{ds} i_{qs} \,, \tag{18}$$

$$P_r = v_{dr} I_{dr} + v_{qr} i_{qr} \,, \tag{19}$$

$$Q_r = v_{qr} i_{dr} - v_{dr} i_{qr} \,, \tag{20}$$

$$P_{total} = P_s + P_r = v_{ds} i_{ds} + v_{qs} i_{qs} + v_{dr} i_{dr} + v_{qr} i_{qr} \,, \tag{21}$$

$$Q_{total} = Q_s + Q_r = v_{qs} i_{ds} - v_{ds} i_{qs} + v_{qr} i_{dr} - v_{dr} i_{qr} \,, \tag{22}$$

where p is the number of poles, I is the current, R is the resistance of the corresponding rotor or stator, ψ is the flux linkage, L_m is the mutual inductance, L_σ is the leakage inductance, H_m is inertia constant of the generator rotor, T_m is the mechanical torque and ω_m is the angular frequency of the generator rotor. The indices s and r indicate stator or rotor side, d and q stand for direct and quadrature components, respectively. P_{total} is the active power fed into the grid by a DFIG. If, however, converter efficiency has to be taken into account the terms with rotor subscript in this expression must be multiplied with converter efficiency to access total power injected into the grid, Q_{total} is the reactive power but is not necessarily the amount fed into the grid because the converters can generate or consume reactive power which thus affects the total amount of reactive power fed into the grid. The algorithm scheme of machine-side converter is illustrated in Fig. 1.

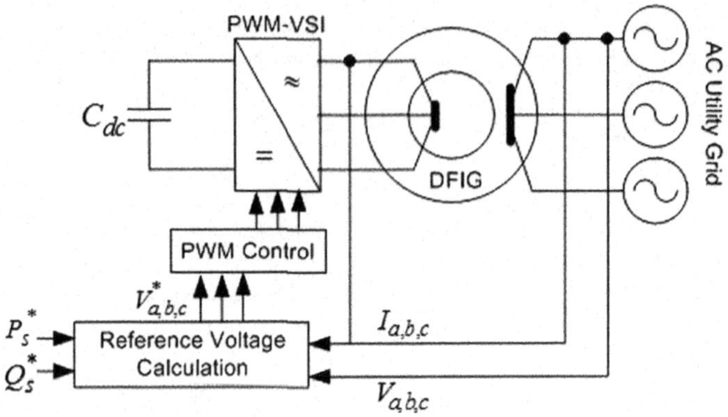

Fig. 1 Algorithm scheme of machine-side converter

3 DFIG based ANN Controller

It is possible to control the rotor speed of the DFIG, the torque as well as the power factor at the stator terminals by means of the converter on the machine side. As mentioned earlier, the machines DFIGs have the capacity of reactive power control and decoupling of P-Q powers. Given that DFIGs can be magnetized by the converter of the rotor circuit and not perforce from the power grid, there are two possibilities of reactive power exchange. The Doubly Fed Induction Generator system will only produce active power and exchange some reactive power with the grid since it is magnetized by the rotor circuit where there is no reactive power exchange with the power system, and also because the machine is connected to a high power power system with a voltage equal, or near to, 1 p.u. On the contrary, the DFIG system can be controlled to produce or absorb a quantity of reactive power, once connected to a weak power system where the voltage can fluctuate. In this case the machine will generate active power and exchange some reactive power with the grid, that is, it has the voltage control ability [18]. The entire system is illustrated in Fig. 1 The PWM three-phase inverter is able to shape and control the three-phase output voltages in magnitude and frequency using sinusoidal pulse-width modulation (SPWM), depending on the reference voltages. The P-Q powers are measured by the stator current and *d*-*q* voltage components, while the reference P-Q powers are calculated by DFIG rotor speed and wind speed υ. The reference stator currents and the reference rotor currents can be calculated from the comparison between the reference and actual P-Q powers and with the stator voltages. At last, the reference rotor voltages are calculated by analyzing the active and reference rotor currents. The complete algorithm blocks are explained in the Fig. 2. The proposed ANN of DFIG machine-side system uses the individual training strategy with fixed weight and supervised models. For this, the neural system is divided into 5 sub-networks namely: reactive and real power measurement sub-networks with dynamic neurons and fixed-weight; reactive calculation and reference real sub-networks with square

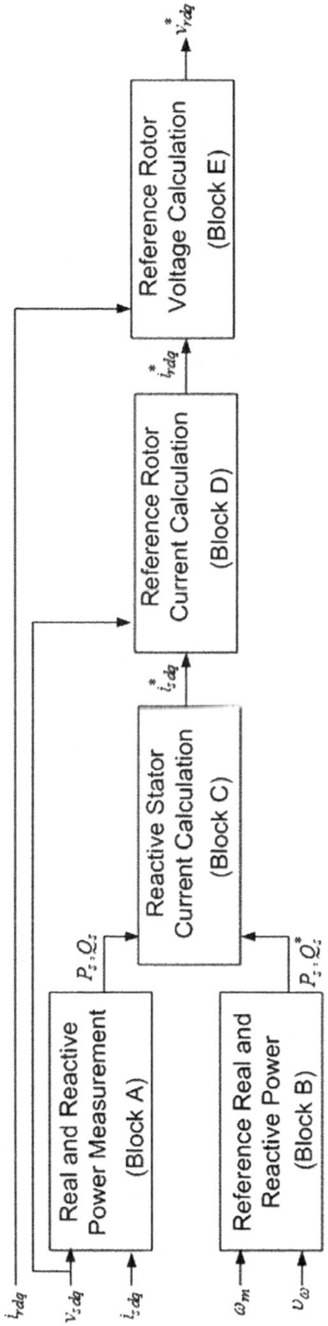

Fig. 2 Block diagram of reference rotor voltage of DFIG machine-side converter

neurons and fixed-weight; reference stator current computation sub-network with logarithm of sigmoid, tan-sigmoid neurons and supervised weight; reference rotor current calculation sub-networks with recurrent neurons and fixed-weight; and reference rotor voltage calculation sub-networks with dynamic neurons and fixed-weight.

3.1 P and Q Power Measurement Sub-networks

The active power at the stator depends on the direct component (*d*) of the stator current, the reactive power at the stator depends on the quadrature component (*q*) of the stator current. The stators-side P and Q powers are given by:

$$P_s = \frac{3}{2} v_{sd} i_{sd} , \tag{23}$$

$$Q_s = -\frac{3}{2} v_{sd} i_{sq} , \tag{24}$$

The P-Q power measurement has been constructed using 2 layer fixed-weight neural network with 8 static neurons composed of a tan-sigmoid and linear-activation function respectively, thus, 4 random inputs of the vector $[i_{sd}, i_{sq}, v_{sd}, v_{sq}]$ generated by *Matlab* are generated, the Eqs. (23) and (24) allows to obtain the target outputs. The convergence can be achieved in relatively few training epochs seen that the network is linear. Figure 3 exhibit the P and Q power measurement sub-network. The weights and biases for the P-Q power measurement have been obtained as follows:

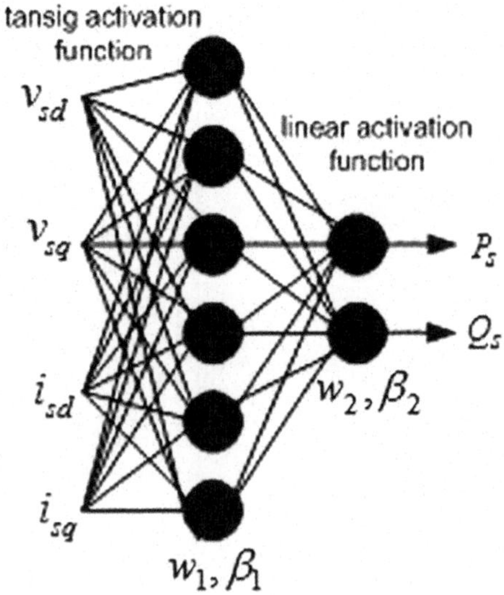

Fig. 3 P and Q power measurement sub-networks, block diagram (A)

$$w_1 = \begin{pmatrix} 0.123 & 0.621 & -0.552 & 1.008 \\ -1.092 & -0.411 & 0.657 & 0.746 \\ 0.209 & -0.341 & 0.881 & 1.091 \\ -1.554 & -0.279 & -1.240 & 1.009 \\ 1.541 & 2.008 & -0.602 & -1.268 \\ 1.560 & -2.776 & 0.554 & -2.781 \end{pmatrix}, \quad (25)$$

$$\beta_1 = \begin{pmatrix} -1.444 \\ 2.091 \\ 0.550 \\ -3.509 \\ -0.333 \\ 1.088 \end{pmatrix}, \quad (26)$$

$$w_2 = \begin{pmatrix} 0.098 & -1.908 & 0.882 & -1.934 & -1.880 & 0.356 \\ 0.077 & 1.504 & -0.615 & 1.033 & -0.773 & -1.761 \end{pmatrix}, \quad (27)$$

$$\beta_2 = \begin{pmatrix} 1.888 \\ -0.973 \end{pmatrix}, \quad (28)$$

3.2 Reference P and Q Calculation Sub-networks

In order to set the values of the reference P and Q power, a technique consists in modifying the generator speed according to the wind velocity υ in order to optimize the output power of the DFIG-WT as indicated in Fig. 4. According to the power system is weak or strong, the power Q is modified to any value (positive, negative or zero), in our case is set to zero. In order to avoid training difficulties, as there are two inputs υ and ω_m the data of all training models will be enormous if high accuracy is required, the fixed weight technique is chosen. A 3 layer fixed weight neural networks is directly associated with 20 neurons as illustrated in Fig. 5. The 1st layer is a tangent-sigmoid active function with the weight is w_1 and bias is β_1, 2nd layer is a square activation function with the weight is w_2 and bias is β_2 while the 3rd layer is a linear active function with the weight is w_3 and bias is β_3.

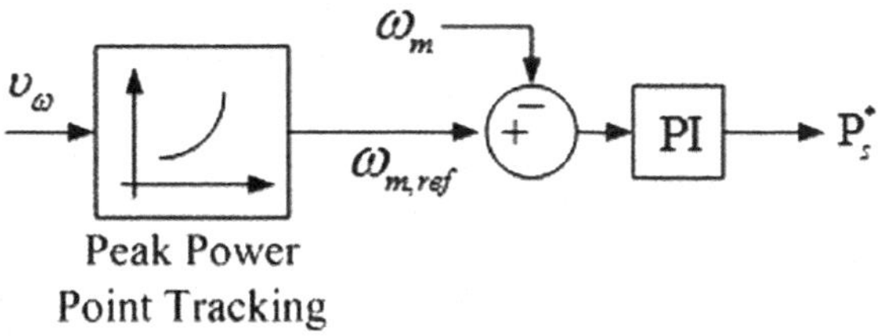

Fig. 4 Speed regulator based reference power calculation

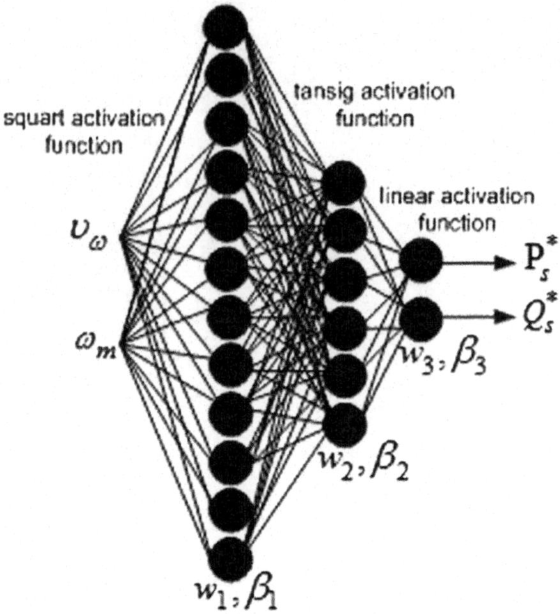

Fig. 5 Reference P and Q calculation sub-networks, block diagram (B)

$$w_1 = \begin{pmatrix} -1.123 & -0.122 & 2.761 & -0.475 & 0.076 & -2.121 & 0.555 & 0.654 & 1,120 & 0.371 & 1.664 & -2.090 \\ 0.089 & -1.332 & 1.729 & -0.113 & 0.774 & 1.101 & -0.443 & 1.934 & -2.109 & 1.554 & -1.030 & 0.886 \end{pmatrix}^t, \tag{29}$$

$$\beta_1 = \left(-0.009\ 0.045\ 0.998\ 0.645\ -1.009\ 0.332\ 1.035\ 0.878\ -0.679\ 1.320\ 0.445\ -0.221 \right)^t, \tag{30}$$

$$w_2 = \begin{pmatrix} -0.339 & 1.634 & 2.092 & -1.360 & 1.564 & -0.561 & -1.113 & 0.915 & 0.007 & 2.245 & -1.799 & 1.228 \\ -2.333 & 0.831 & 0.231 & 0.067 & -2.419 & -1.404 & 0.099 & -2.341 & 2.303 & -0.567 & -0.354 & -0.335 \\ 1.409 & 0.991 & -0.121 & 1.309 & -2.345 & 0.553 & -1.221 & 0.330 & -2.991 & 1.660 & -1.310 & 1.450 \\ 1.231 & -0.241 & 1.538 & -0.440 & 1.440 & -2.001 & 1.556 & 0.508 & 2.599 & -0.341 & 1.495 & 1.224 \\ -2.309 & 3.076 & 0.045 & 2.309 & -0.092 & 1.449 & -1.564 & -0.244 & 2.231 & 0.551 & -0.433 & -2.577 \\ 1.300 & -1.201 & -2.336 & -1.209 & 1.490 & 0.724 & -2.591 & -0.117 & -0.591 & 2.058 & -0.665 & -3.002 \end{pmatrix}, \tag{31}$$

$$\beta_2 = \left(1.129\ 2.265\ -0.774\ -2.487\ 1.204\ -2.300 \right)^t, \tag{32}$$

$$w_3 = \begin{pmatrix} 0.331 & -0.359 & 1.145 & -3.091 & -0.855 & 1.449 \\ -0.993 & -3.322 & 1.302 & -3.334 & 0.089 & -1.454 \end{pmatrix}, \tag{33}$$

$$\beta_3 = \begin{pmatrix} 0.172 \\ 0.389 \end{pmatrix}, \tag{34}$$

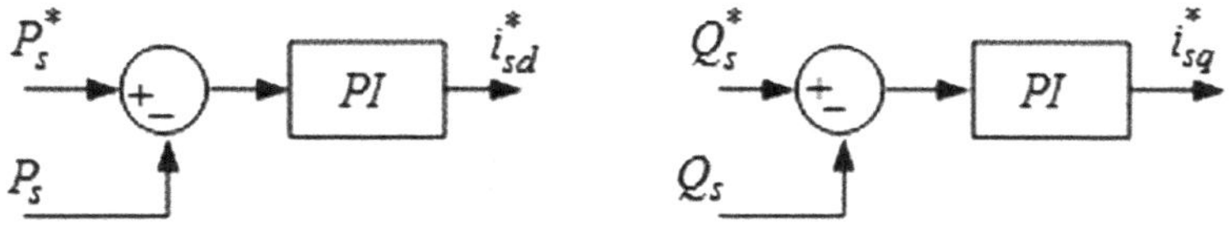

Fig. 6 Reference stator current, block diagram (C)

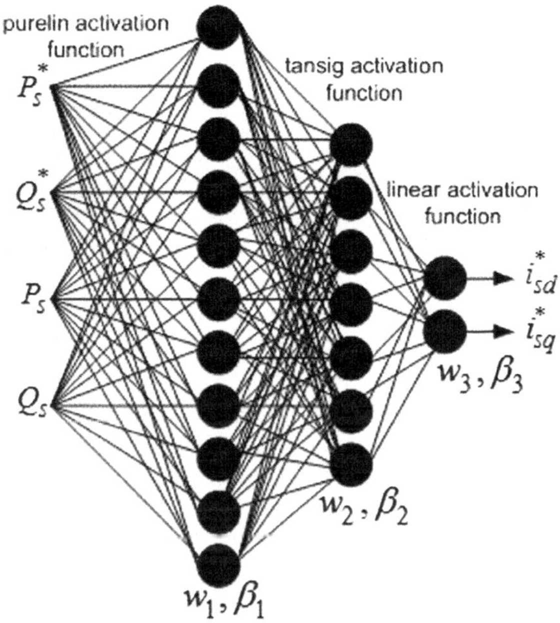

Fig. 7 Reference stator current sub-networks, block diagram (C)

3.3 Reference Stator Current Sub-networks

Knowing that the P-Q control is decoupled, the *d-q* reference components of the stator currents are obtained by the comparison the real and reference powers as shown in Fig. 6. To implement the calculation of the reference stator current we use an ANN constituted of 3 layers with a total of 20 neurons with strict limit as shown in Fig. 7. The first layer has 11 tansig activation function neurons, the 2nd layer has 7 purelin-activation function neurons, and 3rd layer has 2 linear activation function neurons. The 4 random inputs of the reference stator current are $[P_s, P_s^*, Q_s, Q_s^*]$, the ANN is trained by a supervised technique with perception training rule [23, 24]. After 62 training epochs, the sum squared error E converges to the zero value.

$$w_1 = \begin{pmatrix} -1.443 & 1.442 & 2.566 & 1.332 & -2.980 & 1.639 & -0.995 & 0.323 & 0.671 & -1.439 & 3.609 \\ 2.099 & -0.677 & -1.564 & 2.435 & -3.859 & 2.492 & 1.111 & 2.447 & -0.509 & 1.441 & 0.783 \\ -1.322 & 2.768 & -1.342 & -0.543 & 1.548 & -1.099 & 1.555 & -1.344 & -3.678 & 0.712 & 1.951 \\ -0.765 & 1.239 & -3.712 & -2.881 & 0.878 & 1.089 & 0.775 & 1.114 & -1.554 & -2.611 & 0.671 \end{pmatrix}^t, \tag{35}$$

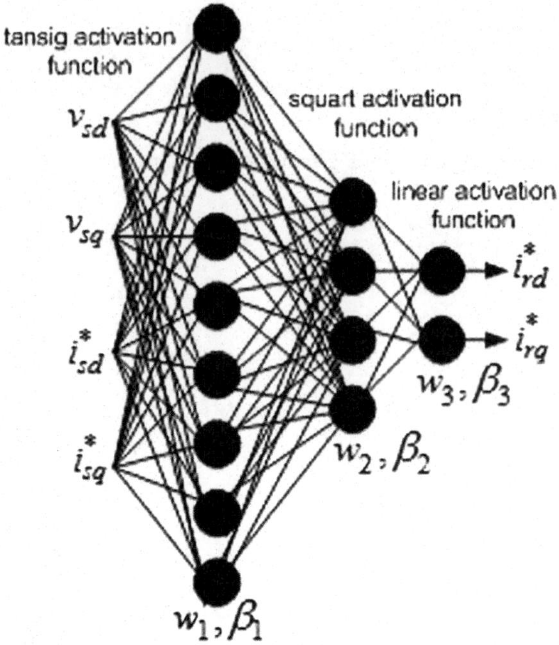

Fig. 8 Reference rotor current sub-networks, block diagram (D)

$$\beta_1 = \begin{pmatrix} 1.331 & -2.850 & -0.403 & 1.392 & -2.554 & -0.655 & -1.098 & 2.756 & -1.443 & 0.675 & -1.093 \end{pmatrix}, \tag{36}$$

$$w_2 = \begin{pmatrix} 0.564 & -2.098 & -1.856 & 1.564 & 0.474 & -2.055 & 1.500 & -1.574 & -0.578 & -3.009 & 0.384 \\ 2.509 & -0.498 & 3.317 & -0.403 & 1.309 & 0.083 & 1.309 & 2.509 & -1.451 & 0.419 & -3.889 \\ 1.092 & -0.009 & -2.880 & 1.709 & -1.248 & 0.980 & -2.678 & 0.577 & -0.990 & 1.234 & 2.456 \\ -3.291 & 1.554 & 0.707 & -1.093 & 2.094 & -1.543 & -1.509 & -1.099 & -0.883 & 1.302 & 0.509 \\ -1.392 & -2.453 & -1.758 & 2.904 & 3.001 & 2.484 & -0.034 & 2.098 & 1.667 & -2.498 & -1.447 \\ 0.449 & 1.449 & 2.904 & -1.495 & -3.494 & 1.495 & -0.433 & -1.598 & 1.007 & -0.443 & -1.303 \\ -1.345 & 1.099 & -3.009 & -1.309 & 2.992 & -0.941 & 1.958 & -1.354 & -2.750 & 1.329 & 2.440 \end{pmatrix}, \tag{37}$$

$$\beta_2 = \begin{pmatrix} -3.344 & 0.009 & 1.503 & -2.812 & -1.808 & -2.109 & 0.221 \end{pmatrix}^t, \tag{38}$$

$$w_3 = \begin{pmatrix} 2.192 & -4.182 & -3.093 & 1.123 & -2.712 & -1.092 & -0.219 \\ -1.298 & 2.001 & 1.271 & 0.271 & -1.929 & -3.022 & 2.832 \end{pmatrix}, \tag{39}$$

$$\beta_3 = \begin{pmatrix} 1.221 \\ 2.034 \end{pmatrix}, \tag{40}$$

3.4 Reference Rotor Current Sub-networks

According to the d-q components of the stator voltages and reference currents we can determine the reference rotor currents from the steady state DFIG system equations. The reference rotor d-q currents equations are provided as follows:

$$i_{rd}^{*} = \frac{-\omega_s L_{ss} i_{sd}^{*} R_s i_{sq}^{*} + v_{sq}}{\omega_s L_m}, \tag{41}$$

$$i_{qr}^{*} = \frac{R_s i_{sd}^{*} \omega_s L_{ss} i_{sq}^{*} - v_{sd}}{\omega_s L_m}, \tag{42}$$

We use a sub-networks consisting of 3 layer fixed-weight with 15 neurons. The 1st layer is a tansig-active function with the weight is w_1 and bias is β_1, the 2nd layer is a square-activation function with the weight is w_2 and bias is β_2 while the 3rd layer is a linear-active function with the weight is w_3 and bias is β_3 as illustrated in Fig. 8. The 4 random inputs of the reference rotor current are $[v_{sd}, v_{sq}, i_{sd}^{*}, i_{sq}^{*}]$, the ANN is trained by a supervised technique with perception training rule [23, 24]. After 12 training epochs, the sum squared error E converges to the zero value.

$$w_1 = \begin{pmatrix} -1.332 & 4.009 & 2.896 & -1.430 \\ -0.853 & 1.665 & -2.155 & 0.344 \\ 3.800 & 2.767 & -1.409 & -2.669 \\ 3.298 & -1.546 & -2.669 & 1.420 \\ 1.935 & 0.513 & -3.981 & 2.773 \\ 0.273 & -1.453 & -2.512 & -1.445 \\ -2.563 & -1.475 & 3.445 & 0.648 \\ 1.009 & 1.475 & -4.551 & -1.342 \\ 0.645 & -3.876 & 1.938 & -2.424 \end{pmatrix}, \tag{43}$$

$$\beta_1 = \left(0.545\ -2.187\ -1.667\ 2.833\ 0.919\ -1.364\ 0.334\ -1.202\ 3.909\right)^t, \tag{44}$$

$$w_2 = \begin{pmatrix} 0.553 & -3.193 & -2.471 & 1.485 & 4.008 & 2.587 & 1.480 & -2.814 & 3.823 \\ -2.556 & 0.764 & -1.495 & 1.366 & 0.445 & 2.565 & -1.342 & 4.551 & 0.568 \\ 1.345 & 0.341 & -3.281 & -2.445 & -0.366 & 1.379 & 0.778 & 1.112 & 0.994 \\ -2.173 & -2.336 & 1.366 & 0.993 & -3.412 & -2.615 & 1.887 & 3.710 & 0.815 \end{pmatrix}, \tag{45}$$

$$\beta_2 = \left(1.665\ -3.288\ 0.981\ 1.282\right)^t, \tag{46}$$

$$w_3 = \begin{pmatrix} 2.199 & -4.120 & -4.773 & 2.188 \\ 1.309 & 1.312 & -3.557 & 1.365 \end{pmatrix}, \tag{47}$$

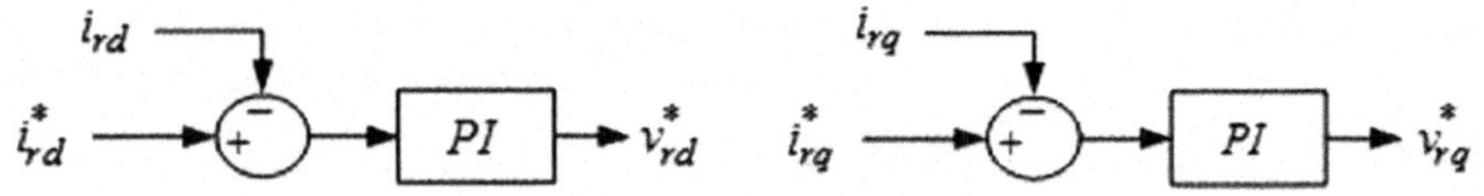

Fig. 9 Reference rotor voltage with PI controller, block diagram (E)

$$\beta_3 = \begin{pmatrix} 1.662 \\ -3.312 \end{pmatrix}, \tag{48}$$

3.5 *Reference Rotor Voltage Sub-networks*

In order to obtain the reference rotor voltages we compared the actual and reference *d-q* rotor voltages as presented in Fig. 9. Owing to the signals noise and harmonic problem, the proportional integral controllers were fine tuned for decrease of the produced noise in the feed-back circuit, ensure stable steady state running, quick and accurate tracking of the reference values and neither steady error. The PI estimator is capable to separate the active and reactive powers and therefore does not need feed forward control. The sub-net of reference rotor voltage computation is built of 2 layer networks with a overall of 12 hard limit neurons as presented in Fig. 10. The first layer has 10 purlin-activation function neurons, and the 2nd layer has 2 linear neurons. The 4 random inputs of the reference rotor voltage calculation are $[i_{rd}, i_{rq}, i^*_{rd}, i^*_{rq}]$, the ANN is trained by a supervised technique with perception training rule [23, 24]. After 38 training epochs, the sum squared error E converges to the zero value.

$$w_1 = \begin{pmatrix} -3.551 & -2.129 & 3.281 & 3.201 & 2.199 & 1.006 & -2.385 & -0.234 & 3.294 & 4.004 \\ 0.574 & 2.509 & -4.384 & 0.467 & -1.354 & 2.471 & 2.496 & 3.299 & -1.357 & -3.699 \\ -2.290 & -0.619 & 3.571 & 0.223 & -4.382 & 1.384 & -3.299 & 2.487 & 0.224 & -3.455 \\ 1.391 & -3.991 & -1.206 & 0.498 & 0.998 & 2.476 & -3.580 & -0.375 & 1.375 & 2.193 \end{pmatrix}^t, \tag{49}$$

$$\beta_1 = \begin{pmatrix} -4.390 & 2.131 & 3.288 & -2.183 & 0.221 & -1.263 & 1.340 & 1.008 & -3.378 & 2.222 \end{pmatrix}^t, \tag{50}$$

$$w_2 = \begin{pmatrix} -1.366 & -4.050 & 2.298 & 0.264 & 0.908 & -3.293 & -2.480 & 3.193 & 3.821 & 1.302 \\ 3.093 & 2.476 & 1.388 & -2.477 & 1.384 & 1.310 & -2.384 & 4.391 & -3.199 & -3.209 \end{pmatrix}, \tag{51}$$

$$\beta_2 = \begin{pmatrix} -1.352 \\ 0.089 \end{pmatrix}, \tag{52}$$

Figure 12 show a comparison of Root Mean Square Error (RMSE) between block (A), (B), (C), (D) and (E) after 150 iterations. For the grid-side estimator, it was built a block diagrams (A-B-C-D-E) ANNs configuration as is illustrate in Fig. 11 with

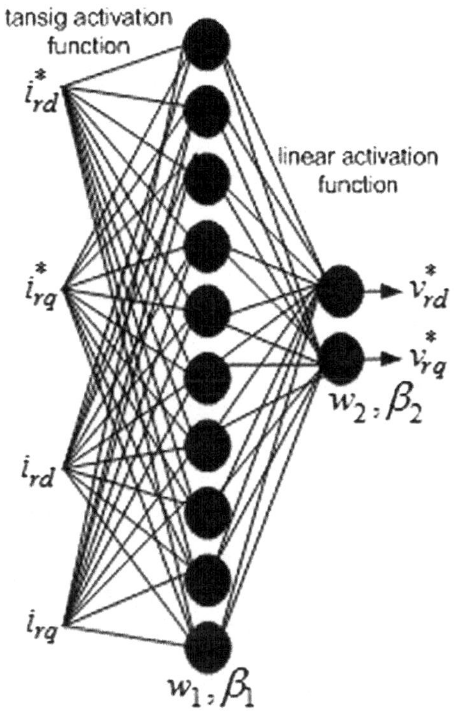

Fig. 10 Reference rotor voltage sub-network, block diagram (E)

the principal parameters in Appendix. To test the claims of the wind speed estimation based on ANN facing different wind speed will be operating at different operating point, the dynamic behavior of a DFIG machine is simulated at two different wind speeds i.e. at below rated wind speed and at rated wind speed. A 3-phase fault is applied at the cable connecting the wind turbine with the grid (infinite bus) at 2.4 s and cleared after 280 ms. A comparison of our approach to that of the Ref. [25] is illustrated in the Figs. 13, 14 and 15, these latter shows the difference in dynamic response of a wind turbine operating at two different operating points when it receives two different wind speeds. In steady state condition, the per unit generator rotor speed estimated by the neural network is slightly higher for the turbine facing the rated wind speed because it is operating at $\omega_{\max}$ whereas at lower wind speed the rotor speed is below $\omega_{\max}$. It can be seen from these figures that the neural rotor speed, neural active power and neural reactive power dynamic response are more stable even when the turbine is facing rated wind speed as compared to when it is facing below rated wind speed.

Another comparison of our approach to that described in the [26] has been analyzed in the Figs. 16 and 17. in fact, in the test of the Neural P-Q decoupled control with the DFIG system, the reference value of V_{dc} is set as 1500 V and the reference value of the reactive power $Q^* = 0.00$ p.u. Figure 16 shows the system response to a step change in which V_{dc}^* is changed from 1500 to 1550 V. The simulation Fig. 16

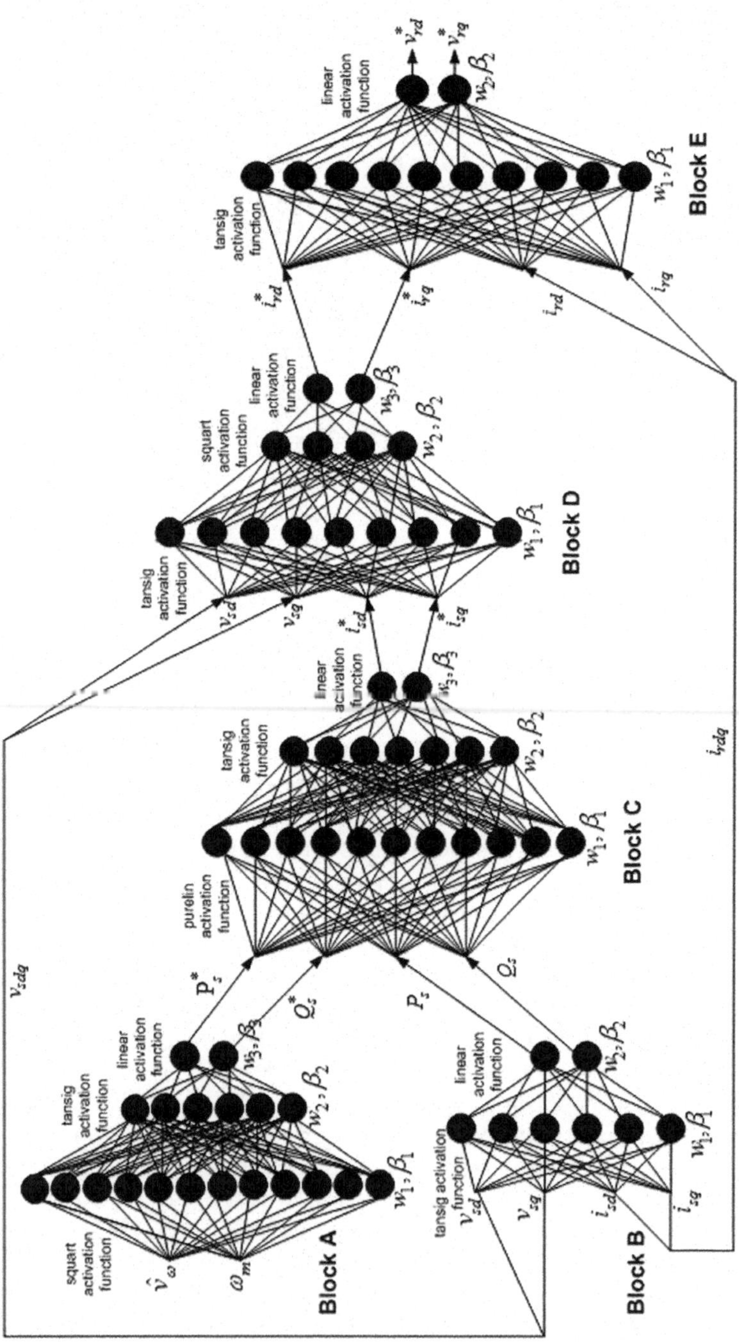

Fig. 11 Wind speed estimation and reference rotor voltage calculation based on neural networks of DFIG machine-side converter

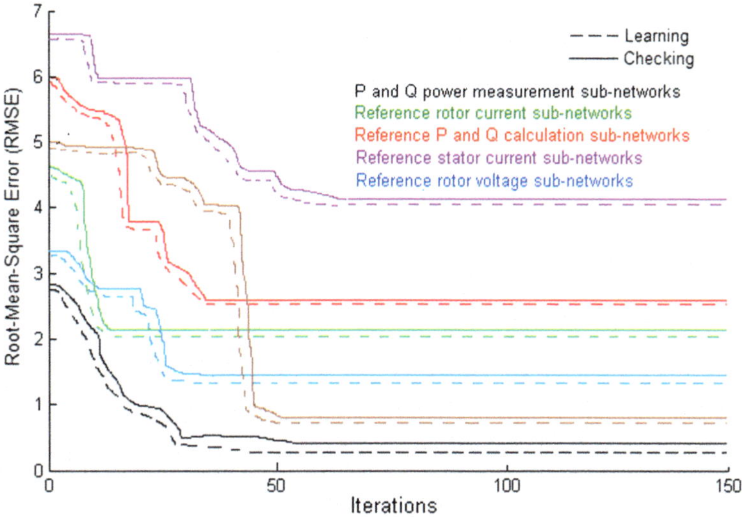

Fig. 12 Representation of the error trend between block (A), (B), (C), (D) and (E) after 150 iterations

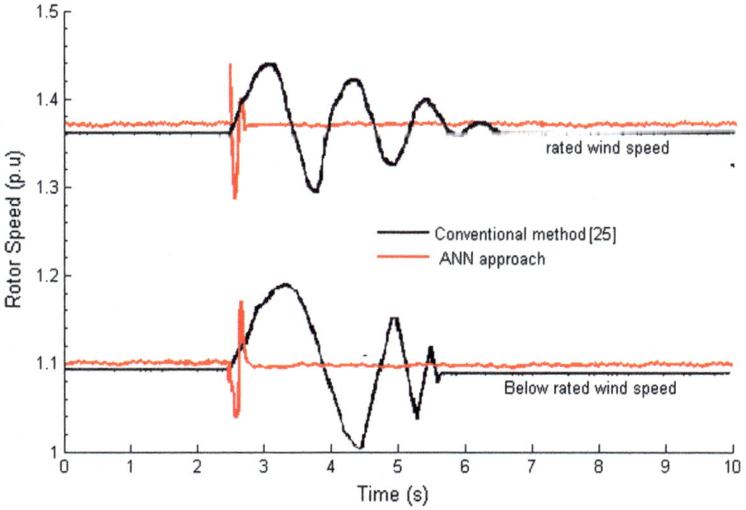

Fig. 13 Response of a DFIG machine under two wind speeds—Generator rotor speed

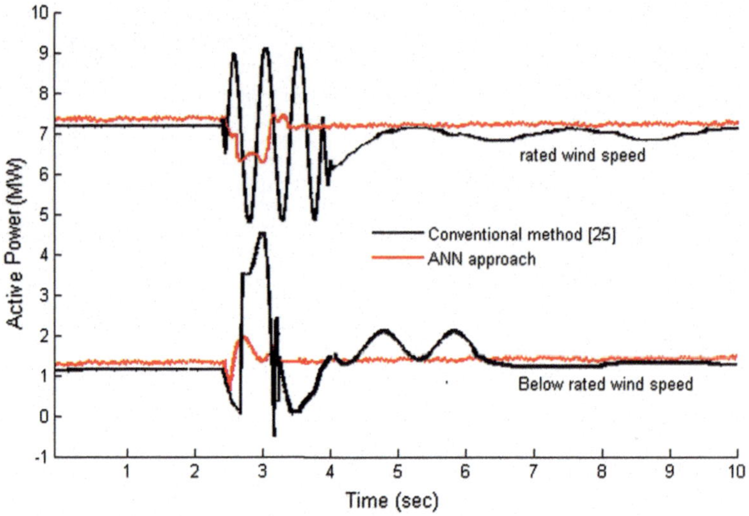

Fig. 14 Response of a DFIG machine under two wind speeds—Active power

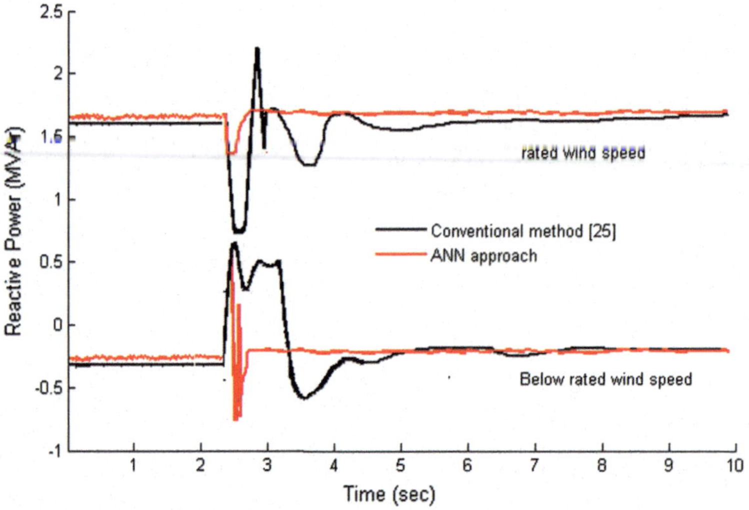

Fig. 15 Response of a DFIG machine under two wind speeds—Reactive power

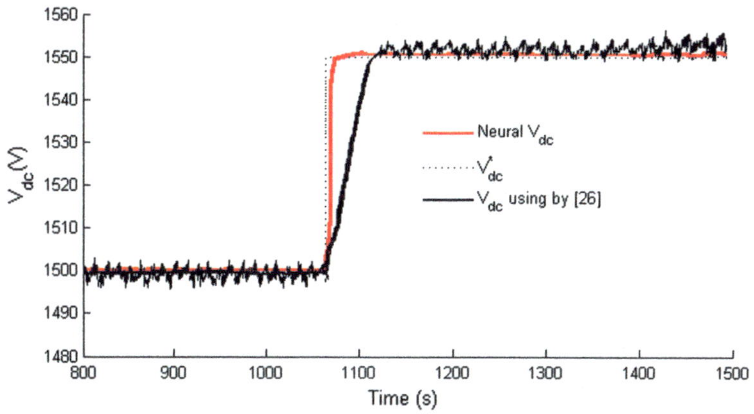

Fig. 16 Step change of V_{dc} from 1500 to 1550 V at $t = 1080$ s

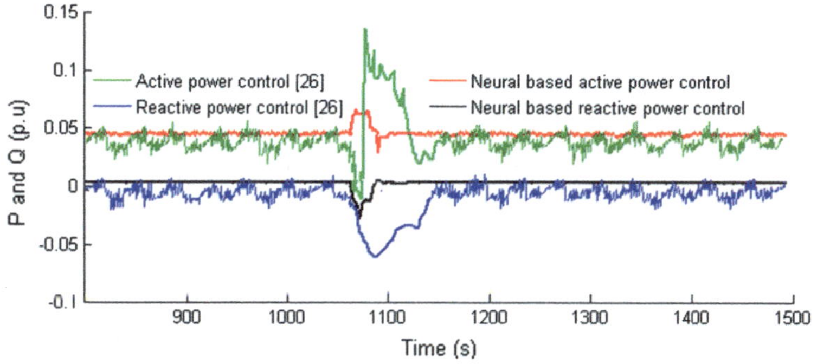

Fig. 17 ANN decoupled P-Q in step change of V_{dc} from 1500 to 1550 V at $t = 1080$ s

shows fast response in V_{dc} that settles to the desired value in 0.01 s. Figure 17 shows the P^* and Q^* during this transient.

The comparison proves that our approach neural can accurately estimate the wind speed, grid-side and rotor-side converters reference voltages under any wind condition. The use of ANN estimator will save simulation time while carrying out the transient stability studies.

4 Conclusion

This chapter has successfully simulated robust decoupled active and reactive control using artificial neural networks as part of a wind turbine driven doubly-fed induction generator, in a situation where the wind velocity is given a variable fluctuation in time, and in the case of changes of machine parameters. Indeed, the simulation and experimental results show there is no steady-state errors for the real and reactive powers compared to their reference values and the control of the real and reactive power are independent (decoupled control). The control remains effective for changes in machine parameters. Under transient conditions, and for step changes of the real and the reactive power references, the DFIG is capable of tracking the references with a response time of less than 1 second. This is fast enough for changes made by the power system operator, and for tracking wind speed variations. Thus, the sensorless measurement of the position is effective in controlling P-Q.

Appendix

Doubly Fed Induction Generator Parameters:
Rated power = 1 MVA
Stator/rotor turns ratio = 1:1
Stator resistance (R_s) = 0.00662 p.u.
Rotor resistance (R_r) = 0.01 p.u.
Stator inductance (L_s) = 3.185 p.u.
Rotor inductance (L_r) = 3.21 p.u.
Mutual inductance (L_m) = 3.1 p.u.
Base impedance (Z_{base}) = 10.98 Ω
Pole pairs (p) = 3
Frequency (f) = 50 Hz

Filter and Grid Parameters:
Inductor $L = 0.005$ H
Resistor $R = 0.25\ \Omega$
Capacitor $C = 4400\ \mu F$

Turbine Parameters:
Radius of the turbine $R_t = 13.5$ m
Gain multiplier $G = 65$
Inertia total moment $J = 10\ \text{Kg m}^2$
Air density $\rho = 1.22\ \text{Kg/m}^2$
Coefficient of viscous friction $f = 0.0001$
Optimal tip speed ration $\lambda_{opt} = 8.5$
Maximal power coefficient $C_{pmax} = 0.5$

References

1. Schmehl R (2017) Airborne wind energy: advances in technology development and research. Springer
2. Ruiz-Cruz R, Sanchez EN, Loukianov A, Ruz-Hernandez JA (2018) Real-time neural inverse optimal control for a wind generator. IEEE Trans Sustain Energy 1(1), Article in Press (2018). https://doi.org/10.1109/tste.2018.2862628
3. Vieto I, Sun J (2018) Sequence impedance modeling and analysis of Type-III wind turbines. IEEE Trans Energy Convers 33(2):537–545. https://doi.org/10.1109/tec.2017.2763585
4. Suppioni VP, Grilo AP, Teixeira JC (2018) Improving network voltage unbalance levels by controlling DFIG wind turbine using a dynamic voltage restorer. Int J Electr Power Energy Syst 96:537–545. https://doi.org/10.1016/j.ijepes.2017.10.002
5. Lodhe PC, Munje RK, Date TN (2015) Sliding mode control for direct power regulation of doubly fed induction generator. In: Paper presented at the 11th IEEE India conference: emerging trends and innovation in technology, INDICON. https://doi.org/10.1109/indicon.2014.7030494
6. Elkington K, Ghandhari M (2013) Non-linear power oscillation damping controllers for doubly fed induction generators in wind farms. IET Renew Power Gener 7(2):172–179. https://doi.org/10.1049/iet-rpg.2011.0145
7. Sguarezi Filho AJ, Filho ER (2012) Model-based predictive control applied to the doubly-fed induction generator direct power control. IEEE Trans Sustain Energy 3(3):398–406. https://doi.org/10.1109/tste.2012.2186834
8. Guo Y, Gao H, Wu Q, stergaard J, Yu D, Shahidehpour M (2019) Distributed coordinated active and reactive power control of wind farms based on model predictive control. Int J Electr Power Energy Syst 104:78–88. https://doi.org/10.1016/j.ijepes.2018.06.043
9. Das S, Subudhi B (2018) H^{∞} robust active and reactive power control scheme for a PMSG-based wind energy conversion system. IEEE Trans Energy Convers 33(3):980–990. https://doi.org/10.1109/TEC.2018.28030673
10. Darvish Falehi A (2014) Optimal design and analysis of NIOFPID-based direct power control to strengthen DFIG power control. J Dyn Syst Measurement Control 140(9):091001. https://doi.org/10.1115/1.4039485
11. Alba E, Mart R (2006) Metaheuristic procedures for training neural networks. Springer
12. Castillo O, Melin P, Kacprzyk J (2018) Fuzzy logic augmentation of neural and optimization algorithms: theoretical aspects and real applications. In: Studies in computational intelligence, vol 749. Springer
13. Ghoudelbourk S, Dib D, Omeiri A (2015) Decoupled control of active and reactive power of a wind turbine based on DFIG and matrix converter. Energy Syst 7(3):483–497. https://doi.org/10.1007/s12667-015-0177-1
14. Jerbi L, Krichen L, Ouali A (2009) A fuzzy logic supervisor for active and reactive power control of a variable speed wind energy conversion system associated to a flywheel storage system. Elect Power Syst Res 79(6):919–925. https://doi.org/10.1016/j.epsr.2008.12.006
15. Rajendran S, Parvathi Sankar DS, Govindarajan U (2014) Active and reactive power regulation in grid connected wind energy systems with permanent magnet synchronous generator and matrix converter. IET Power Electron 7(3):591–603. https://doi.org/10.1049/iet-pel.2013.0058
16. Hore D, Sarma R (2018) Neural network-based improved active and reactive power control of wind-driven double fed induction generator under varying operating conditions. Wind Eng: 0309524X1878040. https://doi.org/10.1177/0309524x18780402
17. Gupta N (2018) Tochastic optimal reactive power planning and active power dispatch with large penetration of wind generation. J Renew Sustaina Energy 10(2):025902. https://doi.org/10.1063/1.5010301
18. Ackermann T (2012) Wind power in power systems. Wiley. https://doi.org/10.1002/9781119941842

19. Monroy A, Alvarez-Icaza L (2006) Real-time identification of wind turbine rotor power coefficient. In: 45th IEEE conference on decision and control, pp 3690–3695. https://doi.org/10.1109/cdc.2006.376895
20. Shamshirband S, Petkovic D, Saboohi H, Anuar NB, Inayat I, Akib S, Cojbaic Z, Nikolic V, Mat Kiah ML, Gani A (2014) Wind turbine power coefficient estimation by soft computing methodologies: comparative study. Energy Convers Manag 81:520–526. https://doi.org/10.1016/j.enconman.2014.02.055
21. Abad G, López J, Rodríguez MA, Marroyo L, Iwanski G (2011) Doubly fed induction machine. Wiley. https://doi.org/10.1002/9781118104965
22. Peresada S, Tilli A, Tonielli A (2004) Power control of a doubly fed induction machine via output feedback. Control Eng Pract 12(1):41–57. https://doi.org/10.1016/S0967-0661(02)00285-X
23. Douiri MR, Belghazi O, Cherkaoui M (2015) Recurrent self-tuning neuro-fuzzy for speed induction motor drive. J Circuits Syst Comput 24(09):1550131. https://doi.org/10.1142/s0218126615501315
24. Douiri MR, Belghazi O, Cherkaoui M (2015) Neuro-fuzzy-based auto-tuning proportional integral controller for induction motor drive. Int J Comput Intell Appl 14(03):1550016. https://doi.org/10.1142/s1469026815500169
25. Xiong L, Wang J, Mi X, Khan MW (2018) Fractional order sliding mode based direct power control of grid-connected DFIG. IEEE Trans Power Syst 33(3):3087–3096. https://doi.org/10.1109/tpwrs.2017.2761815
26. Soares O, Gonalves H, Martins A, Carvalho A (2010) Nonlinear control of the doubly-fed induction generator in wind power systems. Renew Energy 35(8):1662–1670. https://doi.org/10.1016/j.renene.2009.12.008

An Indirect Adaptive Control Paradigm for Wind Generation Systems

Tariq Kamal, Murat Karabacak, Syed Zulqadar Hassan, Luis M. Fernández Ramírez, Indrek Roasto and Laiq Khan

Abstract Globally, there has been a significant evolution in the development of wind energy. Nevertheless, the major difference between the highly stochastic nature of wind speed and the desired demands regarding the electrical energy quality and system stability is the main concern in wind energy system. Hence, wind energy generation according to the standard parameters imposed by the power industry is unachievable without the essential involvement of advanced control technique. In this book chapter, a novel indirect adaptive control for wind energy systems is proposed

T. Kamal (✉)
Faculty of Engineering, Department of Electrical and Electronics Engineering, Sakarya University, Sakarya, Turkey
e-mail: tariq.kamal.pk@ieee.org

T. Kamal
Department of Electrical Engineering, Research Group in Electrical Technologies for Sustainable and Renewable Energy (PAIDI-TEP023), Higher Polytechnic School of Algeciras, University of Cádiz, Cádiz, Spain

M. Karabacak
Department of Electrical and Electronics Engineering, Faculty of Technology, Sakarya University of Applied Sciences, Sakarya, Turkey
e-mail: muratkarabacak@sakarya.edu.tr

S. Z. Hassan
School of Electrical Engineering, Chongqing University, Chongqing, China
e-mail: zulqadar@cqu.edu.cn

L. M. Fernández Ramírez
Department of Electrical Engineering, Research Group in Electrical Technologies for Sustainable and Renewable Energy (PAIDI-TEP-023), University of Cádiz, 11202 EPS Algeciras, Algeciras (Cádiz), Spain
e-mail: luis.fernandez@uca.es

I. Roasto
Department of Electrical Power Engineering and Mechatronics, Tallinn University of Technology (TalTech), Tallinn, Estonia
e-mail: indrek.roasto@ttu.ee

L. Khan
Department of Electrical Engineering, COMSATS University Islamabad, Abbottabad Campus, Khyber Pakhtunkhwa, Pakistan
e-mail: laiq@ciit.net.pk

R.-E. Precup et al. (eds.), *Advanced Control and Optimization Paradigms for Wind Energy Systems*, Power Systems, https://doi.org/10.1007/978-981-13-5995-8_10

considering real load demand and weather parameters. The performance of existing neuro-fuzzy scheme is improved further using a Hermite wavelet in the proposed architecture. The parameters of the controller are trained adaptively online via back-propagation algorithm. The proposed control law adopts the free direct control model which shorten the weight of the lengthy pre-learning, and memory requirements for real time application. Various computer simulation results and performance comparison indexes are given to show that the proposed control law is better in terms of efficiency, output power and steady-state performance over the existing state-of-the-art.

1 Introduction

Globally, the contribution of renewable energies to modern power generation system has increased significantly due to continuous fast depletion of petroleum usage, adversarial environmental effects, and economical benefits [1]. In this context, wind energy has received a substantial jump in the power industry and currently the fastest-growing (about 30% annually) energy source worldwide as compared to other renewable energy sources [2, 3]. This worldwide exploration of wind energy has happened due to the its mature nature, and low maintenance costs [4]. Though, due to the stochastic nature of the primary source (wind speed), its penetration creates volatility, and stability concerns in the system [5].

Due to the stochastic nature of wind, many control schemes are fast growing to increase the efficiency of wind power generation in energy conversion applications [6]. In wind system, the kinetic energy of wind is captured via wind-turbine blades and then convert it into electrical energy using an appropriate generator. Wind turbines are divided into constant and variable speed turbines. Many designers prefer variable speed Wind Turbine (WT) systems due their operation and efficiency (i.e., yield an increase of 38% in efficiency) [7]. Further, Permanent Magnet Synchronous Generator (PMSG) based wind generation systems have high power factor due to the absence of magnetizing current in PMSG. Therefore, the use of PMSG system has a high influence in the development of wind power applications. Since WT is a highly nonlinear and uncertain, and whatever the kind of the WT is used, the control methodology has a major effect on the wind power generation [8, 9]. Consequently, an optimal control methodology is required to harvest the maximum power with maximum dynamic efficiency.

In the literature, many researchers have developed Maximum Power Point Tracking (MPPT) control strategies for wind energy systems which include Optimal Torque Control (OTC) [10], Power Signal Feedback (PSF) [11], Tip Speed Ratio (TSR) [12] and Hill Climbing Search (HCS) [13, 14], and its modifications [15, 16]. The OTC control technique works on the trajectory of the optimum torque, calculated from a quadratic function of turbine speed. Though, the implementation of OTC algorithm is simple but the efficiency obtained from this technique is below the standard due to indirect measurement of the wind speed [10, 17]. For high-scale wind energy systems (e.g., multi megawatt wind turbines) PSF is observed as the most auspicious control algorithm. In PSF method, the control of Maximum Power Point (MPP) is performed through a turbine characteristic curve or a through adjusting electrical power/torque. Nevertheless, in PSF control the extraction of maximum power is difficult due to the dependency of tracking speed on the rotor inertia of a WT [18]. This issue is also well highlighted by many authors in [19–22]. Furthermore, in PSF MPPT control, 5% error usually exists between real wind power coefficient curves and ideal ones as reported by National Renewable Energy Laboratory [23]. This error reduces overall MPPT performances and causes 1–3% energy loss, which cannot be avoided in wind energy industries applications [24]. Some authors have also proposed modified adaptive adjust electrical torques based PSF MPPT control to increase the efficiency of Wind Energy Conversion Systems (WECS) [25, 26].

Anemometers are commonly essential in TSR MPPT scheme to calculate wind speeds [27], but its efficiency highly depends on the performance of sensors. This problem is addressed by researchers using wind speed estimation such as in [28]. Similarly, HCS algorithm is simple in implementation and does not need any prior information of the WT [29]. MPPT is accomplished when dP/dw = 0, through tuning rotor speed or duty cycle of a converter. Though, this technique needs many online computations and, therefore, it is challenging to accomplish MPPT for rapid-fluctuating wind speeds, which significantly reduces its performance. In [30], the authors have developed an improved HCS to capture MPPT procedures. The improvement of the algorithm is that it can instantly search the optimum operating point, therefore, decreasing the operation time. Nevertheless, better performance of the scheme needs anemometer accuracy

Moreover, some researchers have control WECS through non-linear advanced control methodologies such as sliding mode [31, 32], fuzzy control [33, 34], neural control [35], evolutionary computation [36, 37], predictive control [38], and neuro-fuzzy [39] among others. Though, the sliding mode is a popular scheme for high-order nonlinear systems in terms of robustness, but chattering the undesirable phenomenon of oscillations limits its real-world applications [40]. Likewise, some authors have proposed evolutionary computation (i.e., genetic algorithm, particle swarm optimization, etc.), approaches for wind generation, but choice of genetic encoding, convergence speed, and appropriate selection of control parameters are some obstacles regarding these techniques [41]. Predictive control has own concerns such as high computation efforts need to solve online the optimization problem [42]. Similarly, both fuzzy and neural techniques in the literature have exposed their importance in the control of wind. Nevertheless, the implementation of fuzzy needs a-priori-knowledge, while neural networks are computationally intensive to train, hence, their limitations are versatility [43]. Some researchers have preferred neuro-fuzzy control of wind, however, neuro-fuzzy controller fails to differentiate local minima of the search space [44]. In the literature, many researchers have addressed the local minima concern of the neuro-fuzzy structure by wavelets in its consequent part [45]. The integration of wavelets in neuro-fuzzy structure substantially enhances its computational speed [46]. Wavelet transform offers a time-frequency localization of the non-stationary signals. In wavelet transform, the basis functions are used to preserve the maximum energy of the signal [47].

All the above discussed control methodologies are modelled and designed for a certain operating condition. During high fluctuation (i.e., sudden change in load or wind speed) the system constraints may go out of bounds, which may cause instability in the system. An adaptive control paradigm is highly effective to handle non-linearity and unpredictable fluctuations. Roughly, there are types of adaptive control, i.e., direct and indirect [48]. In direct adaptive control, the systems stability sometimes uncertain due to the absence of the explicit identification of the unknown plant [49], while in indirect adaptive control, the closed-loop system stability achieves due the adaptation signal which is directly generated from a predefined control criterion.

In this chapter, Hermite wavelet based modified neuro-fuzzy indirect adaptive control technique is design to acquire maximum power from variable speed wind-turbine. The modeling of the proposed controller starts from the fuzzy control via expert knowledge which initialize the parameters. Then the parameters are updated adaptively online using backpropagation algorithm. Then, the comparison has been made in terms of efficiency, and output power with other maximum power techniques, i.e., Proportional Integral (PI) based HCS MPPT and conventional fuzzy based MPPT. Finally, the superior performance of the proposed controller is shown through simulation.

The layout of this chapter is given as follows. Background of wind energy generation system is given in Sect. 2. Section 3 provides the designing of the proposed controller. Comparison through results is covered in Sect. 4. Finally, conclusions are provided in Sect. 5.

2 Wind Energy Conversion System (WECS)

Typically, WECS consists of a WT, a generator, power electronics converters, interconnection apparatus, filter circuitry and control system. WT is the most important part in WECS which generates electrical energy from the kinetic energy associated with wind speed. The WECS global strategy is given in Fig. 1. The aerodynamic power produced by wind is given as:

$$P_\alpha = 0.5\pi\rho R^2 \Im(\varsigma) \nu_w^3 \tag{1}$$

where $\nu_o > \nu_w > \nu_i$. ς is the tip speed ratio and given as:

$$\varsigma = \frac{R\omega_r}{\nu_w} \tag{2}$$

Fig. 1 Wind energy conversion system

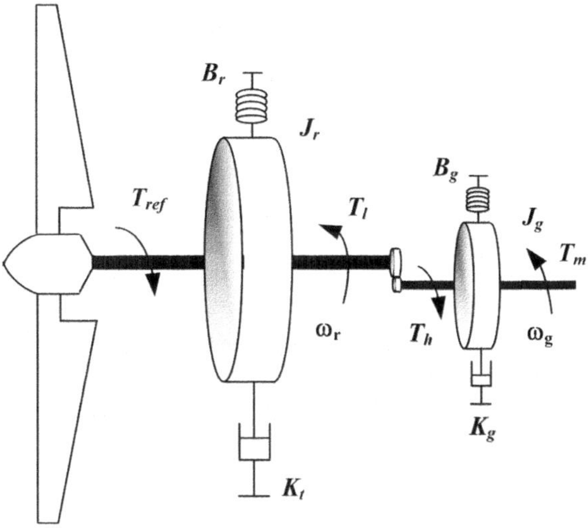

The rotor power and aerodynamic torque (T_α) are calculated as:

$$P_\alpha = \omega_r T_\alpha \tag{3}$$

$$T_\alpha = \frac{\pi R^3 \Im(\varsigma) v_w^2}{2\varsigma} \tag{4}$$

The T_α runs the wind-turbine at the ω_r (speed). T_h is the high speed torque, T_l is the low speed torque, T_{em} is the generator electromagnetic torque. The rotor speed is increased by the gear box. The gear box ratio, η_g to attain the generator speed, ω_g, while the T_l is augmented. For certain wind speed, the output power is calculated as:

$$P_\alpha = z\omega_r^2 \tag{5}$$

where $z = 0.5\rho A\Im(\frac{R}{\varsigma})^3$ and the optimal power obtained by rotational speed is illustrated in Fig. 2. For a certain wind speed, the optimal power is:

$$P_{\alpha,o} = z_o\omega_r^3 \tag{6}$$

where $z_o = 0.5\rho A\Im_o(\frac{R}{\varsigma_o})^3$.

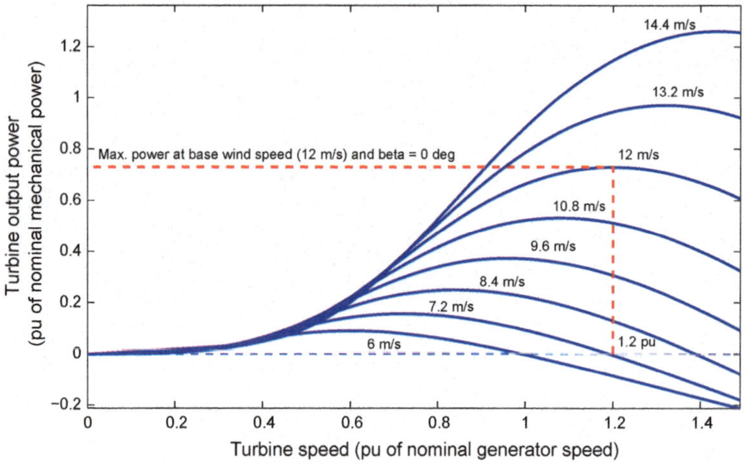

Fig. 2 Wind turbine output power at different speeds

2.1 *Permanent Magnet Synchronous Generator Model*

The rotor dynamics based on the generator inertia are given as follows:

$$\begin{cases} \frac{i_d}{dt} = -\frac{R_a}{L_d} i_d + \frac{L_q}{L_d} P\omega_r i_q + \frac{1}{L_q} v_d \\ \frac{i_d}{dt} = -\frac{R_a}{L_q} i_q - \frac{L_d}{L_q} P\omega_r i_d - \frac{1}{L_q} P\omega_r \varphi_m + \frac{1}{J_q} v_q \\ \frac{d\omega_r}{dt} = \frac{1}{J_t}(T_\alpha - T_g - k_t w_r - B_t \theta_r) \end{cases} \tag{7}$$

The d-axis and q-axis circuits of PMSG are shown in Fig. 3. The gear box ratio is defined as:

$$\eta_g = \frac{\omega_g}{\omega_r} = \frac{T_l}{T_h} \tag{8}$$

where

$$J_t = J_r + \eta_g^2 J_g k_t = k_r + \eta_g^2 k_g B_t = B_r + \eta_g^2 B_g T_g = \eta_g T_e \tag{9}$$

$$T_e = -\frac{3}{2} p_n (L_d - L_q) i_d i_q + \varphi_m i_q \tag{10}$$

where v_d, v_q and i_d, i_q are the stator d-axis and q-axis voltage and current components, respectively. R_a, ω and φ_m represent the stator resistance, frequency and flux linkage, respectively. The torque, T_g of the PMSG is defined as:

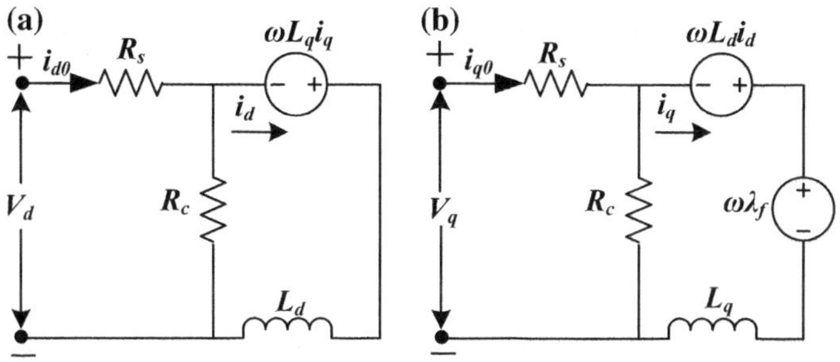

Fig. 3 The d-axis (**a**) and q-axis (**b**) circuits of PMSG

Fig. 4 Drive train as a single lumped mass

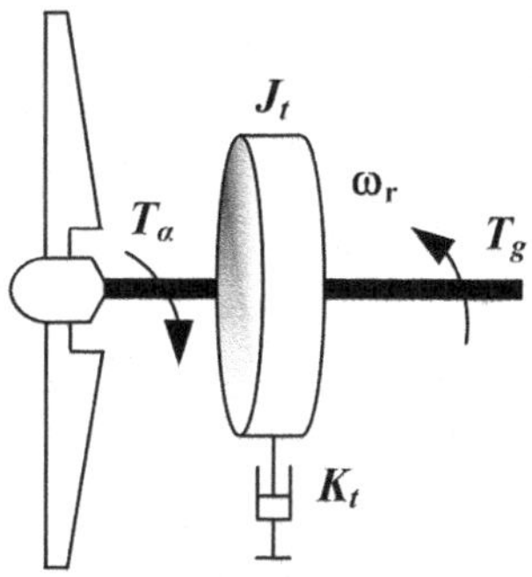

$$T_g = -\frac{3}{2} p_n \eta_g (L_d - L_q) i_d i_q + \varphi_m i_q \tag{11}$$

where p_n represents the number of pole pairs. Finally, the generated power is calculated as:

$$P_g = T_g \omega_r \tag{12}$$

As the external stiffness, B_t is quite small, it can be neglected. So, the drive train can be shown as a single lumped mass as shown in Fig. 4. The q-axis component of the stator current is written in-terms of d-axis component as follows:

$$i_q = \frac{-2T_g}{3 p_n \eta_g (L_d - L_q) i_d i_q + \varphi_m i_q} \tag{13}$$

The maximum efficiency of the PMSG is obtained by reducing the copper losses, P_{cn} and core losses, P_{core} which are calculated as:

$$P_{cn} = R_a (i_d^2 + i_q^2) \tag{14}$$

$$P_{core} = \frac{\omega^2 \{(L_d i_d + \varphi_m)^2 + (L_q i_q)^2\}}{R_c} \tag{15}$$

where R_c represents the core loss component. Therefore, the output power of the PMSG is calculated as:

$$P_{out} = P_a - P_{cn} - P_{core} = T_g\omega_r - R_s(i_d^2 + i_q^2) - \frac{\omega^2\{(L_d i_d + \varphi_m)^2 + (L_q i_q)^2\}}{R_c} \tag{16}$$

The 3ϕ terminal voltages of the PMSG are:

$$\begin{bmatrix} v_a(t) \\ v_b(t) \\ v_c(t) \end{bmatrix} = \frac{1}{\sqrt{3}} \begin{bmatrix} \sqrt{v_d^2 + v_q^2}\cos(\omega t + \phi_a) \\ \sqrt{v_d^2 + v_q^2}\cos(\omega t + \phi_a - \frac{2}{3}\pi) \\ \sqrt{v_d^2 + v_q^2}\cos(\omega t + \phi_a + \frac{2}{3}\pi) \end{bmatrix} = \begin{bmatrix} \sqrt{2}V_{rms}\cos(\omega t + \phi_a) \\ \sqrt{2}V_{rms}\cos(\omega t + \phi_a - \frac{2}{3}\pi) \\ \sqrt{2}V_{rms}\cos(\omega t + \phi_a + \frac{2}{3}\pi) \end{bmatrix} \tag{17}$$

where $V_{rms} = \frac{v_d^2 + v_q^2}{\sqrt{6}}$ and phase angle is $\phi_a = \tan^{-1}(\frac{v_q}{v_d})$.

2.2 3ϕ Diode Bridge Rectifier

Usually, a two-stage conversion (bridge rectifier and boost converter) is used to produce DC from the generated AC of WECS, i.e., AC is processed into DC via diode rectifier and the DC/DC power conversion is performed through boost converter to obtain the required voltage levels. Figure 5 shows 3ϕ bridge rectifier. The 3ϕ voltages at the terminal of bridge rectifier are:

$$v_1 = v_m\cos(\omega_a t) \tag{18}$$

$$v_2 = v_m\cos(\omega_a t - \frac{2\pi}{3}) \tag{19}$$

$$v_3 = v_m\cos(\omega_a t - \frac{4\pi}{3}) \tag{20}$$

Fig. 5 3ϕ diode bridge rectifier

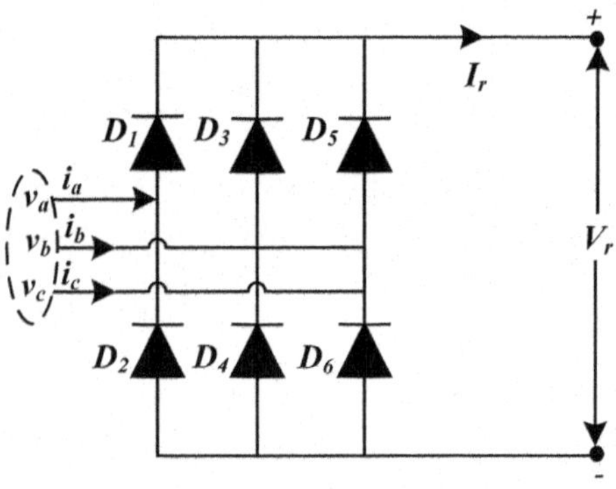

where $v_m = v_{p,rms}\sqrt{2}$. Suppose I_{out} is greater for the entire periods. The maximum phase voltages across two diodes at any instant is written as:

$$v_A = \max(v_a, v_b, v_3) \tag{21}$$

and the minimum phase voltages are calculated as follows:

$$v_B = \min(v_a, v_b, v_3) \tag{22}$$

Fourier series expansion for the output voltage at terminals A and B can be calculated as:

$$v_A = v_{in}\frac{3\sqrt{3}}{\pi}\left[0.5 + \sum_{-\infty}^{+\infty}\frac{(-1)^n + 1}{9n^2 - 1}\cos(3n\omega_o t)\right] \tag{23}$$

$$v_B = v_{in}\frac{3\sqrt{3}}{\pi}\left[-0.5 + \sum_{-\infty}^{+\infty}\frac{1}{9n^2 - 1}\cos(3n\omega_o t)\right] \tag{24}$$

The output voltage of bridge rectifier is calculated as:

$$v_{out} = v_A + v_B \tag{25}$$

2.3 DC–DC Boost Regulator

The boost regulator is illustrated in Fig. 6. The input currents and voltages across inductor and capacitor at any instant can be calculated as:

$$v_i = v_o(t)L\frac{di_L}{dt} \Rightarrow i_L(t) = \frac{1}{L}\int (v_i(t) - v_0(t))dt \tag{26}$$

$$i_c(t) = C\frac{dv_0(t)}{dt} \Rightarrow v_0(t) = \frac{1}{C}\int i_C(t)dt \tag{27}$$

$$v_L(t) = v_i(t) \times F + ((v_i(t) - v_0(t)) \times \overline{\mathrm{F}} \times sign(i_L(t)) \tag{28}$$

$$i_C(t) = -i_0(t) \times F + (i_L(t)) \times \overline{\mathrm{F}} \times sign(i_L(t)) = C\frac{dv_0}{dt} \tag{29}$$

The voltage at the output terminal of boost regulator is written as:

$$v_o(t) = \frac{1}{C}\int (i_c(t)dt) = \frac{1}{C}\int (-i_0(t) \times F + i_L(t) \times \overline{\mathrm{F}} \times sign(i_L(t))dt \tag{30}$$

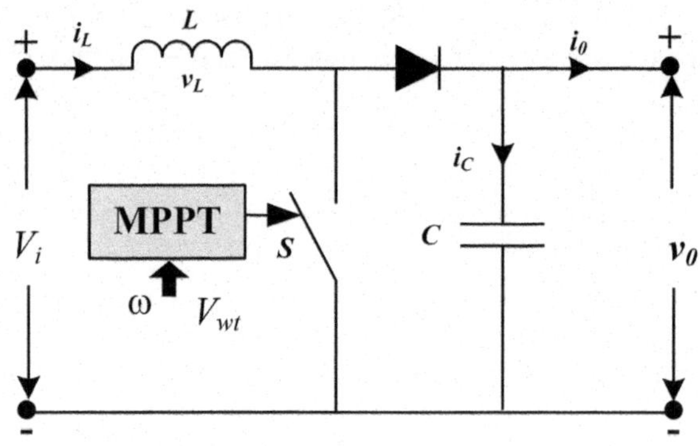

Fig. 6 Boost converter

Finally, the output voltage is simplified through Fourier series expansion as follows:

$$v_0 = v_{in}\frac{3\sqrt{3}}{3}\left[1 - \sum_{k=1}^{+\infty}\frac{2}{36k^2 - 1}\cos(6k\omega_0 t)\right] \tag{31}$$

The DC component voltage is calculated as:

$$v_{0,dc} = v_m\frac{3\sqrt{3}}{\pi} \approx 1.65v_m \approx 2.34v_{p,rms} \tag{32}$$

Similarly, the voltage AC component is calculated as:

$$v_{o,ac} = v_m\frac{3\sqrt{3}}{\pi}\left[1 - \sum_{k=1}^{+\infty}\frac{2}{36k^2 - 1}\cos(6k\omega_o t)\right] \tag{33}$$

The voltage across capacitor is calculated as:

$$v_C = -v_A - v_B \tag{34}$$

Fourier series expansion for v_C is given as:

$$v_C = v_m\frac{3\sqrt{3}}{\pi}\sum_{k=1}^{+\infty}\frac{2}{(6k-3)^2 - 1}\cos((6k-3)\omega_o t) \tag{35}$$

The currents of the diodes are:

$$i_{Dk} = d_k(\omega_o t)I_o \quad for \quad k \in \{1, 2, 3, 4, 5, 6\} \tag{36}$$

$$I_D = \frac{1}{3}I_o \tag{37}$$

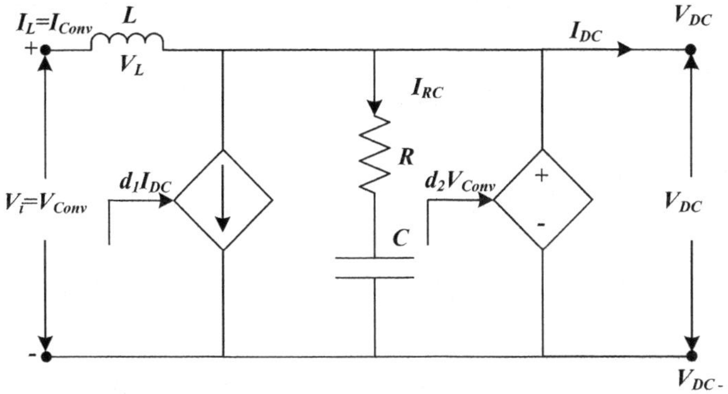

Fig. 7 Average model of the DC voltage regulator

The output power of rectifier is calculated as follows:

$$P_o = V_o I_o = v_m I_o \frac{3\sqrt{3}}{\pi} = P_{in} \tag{38}$$

2.4 Average Model of DC–DC Boost Regulator

The average model of DC voltage regulator is illustrated in Fig. 7. The regulator output voltage, V_{DC} is calculated as:

$$V_{DC} = d_2 V_{Conv} \tag{39}$$

$$I_{RC} = I_L - d_1 I_{DC} \tag{40}$$

$$V_{DC} = d_2 V_{Conv} \tag{41}$$

where d_1 and d_2 are duty cycles which can be calculated using Kirchhoff's voltage law:

$$V_{Conv} - V_L = I_{RC} R + \frac{1}{C}\int I_{RC} dt \Rightarrow V_{Conv} = V_L + I_{RC}\left(R + \frac{1}{C}\int dt\right) \tag{42}$$

Putting V_{Conv} in (39), the value of V_{DC} is calculated as:

$$V_{DC} = d_2\left\{V_L + I_{RC}\left(R + \frac{1}{C}\int dt\right)\right\} \tag{43}$$

From (40) put the value of I_{RC} in (43), we get the final value of V_{DC} as follows:

$$V_{DC} = d_2 \left\{ V_L + I_L - I_{DC} d_1 \left(R + \frac{1}{C} \int dt \right) \right\} \tag{44}$$

The output power at the DC bus is:

$$P_{DC} = V_{DC} \times I_{DC} \tag{45}$$

Finally, the maximum power injected by WECS is calculated as:

$$P_{mpp} = V_{mpp} I_{mpp} \tag{46}$$

where $P_{mpp} = V_{DC}$ and $V_{mpp} = I_{DC}$ are the voltage and current at MPP.

3 Indirect Adaptive Hermite-Wavelet Based Controller (IAHWBC)

For WECS, the maximum power of point extraction is carried out via an indirect adaptive Hermite-wavelet based algorithm. The WECS system model $f_{IAHWBC-WT}$ $(\mho(n))$ is first identified online as shown in Fig. 8, where $\mho(n) = y(n-1), \ldots, y(n-p), u(k-1), \ldots, u(k-q)$. The nonlinear subsystem dynamic model, $f_{IAHWBC-WT}$ $(\mho(n))$ for WECS is identified by using the following general objective function:

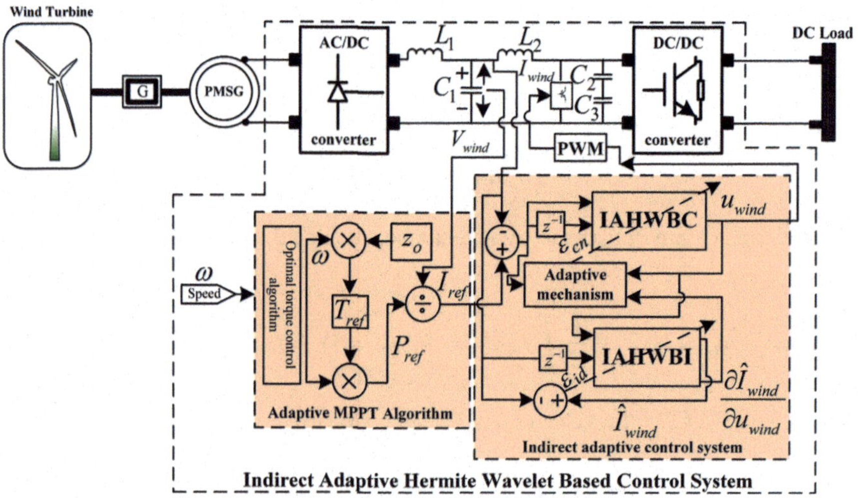

Fig. 8 WECS-PMSG subsystem closed-loop and adaptive control model

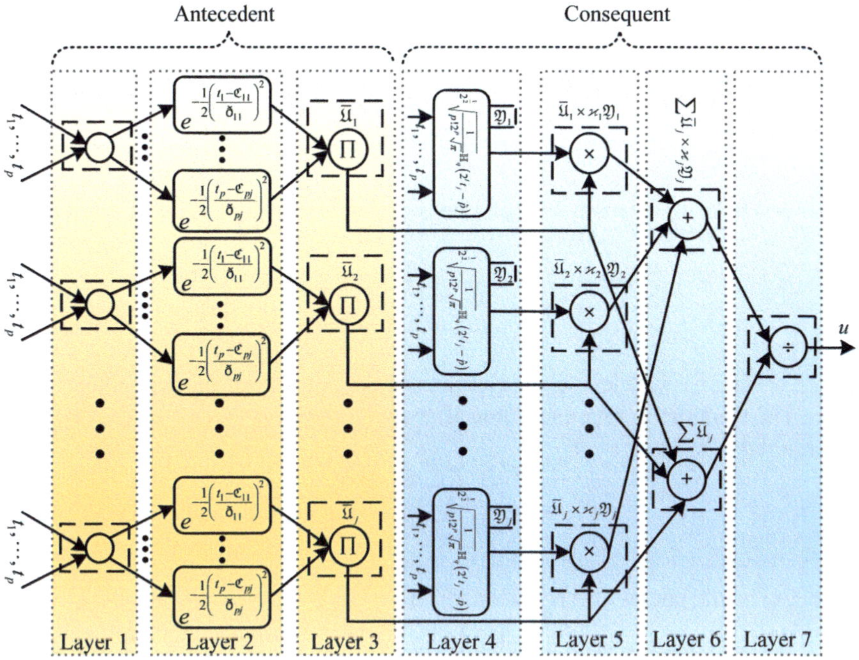

Fig. 9 Neurofuzzy structure of the proposed controller

$$Min\ \mathfrak{A}_{wind-id} = \frac{1}{2}\left[I_{wind}(n) - \hat{I}_{wind}(n)\right] \tag{47}$$

s.t.

$$P_{wind}(n) = \begin{cases} 0\,(WT - stall) & if \quad \nu_w < \nu_i \quad or \quad \nu_w > \nu_o \\ \mathfrak{Z}_{gr}\mathfrak{Z}_{gn}k_{wind}\Im(\varsigma, \mathfrak{B})\frac{\rho A}{2}\nu_w^3 & if \quad \nu_i < \nu_w \nu_r \\ \mathfrak{Z}_{gr}\mathfrak{Z}_{gn}k_{wind}P_r & if \quad \nu_r < \nu_w \nu_o \end{cases}$$

where ν_r, ν_o, ν_i and ν_w are the reference, cut-out, cut-in and current wind speed, respectively. $\mathfrak{Z}_{gn}$ and $\mathfrak{Z}_{gr}$ are the generator and gearbox efficiencies, respectively. Similarly, A and $\mathfrak{B}$ are the turbine swept area and blade pitch angle, respectively. There are seven layers in the proposed structure as shown in the Fig. 9. IF-THEN rules are defined as follows:

$$R^i : \text{if } t_1 \text{ is } \mathfrak{U}_1^i,\ t_2 \text{ is } \mathfrak{U}_2^i \text{ and } t_p^i \text{ is } \mathfrak{U}_p^i,\ \text{ then } y = \mathfrak{Y}_{pq}$$

These seven layers are described as follows:

Layer 1: This is input layer which collects the input from the plant. These inputs are directly sent to the fuzzification layer (i.e., layer 2) using nodes.

Layer 2: Here the linguistic terms and their degree of membership are allocated to each input. The linguistic terms using Gaussian membership function for each input as follows:

$$\mathfrak{U}_j^i = e^{\frac{-1}{2}\left[\frac{t_j - \mathfrak{C}_j^i}{\mathfrak{d}_j^i}\right]^2} \tag{48}$$

Layer 3: The firing intensity of each rule is determined using product T-norm in this layer.

$$\bar{\mathfrak{U}}_j^i = \prod_{i=1}^{q} \mathfrak{U}_j^i \tag{49}$$

Layer 4: This is wavelets layer. Hermite wavelet functions are integrated in this layer. The weighted consequent value after applying Hermite wavelet of each rule is determined as follows:

$$\wp^i = \sum_{j=1}^{2^{n-1}} \varkappa_j^i \mathfrak{Y}_{pq}(t) \tag{50}$$

where, $\wp^i$ is the output of this layer and $\varkappa$ represents weighting factor.

Layer 5: The output of layer 3 which is the last layer of antecedent part and layer 4 which is first layer of consequent part are multiplied for each input and then they are sum.

Layer 6: This layer provides the sum of output of layer 3.

Layer 7: The final output of neurofuzzy structure is calculated and given as:

$$u(t) = \frac{\sum_{j=1}^{p} \bar{\mathfrak{U}}_j^i \wp^i}{\sum_{j=1}^{p} \bar{\mathfrak{U}}_j^i} \tag{51}$$

Where $i = 1, \ldots, q$ and $j = 1, \ldots, p$.

The above derived wavelet based neurofuzzy stricture is proven as a universal approximator for continuous functions over compact sets. For optimizing adaptive function $f_{IAHWBC-WT}(\mho(n))$, it needs the adaptation of several parameters, $\aleph_j^i \in \{\mathfrak{C}_j^i, \mathfrak{d}_j^i, \varkappa_j^i\}$ which is based on $\mathfrak{A}_{wind-id}$. Where $\mathfrak{C}_j^i$ is the mean of the Gaussian membership function for jth input, ith rule, $\mathfrak{d}_j^i$ is the variance of the Gaussian membership function for jth input, ith rule and $\varkappa_j^i$ is the weight of the Gaussian membership function for jth input, ith rule. Parameters for training algorithm are $\mathfrak{C}_j^i$, $\mathfrak{d}_j^i$ and $\varkappa_j^i$. Gradient descent method is used to tune the parameters, the generalized update equation is written as follows:

$$\aleph_j^i(n+1) = \aleph_j^i(n) + \Upsilon_{wind}\varepsilon_{id}\frac{\partial \hat{I}_{wind}(n)}{\partial \aleph_j^i(n)} \tag{52}$$

where Υ_{wind} and $\varepsilon_{id} = I_{wind}(n) - \hat{I}_{wind}(n)$ denotes the learning rate and identification error. The update equations for all parameters are:

$$\mathfrak{C}_j^i(n+1) = \mathfrak{C}_j^i(n) + \Upsilon_{wind}\varepsilon_{id}\left[\left\{\frac{\wp_j - \hat{I}_{wind}(n)}{\sum_{j=1}^{p}\mathfrak{U}_j}\right\}\mathfrak{U}_j\left\{\frac{t_j - \mathfrak{C}_j^i}{\mathfrak{d}_j^{i^2}}\right\}\right] \tag{53}$$

$$\mathfrak{d}_j^i(n+1) = \mathfrak{d}_j^i(n) + \Upsilon_{wind}\varepsilon_{id}\left[\left\{\frac{\wp_j - \hat{I}_{wind}(n)}{\sum_{j=1}^{p}\mathfrak{U}_j}\right\}\mathfrak{U}_j\left\{\frac{(t_j - \mathfrak{C}_j^i)^2}{\mathfrak{d}_j^{i^3}}\right\}\right] \tag{54}$$

$$\varkappa_j^i(n+1) = \varkappa_j^i(n) + \Upsilon_{wind}\varepsilon_{id}\left[\left\{\frac{\mathfrak{U}_j}{\sum_{j=1}^{p}\mathfrak{U}_j}\right\}\mathfrak{Y}_{pq}(t_j)\right] \tag{55}$$

where t_j is the jth input, $\wp_j = \varkappa_j^i \times \mathfrak{Y}_{pq}(t_j)$ and $\mathfrak{Y}_{pq}(t_j)$ is the Hermite wavelet which is taken from the interval, [0, 1] as follows [50]:

$$\mathfrak{Y}_{pq}(t_j) = \begin{cases} 2^{\frac{l}{2}}\sqrt{\frac{1}{p!2^p\sqrt{\pi}}}\mathbb{H}_q(2^l t_j - \hat{p}), & \frac{\hat{p}-1}{2^l} \le t_j \le \frac{\hat{p}+1}{2^l} \\ 0 & \text{Otherwise} \end{cases} \tag{56}$$

where $p = 1, 2, \ldots, 2^{l-1}, \hat{p} = 2p - 1$ which explains the translation parameter and $q = 0, 1, \ldots, Q - 1$ defines the order of polynomial. Where $\mathbb{H}_q$ denotes an orthogonal related to the weight function as:

$$\int_{-\infty}^{\infty} e^{-t^2}\mathbb{H}_q\mathbb{H}_p = \begin{cases} 0, & q \ne p \\ p!2^p\sqrt{\pi} & q = p \end{cases} \tag{57}$$

where $\mathbb{H}_q$ are Hermite polynomials which can be calculated as:

$$\mathbb{H}_0 = 1, \mathbb{H}_1 = 2t_j \text{ and } \mathbb{H}_{q+1} = 2t_j\mathbb{H}_q - 2q\mathbb{H}_{q-1} \tag{58}$$

The cost function is given as:

$$Min\ \mathfrak{A}_{wind-cn} = \frac{1}{2}[I_{wind}(n) - I_{wind-ref}(n)] \tag{59}$$

The control law $u(n) = u_{wind}(n)$ for WECS is:

$$u_{wind}(n) = \frac{\sum_{j=1}^{p}\mathfrak{U}_j^i\wp^i}{\sum_{j=1}^{p}\mathfrak{U}_j^i} = \frac{\sum_{j=1}^{p}\left(\left[\prod_{j=1}^{p}\exp\left(\frac{-1}{2}\left[\frac{t_j-\mathfrak{C}_j^i}{\mathfrak{d}_j^i}\right]^2\right)\right]\times[\varkappa_j^i \times \mathfrak{Y}_{pq}(t_j)]\right)}{\sum_{j=1}^{p}\left[\prod_{j=1}^{p}\exp\left(\frac{-1}{2}\left[\frac{t_j-\mathfrak{C}_j^i}{\mathfrak{d}_j^i}\right]^2\right)\right]} \tag{60}$$

In general the update equation for control law $u_{wind}(n)$ is written as:

$$\mathfrak{K}_j^i(n+1) = \mathfrak{K}_j^i(n) + \zeta_{wind} \frac{\partial \Lambda_{wind}(n)}{\partial \mathfrak{K}_j^i(n)} + \zeta_{wind} \Delta\varepsilon_{cn}(n+1) \tag{61}$$

The $\varepsilon_{cn} = I_{wind}(n) - I_{wind-ref}(n)$ is used to tune the parameters $\mathfrak{K}_j^i \in \{\hbar_j^i, \varrho_j^i, \tau_j^i\}$ of the controller. Where $\Delta\varepsilon_{cn}(n+1) = \varepsilon_{cn}(n) - \varepsilon_{cn}(n-1)$, ζ_{wind} is the learning rate, $\hbar_j^i$ is the mean of the Gaussian fuzzy function for jth input, ith rule, ϱ_j^i is the variance of the Gaussian fuzzy function for jth input, ith rule and τ_j^i is the weight of the Gaussian membership function for jth input, ith rule. The term, $\Lambda_{wind}(n)$ can be determined as:

$$\Lambda_{wind}(n) = \frac{1}{2}\left[\varepsilon_{cn}(n) + \zeta_{wind} u_{wind}^2(n)\right] \tag{62}$$

where $\frac{\partial \Lambda_{wind}(n)}{\partial \mathfrak{K}_j^i(n)}$ can be simplified as:

$$\frac{\partial \Lambda_{wind}(n)}{\partial \mathfrak{K}_j^i(n)} = \left[\varepsilon_{cn}(n) \frac{\partial \hat{I}_{wind}(n)}{\partial u_{wind}(n)} - \zeta_{wind} u_{wind}(n)\right] \frac{\partial u_{wind}(n)}{\partial \mathfrak{K}_j^i(n)} \tag{63}$$

The term, $\frac{\partial \hat{I}_{wind}(n)}{\partial u_{wind}(n)}$ can be simplified as:

$$\frac{\partial \hat{I}_{wind}(n)}{\partial u_{wind}(n)} = \frac{\sum_{j=1}^{p} \mathfrak{U}_j \left[-\left(\frac{u_{wind}(n) - \mathfrak{C}_j^i}{\mathfrak{F}_j^{i2}} \right) (\wp_j - \hat{I}_{wind}(n)) + 2\sqrt{\frac{2}{\pi}} \{8 r_{11}^j + r_{12}^j (128 u_{wind}(n) - 0.5)\} \right]}{\sum_{j=1}^{p} \mathfrak{U}_j} \tag{64}$$

where r_{11}^j and r_{12}^j are wavelet coefficients for jth input. The update equations for control law, $u_{wind}(n)$ are:

$$\begin{aligned} \hbar_j^i(n+1) = {} & \hbar_j^i(n) + \zeta_{wind}\left(\varepsilon_{cn}(n) \frac{\partial \hat{I}_{wind}}{\partial u_{wind}(n)} - \zeta_{wind} u_{wind}(n)\right) \\ & \left[\frac{\wp_j - u_{wind}(n)}{\sum_{j=1}^{p} \mathfrak{U}_j}\right] \mathfrak{U}_i \frac{(t_j - \hbar_j^i)}{\varrho_j^{i2}} \end{aligned} \tag{65}$$

$$\begin{aligned} \varrho_j^i(n+1) = {} & \varrho_j^i(n) + \zeta_{wind}\left(\varepsilon_{cn}(n) \frac{\partial \hat{I}_{wind}}{\partial u_{wind}(n)} - \zeta_{wind} u_{wind}(n)\right) \\ & \left[\frac{\wp_j - u_{wind}(n)}{\sum_{j=1}^{p} \mathfrak{U}_j}\right] \mathfrak{U}_i \frac{(t_j - \hbar_j^i)}{\varrho_j^{i3}} \end{aligned} \tag{66}$$

$$\tau_j^i(n+1) = \tau_j^i(n) + \zeta_{wind}\left(\varepsilon_{cn}(n) \frac{\partial \hat{I}_{wind}}{\partial u_{wind}(n)} - \zeta_{wind} u_{wind}(n)\right) \left[\frac{\mathfrak{U}_j}{\sum_{j=1}^{p} \mathfrak{U}_j} \mathfrak{Y}_{pq}(t_j)\right] \tag{67}$$

Table 1 System modelling parameters

Name	Values
Working mode	Network connection
Sweep area	$1810\,m^2$
Cut in wind speed	3 m/s
Nominal wind speed	12 m/s
Cut out wind speed	25 m/s
Rotor diameter	48 m
Nominal output power	500 kW
Generator used	PMSG
Nominal rotor speed	30 rpm
Rotor speed range	10–30 rpm
Generator nominal output	250 kW
Generator nominal cycle	1,500 rpm
Gear box rate	01:50
Blade number	3
Generator number	2
Generator voltage	690 V

4 Simulation Results

To demonstrate the accuracy of the proposed control method (denoted as IAHWBC in the figures) over other existing methods, PI, Fuzzy Logic Controller (FLC) and IAHWBC are dynamically modelled in MATLAB/Simulink. Then detailed simulations are performed for each controller under the same operating conditions (i.e., same type and number of input variable such as different wind speed and load parameters). Different parameters which are adopted during the modelling of wind model are given in Table 1.

The real time wind speed for Islamabad, Pakistan is recorded for 12 July, 2018 and used for simulation. The wind speed in m/s is shown in Fig. 10a. In Fig. 10b–d, the output power with PI, FLC and proposed IAHWBC is shown. From Fig. 10, it is observed that the maximum overshoots reaches 760 kW in case of PI controller, 619 kW using FLC and 528 kW with proposed controller. The zoomed figures of output power comparison at different intervals are shown in the Fig. 11.

In order to compare the performance of proposed IAHWBC, two different control schemes, i.e., PI, and FLC are used. Figure 12 shows that the spider plot for efficiency, overshoot, undershoot, steady state error and MPPT error with proposed and all other control schemes. The overshoot, MPPT error and percentage overshoot of PI controller is very high followed by FLC. The steady state error of FLC is also very high. The proposed IAHWBC shows best performance in all aspects over other controllers.

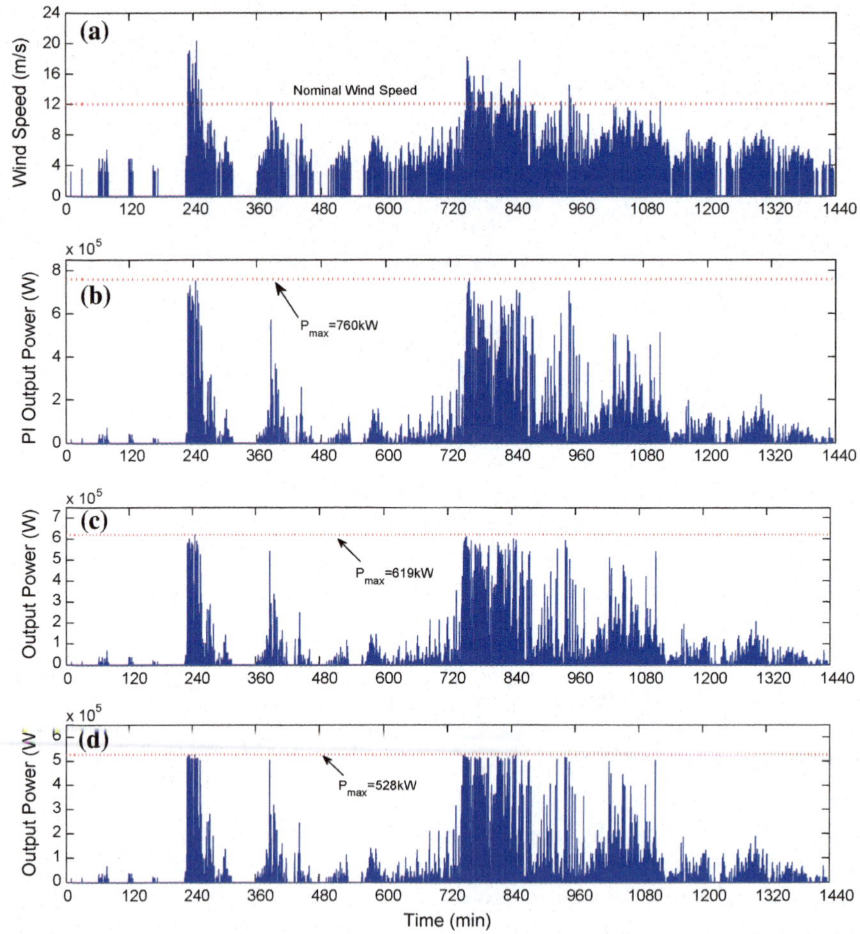

Fig. 10 **a** Wind speed, **b** PI output power, **c** FLC output power, **d** IAHWBC output power

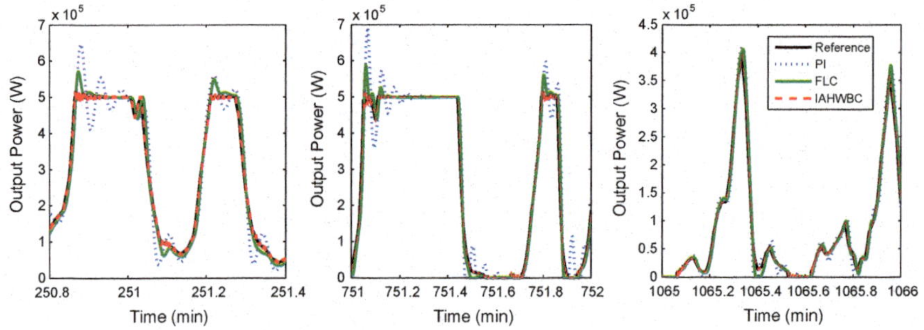

Fig. 11 Output power comparison using different controllers

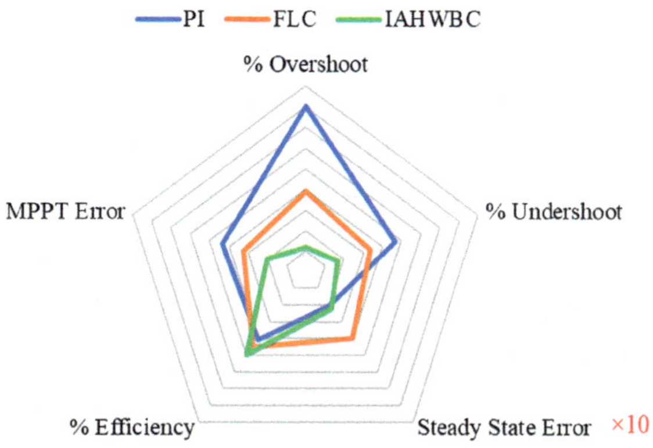

Fig. 12 Spider chart (efficiency, overshoot, undershoot, steady state error, MPPT error) for various controllers

To test further the dynamic performance of all three controllers, various performance indexes, i.e., Mean Relative Error (MRE), Integral Squared Error (ISE), Integral Time-weighted Absolute Error (ITAE), Integral Absolute Error (IAE), and Integral Time-weighted Squared Error (ITSE) are calculated below and are shown in Fig. 13.

$$ITAE = \int_0^t |te_{wind}|dt \tag{68}$$

$$IAE = \int_0^t |e_{wind}|dt \tag{69}$$

$$ISE = \int_0^t e_{wind}^2 dt \tag{70}$$

$$ITSE = \int_0^t te_{wind}^2 dt \tag{71}$$

where

$$e_{wind}(t) = P_{wind-ref} - P_{wind} \tag{72}$$

$$MRE = \frac{1}{T}\sum_{t=1}^{T} \frac{P_{wind-ref} - P_{wind}}{P_{wind}} \tag{73}$$

It is seen from Fig. 13 that all index values are quite small with IAHWBC as compared to other two control methods.

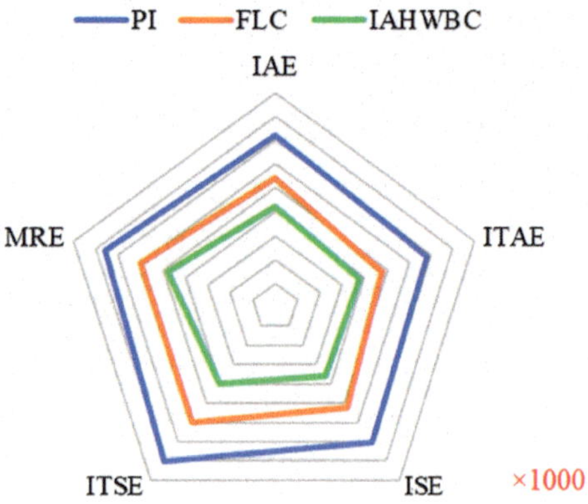

Fig. 13 Spider chart (MRE, ITAE, IAE, ISE, ITSE) for various controllers

5 Conclusion

A wind system was controlled considering real load and weather parameters via a modified neural-fuzzy controller in this book chapter. The performance of wind system in terms of efficiency, output power and steady-state characteristics (i.e., small overshoot and short settling time) was considerably improved by the hybridization of wavelet (Hermite) in the conventional neural-fuzzy structure. The new developed controller has maintained its self adaptively behavior under uncertainties generating from various load disturbances and wind speed variation. The approximating capability of Hermite wavelet decomposition enhanced the adaptive capability further of the IAHWBC. Simulation results were given to demonstrate the superiority over PI/fuzzy controllers under similar operating conditions. In future work, the application of proposed controller will modify to be applied to inverter to raise the reliability of the grid connected system.

Acknowledgements The authors gratefully thank to Tallinn University of Technology and Archimedes Foundation for providing Dora Plus grant in the frame of the European Regional Development Funds Doctoral Studies and Internationalisation Programme.

References

1. Bradai R, Boukenoui R, Kheldoun A et al (2017) Experimental assessment of new fast MPPT algorithm for PV systems under non-uniform irradiance conditions. Appl Energy 199:416–429. https://doi.org/10.1016/J.APENERGY.2017.05.045
2. Osório GJ, Matias JCO, Catalão JPS (2015) Short-term wind power forecasting using adaptive neuro-fuzzy inference system combined with evolutionary particle swarm optimization, wavelet transform and mutual information. Renew Energy 75:301–307. https://doi.org/10.1016/J.RENENE.2014.09.058
3. Amjady N, Keynia F, Zareipour H (2011) Short-term wind power forecasting using ridgelet neural network. Electr Power Syst Res 81:2099–2107. https://doi.org/10.1016/J.EPSR.2011.08.007

4. Khalid M, Savkin AV (2012) A method for short-term wind power prediction with multiple observation points. IEEE Trans Power Syst 27:579–586. https://doi.org/10.1109/TPWRS.2011.2160295
5. Xia S, Zhang Q, Hussain ST et al (2018) Impacts of integration of wind farms on power system transient stability. Appl Sci 8. https://doi.org/10.3390/app8081289
6. Petković D, Ćojbašić Ž, Nikolić V et al (2014) Adaptive neuro-fuzzy maximal power extraction of wind turbine with continuously variable transmission. Energy 64:868–874. https://doi.org/10.1016/J.ENERGY.2013.10.094
7. Tan K, Islam S (2004) Optimum control strategies in energy conversion of PMSG wind turbine system without mechanical sensors. IEEE Trans Energy Convers 19:392–399. https://doi.org/10.1109/TEC.2004.827038
8. Boukhezzar B, Siguerdidjane H (2009) Nonlinear control with wind estimation of a DFIG variable speed wind turbine for power capture optimization. Energy Convers Manag 50:885–892. https://doi.org/10.1016/J.ENCONMAN.2009.01.011
9. Lin C-H, (2014) Recurrent wavelet neural network control of a PMSG system based on a PMSM wind turbine emulator. TURKISH J Electr Eng Comput Sci 22:795–824. https://doi.org/10.3906/elk-1208-3
10. Yin M, Li W, Chung CY et al (2017) Optimal torque control based on effective tracking range for maximum power point tracking of wind turbines under varying wind conditions. IET Renew Power Gener 11:501–510. https://doi.org/10.1049/iet-rpg.2016.0635
11. Taveiros FEV, Barros LS, Costa FB (2015) Back-to-back converter state-feedback control of DFIG (doubly-fed induction generator)-based wind turbines. Energy 89:896–906. https://doi.org/10.1016/J.ENERGY.2015.06.027
12. Eriksson S, Kjellin J, Bernhoff H (2013) Tip speed ratio control of a 200 kW VAWT with synchronous generator and variable DC voltage. Energy Sci Eng 1:135–143. https://doi.org/10.1002/ese3.23
13. Zhong Q-H, Ruan Y, Zhao M-H, Tan L (2013) Application of variable-step hill climbing searching in maximum power point tracking for DFIG wind power generation system. Power Syst Prot Control 41:67–73
14. Lalouni S, Rekioua D, Idjdarene K, Tounzi A (2015) Maximum power point tracking based hybrid hill-climb search method applied to wind energy conversion system. Electr Power Components Syst 43:1028–1038. https://doi.org/10.1080/15325008.2014.999143
15. Harrag A, Messalti S (2015) Variable step size modified P&O MPPT algorithm using GA-based hybrid offline/online PID controller. Renew Sustain Energy Rev 49:1247–1260. https://doi.org/10.1016/J.RSER.2015.05.003
16. Jiang L (2015) An improved hybrid hill climb searching control for MPPT of wind power generation systems under fast varying wind speed. IET Conf Proc 1–6. https://doi.org/10.1049/cp.2015.0493
17. Rezaei MM (2018) A nonlinear maximum power point tracking technique for DFIG-based wind energy conversion systems. Eng Sci Technol an Int J 21:901–908. https://doi.org/10.1016/J.JESTCH.2018.07.005
18. Li B, Tang W, Xiahou K, Wu Q (2017) Development of novel robust regulator for maximum wind energy extraction based upon perturbation and observation. Energies 10. https://doi.org/10.3390/en10040569
19. Kazmi SMR, Goto H, Guo H, Ichinokura O (2011) A novel algorithm for fast and efficient speed-sensorless maximum power point tracking in wind energy conversion systems. IEEE Trans Ind Electron 58:29–36. https://doi.org/10.1109/TIE.2010.2044732
20. Huang C, Li F, Jin Z (2015) Maximum power point tracking strategy for large-scale wind generation systems considering wind turbine dynamics. IEEE Trans Ind Electron 62:2530–2539. https://doi.org/10.1109/TIE.2015.2395384
21. Tang C, Soong WL, Freere P et al (2012) Dynamic wind turbine output power reduction under varying wind speed conditions due to inertia. Wind Energy 16:561–573. https://doi.org/10.1002/we.1507

22. Kim K, Van TL, Lee D et al (2013) Maximum output power tracking control in variable-speed wind turbine systems considering rotor inertial power. IEEE Trans Ind Electron 60:3207–3217. https://doi.org/10.1109/TIE.2012.2200210
23. Zhang X, Huang C, Hao S et al (2016) An improved adaptive-torque-gain MPPT control for direct-driven PMSG wind turbines considering wind farm turbulences. Energies 9. https://doi.org/10.3390/en9110977
24. Johnson KE, Pao LY, Balas MJ, Fingersh LJ (2006) Control of variable-speed wind turbines: standard and adaptive techniques for maximizing energy capture. IEEE Control Syst Mag 26:70–81. https://doi.org/10.1109/MCS.2006.1636311
25. Xia Y, Ahmed KH, Williams BW (2011) A new maximum power point tracking technique for permanent magnet synchronous generator based wind energy conversion system. IEEE Trans Power Electron 26:3609–3620. https://doi.org/10.1109/TPEL.2011.2162251
26. Xia Y, Ahmed KH, Williams BW (2013) Wind turbine power coefficient analysis of a new maximum power point tracking technique. IEEE Trans Ind Electron 60:1122–1132. https://doi.org/10.1109/TIE.2012.2206332
27. Satpathy AS, Kishore NK, Kastha D, Sahoo NC (2014) Control scheme for a stand-alone wind energy conversion system. IEEE Trans Energy Convers 29:418–425. https://doi.org/10.1109/TEC.2014.2303203
28. Zhao H, Wu Q, Rasmussen CN, Blanke M (2014) L_1 adaptive speed control of a small wind energy conversion system for maximum power point tracking. IEEE Trans Energy Convers 29:576–584. https://doi.org/10.1109/TEC.2014.2312978
29. Koutroulis E, Kalaitzakis K (2006) Design of a maximum power tracking system for wind-energy-conversion applications. IEEE Trans Ind Electron 53:486–494. https://doi.org/10.1109/TIE.2006.870658
30. Heo SY, Kim MK, Choi JW (2015) Hybrid intelligent control method to improve the frequency support capability of wind energy conversion systems. Energies 8:11430–11451. https://doi.org/10.3390/en81011430
31. Martinez MI, Susperregui A, Tapia G (2017) Second-order sliding-mode-based global control scheme for wind turbine-driven DFIGs subject to unbalanced and distorted grid voltage. IET Electr Power Appl 11:1013–1022. https://doi.org/10.1049/iet-epa.2016.0711
32. Martinez MI, Susperregui A, Tapia G, Xu L (2013) Sliding-mode control of a wind turbine-driven double-fed induction generator under non-ideal grid voltages. IET Renew Power Gener 7:370–379. https://doi.org/10.1049/iet-rpg.2012.0172
33. Belmokhtar K, Doumbia ML, Agbossou K (2014) Novel fuzzy logic based sensorless maximum power point tracking strategy for wind turbine systems driven DFIG (doubly-fed induction generator). Energy 76:679–693. https://doi.org/10.1016/j.energy.2014.08.066
34. Hassan SZ, Li H, Kamal T et al (2017) An intelligent pitch angle control of wind turbine. In: 2017 international symposium on recent advances in electrical engineering (RAEE). https://doi.org/10.1109/RAEE.2017.8246144
35. Pucci M, Cirrincione M (2011) Neural MPPT control of wind generators with induction machines without speed sensors. IEEE Trans Ind Electron 58:37–47. https://doi.org/10.1109/TIE.2010.2043043
36. Khanali M, Ahmadzadegan S, Omid M et al (2018) Optimizing layout of wind farm turbines using genetic algorithms in Tehran province, Iran. Int J Energy Environ Eng 9:399–411. https://doi.org/10.1007/s40095-018-0280-x
37. Chang TP (2011) Wind energy assessment incorporating particle swarm optimization method. Energy Convers Manag 52:1630–1637. https://doi.org/10.1016/J.ENCONMAN.2010.10.024
38. Yang X, Liu G, Li A, Van Dai L (2017) A predictive power control strategy for DFIGs based on a wind energy converter system. Energies 10. https://doi.org/10.3390/en10081098
39. Medjber A, Guessoum A, Belmili H, Mellit A (2016) New neural network and fuzzy logic controllers to monitor maximum power for wind energy conversion system. Energy 106:137–146. https://doi.org/10.1016/J.ENERGY.2016.03.026
40. Kamal S, Bandyopadhyay B (2014) Higher order sliding mode control: a control lyapunov function based approach. WSEAS Trans Syst Control 9:38–46

41. Selvi V, Umarani DR (2010) Comparative analysis of ant colony and particle swarm optimization techniques. Int J Comput Appl 5:4
42. Vzquez Prez S, Len Galvn JI, Garca Franquelo L et al (2009) Model predictive control with constant switching frequency using a discrete space vector modulation with virtual state vectors. In: International conference on industrial technology, Gippsland, Victoria, Australia. IEEE, pp 1–6
43. Kamal T, Karabacak M, Hassan SZ et al (2018) A robust online adaptive B-spline MPPT control of three-phase grid-coupled photovoltaic systems under real partial shading condition. IEEE Trans Energy Convers 1. https://doi.org/10.1109/TEC.2018.2878358
44. Atakulreka A, Sutivong D (2007) Avoiding local minima in feedforward neural networks by simultaneous learning BT. In: Orgun MA, Thornton J (eds) AI 2007: advances in artificial intelligence. Springer, Heidelberg, Berlin, pp 100–109
45. Hassan SZ, Li H, Kamal T et al (2017) Neuro-fuzzy wavelet based adaptive MPPT algorithm for photovoltaic systems. Energies 10:394. https://doi.org/10.3390/en10030394
46. Abiyev RH, Kaynak O (2008) Identification and control of dynamic plants using fuzzy wavelet neural networks. In: 2008 IEEE international symposium on intelligent control. IEEE. https://doi.org/10.1109/ISIC.2008.4635940
47. Badar R, Khan L (2013) Hybrid neuro-fuzzy legendre-based adaptive control algorithm for static synchronous series compensator. Electr Power Compon Syst. https://doi.org/10.1080/15325008.2013.792882
48. Cao C, Ma L, Xu Y (2012) Adaptive control theory and applications. J Control Sci Eng 2012:2
49. Tao G (2003) Adaptive control design and analysis. Wiley
50. Mumtaz S, Khan L, Ahmed S, Bader R (2017) Indirect adaptive soft computing based wavelet-embedded control paradigms for WT/PV/SOFC in a grid/charging station connected hybrid power system. PLoS One 12:e0183750. https://doi.org/10.1371/journal.pone.0183750

Printed by Printforce, the Netherlands